Communications
in Computer and Information Science   357

Joaquim Gabriel   Jan Schier
Sabine Van Huffel   Emmanuel Conchon
Carlos Correia   Ana Fred   Hugo Gamboa (Eds.)

Biomedical Engineering Systems and Technologies

5th International Joint Conference, BIOSTEC 2012
Vilamoura, Portugal, February 1-4, 2012
Revised Selected Papers

 Springer

Volume Editors

Joaquim Gabriel
University of Porto, Portugal
E-mail: jgabriel@fe.up.pt

Jan Schier
Institute of Information Theory and Automation of the ASCR
Prague, Czech Republic
E-mail: schier@utia.cas.cz

Sabine Van Huffel
Katholieke Universiteit Leuven, Heverlee, Belgium
E-mail: sabine.vanhuffel@esat.kuleuven.be

Emmanuel Conchon
University of Toulouse, France
E-mail: emmanuel.conchon@univ-jfc.fr

Carlos Correia
University of Coimbra, Portugal
E-mail: correia@fis.uc.pt

Ana Fred
IST - Technical University of Lisbon, Portugal
E-mail: afred@lx.it.pt

Hugo Gamboa
Institute of Telecommunications, Lisbon, Portugal
E-mail: hgamboa@gmail.com

ISSN 1865-0929   e-ISSN 1865-0937
ISBN 978-3-642-38255-0   e-ISBN 978-3-642-38256-7
DOI 10.1007/978-3-642-38256-7
Springer Heidelberg Dordrecht London New York

Library of Congress Control Number: 2013937509

CR Subject Classification (1998): J.3, K.4.1-2, H.2.8, I.5.4, I.4.9, H.5.2-3

Typesetting: Camera-ready by author, data conversion by Scientific Publishing Services, Chennai, India

Printed on acid-free paper

Springer is part of Springer Science+Business Media (www.springer.com)

Preface

The present book includes extended and revised versions of a set of selected papers from the 5th International Joint Conference on Biomedical Engineering Systems and Technologies (BIOSTEC 2012), held in Vilamoura, Algarve, Portugal, during February 1–4, 2012.

BIOSTEC was sponsored by the Institute for Systems and Technologies of Information, Control and Communication (INSTICC), in cooperation with the Association for the Advancement of Artificial Intelligence (AAAI) and technically co-sponsored by the European Society for Engineering and Medicine (ESEM), Biomedical Engineering Society (BMES), IEEE Engineering in Medicine and Biology Society (IEEE EMBS) and IEEE Portugal EMBS Chapter.

The purpose of this conference is to bring together researchers and practitioners, including engineers, biologists, health professionals and informatics/computer scientists, interested in both theoretical advances and applications of information systems, artificial intelligence, signal processing, electronics, and other engineering tools in knowledge areas related to biology and medicine.

BIOSTEC is composed of four co-located conferences; each specializes in one of the aforementioned main knowledge areas, namely:

- BIODEVICES (International Conference on Biomedical Electronics and Devices) focuses on aspects related to electronics and mechanical engineering, especially equipment and material inspired from biological systems and/or addressing biological requirements. Monitoring devices, instrumentation sensors and systems, biorobotics, micro-nanotechnologies, and biomaterials are some of the technologies addressed at this conference.
- BIOINFORMATICS (International Conference on Bioinformatics Models, Methods and Algorithms) focuses on the application of computational systems and information technologies to the field of molecular biology, including, for example, the use of statistics and algorithms to understanding biological processes and systems, with a focus on new developments in genome bioinformatics and computational biology.
- BIOSIGNALS (International Conference on Bio-inspired Systems and Signal Processing) is a forum for those studying and using models and techniques inspired from or applied to biological systems. A diversity of signal types can be found in this area, including image, audio, and other biological sources of information. The analysis and use of these signals is a multidisciplinary area including signal processing, pattern recognition, and computational intelligence techniques, amongst others.
- HEALTHINF (International Conference on Health Informatics) promotes research and development in the application of information and communication technologies (ICT) to healthcare and medicine in general and to the specialized support to persons with special needs in particular. Databases, network-

ing, graphical interfaces, intelligent decision support systems, and specialized programming languages are just a few of the technologies currently used in medical informatics. Mobility and ubiquity in healthcare systems, standardization of technologies and procedures, certification, privacy are some of the issues that medical informatics professionals and the ICT industry in general need to address in order to further promote ICT in healthcare.

The joint conference, BIOSTEC, received 522 paper submissions from 66 countries in all continents. To evaluate each submission, a double-blind paper review was performed by the Program Committee. After a stringent selection process, 65 papers were published and presented as full papers, i.e., completed work (10 pages/30-min oral presentation), 128 papers reflecting work-in-progress or position papers were accepted for short presentation, and another 110 contributions were accepted for poster presentation. These numbers, leading to a "full-paper" acceptance ratio of about 12% and a total oral paper presentations acceptance ratio close to 37%, show the intention of preserving a high-quality forum for the next editions of this conference.

The conference included a panel and six invited talks delivered by internationally distinguished speakers, namely: José C. Príncipe, Richard Bayford, Jan Cabri, Mamede de Carvalho, Franco Docchio, and Miguel Castelo Branco.

We would like to thank the authors, whose research and development efforts are recorded here for future generations. We are also grateful to the keynote speakers for their very interesting talks and invaluable contribution. Finally, special thanks to all the members of the INSTICC team, whose collaboration was fundamental for the success of this conference.

December 2012

Joaquim Gabriel
Jan Schier
Sabine Van Huffel
Emmanuel Conchon
Carlos Correia
Ana Fred
Hugo Gamboa

Organization

Conference Co-chairs

Carlos Correia	University of Coimbra, Portugal
Ana Fred	Technical University of Lisbon / IT, Portugal
Hugo Gamboa	CEFITEC / FCT - New University of Lisbon, Portugal

Program Co-chairs

BIODEVICES

Joaquim Gabriel	FEUP, Portugal

BIOINFORMATICS

Jan Schier	Institute of Information Theory and Automation of the ASCR, Czech Republic

BIOSIGNALS

Sabine Van Huffel	Katholieke Universiteit Leuven, Belgium

HEALTHINF

Emmanuel Conchon	University of Toulouse, IRIT/ISIS, France

Organizing Committee

Sérgio Brissos	INSTICC, Portugal
Helder Coelhas	INSTICC, Portugal
Vera Coelho	INSTICC, Portugal
Patrícia Duarte	INSTICC, Portugal
Liliana Medina	INSTICC, Portugal
Carla Mota	INSTICC, Portugal
Raquel Pedrosa	INSTICC, Portugal
Vitor Pedrosa	INSTICC, Portugal
Daniel Pereira	INSTICC, Portugal
Cláudia Pinto	INSTICC, Portugal
José Varela	INSTICC, Portugal
Pedro Varela	INSTICC, Portugal

BIODEVICES Program Committee

Omar Abdallah, Germany
W. Andrew Berger, USA
Martin Bogdan, Germany
Luciano Boquete, Spain
Carlos Correia, Portugal
Fernando Cruz, Portugal
Maeve Duffy, Ireland
G.S. Dulikravich, USA
Michele Folgheraiter, Germany
Hugo Gamboa, Portugal
Maki Habib, Egypt
Clemens Heitzinger, UK
Toshiyuki Horiuchi, Japan
Leonid Hrebien, USA
Takuji Ishikawa, Japan
Sandeep Jha, Korea, Republic of
Walter Karlen, Canada
Frank Kirchner, Germany
Ondrej Krejcar, Czech Republic
Xiao Liu, UK
Mai Mohammed Said Mabrouk, Egypt
Jordi Madrenas, Spain
DanMandru, Romania

Joseph Mizrahi, Israel
Raimes Moraes, Brazil
Raul Morais, Portugal
Umberto Morbiducci, Italy
Alexandru Morega, Romania
Toshiro Ohashi, Japan
Kazuhiro Oiwa, Japan
Mónica Oliveira, Portugal
Abraham Otero, Spain
Nathalia Peixoto, USA
Seonghan Ryu, Korea, Republic of
Mario Sarcinelli-Filho, Brazil
Chutham Sawigun, The Netherlands
Fernando di Sciascio, Argentina
Federico Vicentini, Italy
Bruno Wacogne, France
Shigeo Wada, Japan
Sen Xu, USA
Jing Yang, China
Aladin Zayegh, Australia
John X. J. Zhang, USA

BIODEVICES Auxiliary Reviewers

Tiago Araújo, Portugal
Shan Cheng, USA
Matthias Görges, Canada

Joana Sousa, Portugal

BIOINFORMATICS Program Committee

Tatsuya Akutsu, Japan
Charles Auffray, France
Rolf Backofen, Germany
Tim Beissbarth, Germany
Inanc Birol, Canada
Rainer Breitling, UK
Carlos Brizuela, Mexico
Egon L. van den Broek,
    The Netherlands
Chris Bystroff, USA

Stefan Canzar, The Netherlands
Kun-Mao Chao, Taiwan
Antoine Danchin, France
Thomas Dandekar, Germany
Richard Edwards, UK
Nadia El-Mabrouk, Canada
Elisa Ficarra, Italy
Liliana Florea, USA
Gianluigi Folino, Italy
Andrew French, UK

Weisen Guo, Japan
Reinhard Guthke, Germany
Sion Hannuna, UK
Jiankui He, USA
Liisa Holm, Finland
Yongsheng Huang, USA
Bo Jin, USA
Giuseppe Jurman, Italy
Inyoung Kim, USA
Jirí Kléma, Czech Republic
Andrzej Kloczkowski, USA
Ina Koch, Germany
Andrzej Kolinski, Poland
Malgorzata Kotulska, Poland
Bohumil Kovár, Czech Republic
Lukasz Kurgan, Canada
Zoe Lacroix, USA
Yinglei Lai, USA
Matej Lexa, Czech Republic
Nianjun Liu, USA
Stefano Lonardi, USA
Bin Ma, Canada
Shuangge Ma, USA
Veli Mäkinen, Finland
Elena Marchiori, The Netherlands
Majid Masso, USA
FrancescoMasulli, Italy
Pavel Matula, Czech Republic
Petr Matula, Czech Republic
Imtraud Meyer, Canada
Yves Moreau, Belgium
Burkhard Morgenstern, Germany
Hunter Moseley, USA
Vincent Moulton, UK
Chad Myers, USA
Radhakrishnan Nagarajan, USA
Jean-Christophe Nebel, UK
José Luis Oliveira, Portugal
Matteo Pellegrini, USA

Michael R. Peterson, USA
Esa Pitkänen, Finland
Xiaoning Qian, USA
Jaques Reifman, USA
Marylyn Ritchie, USA
Eric Rivals, France
Miguel Rocha, Portugal
David Rocke, USA
Simona E. Rombo, Italy
Chiara Romualdi, Italy
Juho Rousu, Finland
Andrey Rzhetsky, USA
Jan Schier, Czech Republic
Thomas Schlitt, UK
Reinhard Schneider, Luxembourg
Mark Segal, USA
Hamid Reza Shahbazkia, Portugal
Christine Sinoquet, France
Peter F. Stadler, Germany
Ruixiang Sun, China
Yanni Sun, USA
David Svoboda, Czech Republic
Sandor Szedmak, Austria
Silvio C. E. Tosatto, Italy
Jyh-Jong Tsay, Taiwan
Alexander Tsouknidas, Greece
Massimo Vergassola, France
Bing Wu, Ireland
Dong Xu, USA
Tangsheng Yi, USA
Yanbin Yin, USA
Jingkai Yu, China
Qingfeng Yu, USA
Filip Zelezny, Czech Republic
Erliang Zeng, USA
Jie Zheng, Singapore
Leming Zhou, USA

BIOINFORMATICS Auxiliary Reviewers

Qi Liu, China
Qin Ma, USA
Marc Parisien, USA
Dusan Popovic, Belgium

Krister Swenson, Canada
Leon-Charles Tranchevent, Belgium

BIOSIGNALS Program Committee

Oliver Amft, The Netherlands
Peter Bentley, UK
Egon L. van den Broek,
  The Netherlands
Tolga Can, Turkey
M. Emre Celebi, USA
Jan Cornelis, Belgium
Carlos Correia, Portugal
Rezarta Islamaj Dogan, USA
Jose-Jesus Fernandez, Spain
Hugo Gamboa, Portugal
Juan I. Godino-Llorente, Spain
Verena Hafner, Germany
Thomas Hinze, Germany
Sabine Van Huffel, Belgium
Bart Jansen, Belgium
Pasi Karjalainen, Finland
Georgios Kontaxakis, Spain
Vaclav Kremen, Czech Republic
KiYoung Lee, Korea, Republic of
Alexandre Luís Magalhães Levada,
  Brazil
Lenka Lhotska, Czech Republic
Hari KrishnaMaganti, Italy
Emanuela Marasco, USA
KostasMarias, Greece
G. K. Matsopoulos, Greece
Paul Meehan, Australia
Mihaela Morega, Romania
Kayvan Najarian, USA
Nicoletta Nicolaou, Cyprus
Kazuhiro Oiwa, Japan
Krzysztof Pancerz, Poland
George Panoutsos, UK
Joao Papa, Brazil

Sever Pasca, Romania
Shahram Payandeh, Canada
Gennaro Percannella, Italy
Ernesto Pereda, Spain
Octavian Postolache, Portugal
Ales Prochazka, Czech Republic
Dick de Ridder, The Netherlands
José Joaquín Rieta, Spain
Marcos Rodrigues, UK
Heather Ruskin, Ireland
Carlo Sansone, Italy
Andres Santos, Spain
Gerald Schaefer, UK
Emanuele Schiavi, Spain
Tapio Seppänen, Finland
Jordi Solé-Casals, Spain
Olga Sourina, Singapore
Daby Sow, USA
Asser Tantawi, USA
Wallapak Tavanapong, USA
Petr Tichavsky, Czech Republic
Duygu Tosun, USA
Carlos M. Travieso, Spain
Mahdi Triki, The Netherlands
Bart Vanrumste, Belgium
Aniket Vartak, USA
Michal Vavrecka, Czech Republic
Eric Wade, USA
Yuanyuan Wang, China
Takashi Watanabe, Japan
Pew-Thian Yap, USA
Huiyu Zhou, UK
Li Zhuo, China

BIOSIGNALS Auxiliary Reviewers

Vânia Almeida, Portugal
Tiago Araújo, Portugal
Vanya Van Belle, Belgium
Bert Bonroy, Belgium

Manuel Cabeleira, Portugal
Alexander Caicedo, Belgium
João Cardoso, Portugal
Anca Croitor, Belgium

Kris Cuppens, Belgium
Daniela D'Auria, Italy
Glen Debard, Belgium
Maria Isabel Osorio Garcia, Belgium
Eduardo Iáñez, Spain
Peter Karsmakers, Belgium
Michal Kawulok, Poland
Jan Luts, Belgium
Marc Mertens, Belgium

Bogdan Mijovic, Belgium
Neuza Nunes, Portugal
Tânia Pereira, Portugal
Sharad Shandilya, USA
Andre Spadotto, Brazil
Katrien Vanderperren, Belgium
Maarten De Vos, Belgium
Qi Zhu, Belgium

HEALTHINF Program Committee

Sergio Alvarez, USA
Francois Andry, USA
Philip Azariadis, Greece
Adrian Barb, USA
Rémi Bastide, France
Bert-Jan van Beijnum,
*4mmThe Netherlands
Egon L. van den Broek,
*4mmThe Netherlands
Eric Campo, France
James Cimino, USA
Miguel Coimbra, Porugal
Emmanuel Conchon, France
Carlos Costa, Portugal
Donald Craig, Canada
Ricardo João Cruz-Correia, Portugal
Stephan Dreiseitl, Austria
O. Ferrer-Roca, Spain
José Fonseca, Portugal
Christoph M. Friedrich, Germany
Ioannis Fudos, Greece
Hugo Gamboa, Portugal
Jonathan Garibaldi, UK
Alfredo Goñi, Spain
David Greenhalgh, UK
Nicolas Guelfi, Luxembourg
Alexander Hörbst, Austria
Chun-Hsi Huang, USA
Ivan Evgeniev Ivanov, Bulgaria
Anastasia Kastania, Greece
Andreas Kerren, Sweden
Georgios Kontaxakis, Spain
Baoxin Li, USA

Giuseppe Liotta, Italy
Guillaume Lopez, Japan
Martin Lopez-Nores, Spain
Michele Luglio, Italy
Emilio Luque, Spain
Paloma Martínez, Spain
Alice Maynard, UK
Rob van der Mei, The Netherlands
Gerrit Meixner, Germany
M. Mohyuddin, Saudi Arabia
Sai Moturu, USA
Radhakrishnan Nagarajan, USA
Goran Nenadic, UK
Shane O'Hanlon, Ireland
José Luis Oliveira, Portugal
Rui Pedro Paiva, Portugal
Chaoyi Pang, Australia
Danilo Pani, Italy
José J. Pazos-arias, Spain
Carlos Eduardo Pereira, Brazil
Rosario Pugliese, Italy
Juha Puustjärvi, Finland
Arkalgud Ramaprasad, USA
Marcos Rodrigues, UK
George Sakellaropoulos, Greece
Nickolas S. Sapidis, Greece
Akio Sashima, Japan
Bettina Schnor, Germany
Arash Shaban-Nejad, Canada
Kulwinder Singh, Canada
Jan Stage, Denmark
Zoran Stevic, Serbia
Andrzej Swierniak, Poland

Kåre Synnes, Sweden
Francesco Tiezzi, Italy
Yulia Trusova, Russian Federation
Alexey Tsymbal, Germany
Aristides Vagelatos, Greece

Francisco Veredas, Spain
Lixia Yao, USA
Vera Yashina, Russian Federation
André Zúquete, Portugal

HEALTHINF Auxiliary Reviewers

Eduardo C. Cabrera, Spain
Adrien Defossez, France
Lourdes Moreno, Spain
Neuza Nunes, Portugal
Nicolas Singer, France

Nathalie Souf, France
Joana Sousa, Portugal
Manel Taboada, Spain

Invited Speakers

José C. Príncipe   University of Florida, USA
Richard Bayford   Middlesex University, UK
Jan Cabri   Norwegian School of Sport Sciences, Norway
Mamede de Carvalho   Institute of Molecular Medicine - University of
                           Lisbon, Portugal
Franco Docchio   Università degli studi di Brescia, Italy
Miguel Castelo-Branco   University of Coimbra, Portugal

Table of Contents

Invited Paper

Part I: Biomedical Electronics and Devices

Part II: Bioinformatics Models, Methods and Algorithms

Part III: Bio-inspired Systems and Signal Processing

Part IV: Health Informatics

Invited Paper

Biomedical 2D and 3D Imaging:
State of Art and Future Perspectives

Giovanna Sansoni[1] and Franco Docchio[2]

[1] Department of Information Engineering, University of Brescia, Italy
[2] Department of Mechanical and Industrial Engineering, University of Brescia, Italy
{giovanna.sansoni,franco.docchio}@ing.unibs.it

Abstract. The increasing and rapidly evolving role of 2D and 3D vision in biomedical science and technology is herein presented, based on the experience of our Laboratory and of its start-ups in recent years. Applications to ophthalmology, dentistry, forensic science and prosthetic technology are discussed.

Keywords: 2D Vision, 3D Vision, Ophthalmology, Forensic Science, Dentistry, Prosthetic Science, Lasers, Optics, Cameras, Start-ups.

1   Introduction

The continuously increasing impact of vision science and vision-related techniques is one of the most remarkable aspects in the early years of this century [1]. None of the fields of our everyday life is immune to this rapid penetration of vision technology. Suffice to consider the huge amount of cameras continuously monitoring buildings, crossings, streets; the presence of a 2D or 3D camera within almost every PC and every smartphone; the impressive use of virtual reality techniques (based on the use of 2D and 3D cameras) to „place" our avatars in social spaces, and the huge amount of vision techniques to assist, monitor and automate the industrial production lines.

Along with the above evolution, biomedicine is another key target for the development of vision-based techniques and instruments. Biomedicine, indeed, is worldwide reputated to be the key enabling technology of the third millennium, the one where most of R&D investments will converge, given its tremendous impact on the GRD of mature and emerging countries, and given the progressive increase in the population's life expectation. This explains the increasing number of laboratories and small hi-tech start-ups focusing on biomedical applications of mature and emerging technologies.

The Laboratory of Optoelectronics of the University of Brescia has been active for years in this, as well as in other domains of application of 2D and 3D vision [2]. Its research areas span from ophthalmology to dermatology, from blood serum analysis using optical sensors to dentistry and orthodontics, from oxymetry to maxillo-facial prosthetics, and to forensic medicine. Optolab has generated, over the years, a remarkable number of start-up companies, most of which operate in one or more of the above domains, and which represent an optoelectronic technical pole in the province of Brescia [3].

J. Gabriel et al. (Eds.): BIOSTEC 2012, CCIS 357, pp. 3–19, 2013.

In this chapter, some of the relevant research topics of the Laboratory in the recent past and in the present will be described, which represent the state of art of science and technology in the respective fields. Some examples of technology transfer to the market via our start-ups will be given for some of the applications.

In ophthalmology, optical coherence tomography (OCT) is a diagnostic method based on low-coherence interferometry that opened new perspectives in the 3D examination of the ocular fundus and the retina, allowing the study of ocular pathologies and in particular the macular ones. However, up to present the interpretation of OCT scans is still subjective, in the absence of imaging elaboration tools able to quantify the morphology of the retinal tissue. Our research was focused on the development of a user-friendly software tool, easy to use, portable and reliable, to be used by the physician as an aid to obtain quantitative information related to the pathology under examination (Section 2.1).

Another interesting domain of interest for vision in ophthalmology is the study of the correlation of some morphologic quantities of retinal blood vessels with the degree of hypertension. This is a so-called „indirect" ophthalmic application: indeed the goal of the investigation is not the diagnosis of an ocular pathology, rather a non-invasive diagnostic procedure (using the eye) to predict pressure-related alterations in blood vessels, and therefore to early diagnose at risk populations (Section 2.2).

In dentistry, the use of 2D and 3D vision techniques have rapidly evolved over the years. Many investigators have developed 3D acquisition instruments to scan oral cavities as an alternative to unpleasant oral plasters, to make models for subsequent orthodontic, prosthetic, implant devices. Our studies on 3D acquisition instruments have been put to market by one of our start-ups (Open Technologies s.r.l.), to develop a widely marketed 3D scanner that enables reverse engineering of oral cavities in a very fast and efficient way (Section 3).

In forensic pathology and anthropology, a correct analysis of lesions on soft tissues and bones is of utmost importance in order to verify the cause of death and the modality of events. Bone structure injuries allow a precise reconstruction of tool and origin; however, the same lesions on soft tissues are less informative because of the higher elasticity degree. Moreover, decomposition of soft tissues makes morphological analysis more difficult to be assessed, and changes in lesion appearance with time. In this context, 3D imaging and suitable reconstruction, CAD modelling and reverse engineering procedures are a valuable alternative for the non-contact acquisition of the morphology of even soft tissues, and for the subsequent analysis of the interaction of the offensive tool with the corpse or with its wounds (Section 4).

Finally, in maxillofacial prosthetic restoration of facial defects, the use of 3D optical sensors have been recently proposed, to safely acquire patient's face segments and to model and prototype prosthetic elements. This trend is justified by (i) their non-invasiveness for the patient, due to the pure-reflective approach to the measurement, (ii) their acquisition speed, which increases the patient comfort and guarantees the accuracy of the measurements despite the unavoidable patient movements, (iii) their market availability at by far lower costs with respect to CT/MRI systems, and (iv) their performances in terms of data quality, system portability and ruggedness (Section 5).

2   Applications to Ophthalmology

2.1   Quantitative Image Analysis Applied to OCT Scans

Optical Coherence tomography (OCT) is a well-known imaging diagnostic tool, which is considered as a new era in ophthalmic diagnosis, allowing the study of the main ocular pathologies, of the fundus as well as of the anterior chamber. From the optical and hardware viewpoints, a considerable amount of research has been, and is presently carried out to increase the axial and lateral resolution of the technique, to add spectral information and data presentation [5]. In addition, most of the commercial instruments are equipped with software tools for the interpretation of the OCT images. The still present drawback, however, is an excessive subjectivity in the interpretation of the images, in the absence of metric tools to quantitatively interpret the images.

Our Laboratory, in partnership with the Ophthalmic Clinics of the University of Insubria (Varese) was active in the development of a new software tool which could, in a user friendly way, favour the diagnostic work by providing quantitative data connected to the images obtained by the OCT [5].

The OCT frame elaboration (Fig. 1) is based on the search, within the macular area, of a Region of Interest (ROI). Within each region, measurement information based on the presence of low-reflectance areas within the scan is obtained by image elaboration techniques [6].

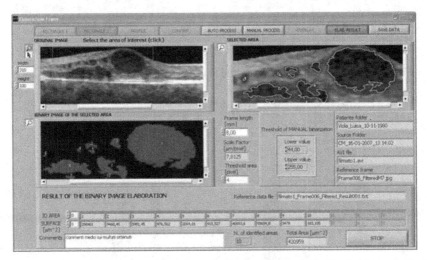

Fig. 1. Selection of the macular edema and measurement of the area of detected hyporeflective regions

The software tool shown in Fig. 1 allows the identification of the intra-retinal region, to identify low-reflectance areas, and, using blob analysis algorithms, allows the collection of measurement information such as perimeter and area of those low reflectance areas. The software tool proved to be ideal for patient follow up, e.g. after surgical or drug procedures: Fig. 2 shows equal ROIs of the same patient at two different times, showing the corresponding quantitative data in the two situations.

The system has been initially tested on three groups of patients affected by three important macular pathologies, i.e., macular edema, macular pucker, and diabetic macular edema. As an example, Fig. 3 shows a quantitative comparison between the areas of the low-reflection regions of the macula (liquid) pre- and post drug treatment in the case of pucker suffering patients. From the figure, reduction of low reflection areas between 27% and 89% after treatment is evident, and this is a quantitative proof of the effectiveness of the treatment.

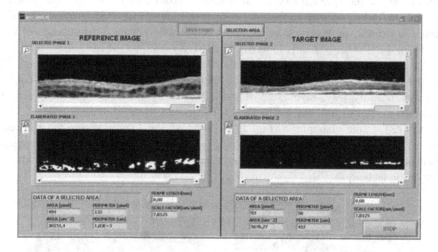

Fig. 2. Example of the software interface developed to perform patient's follow-up

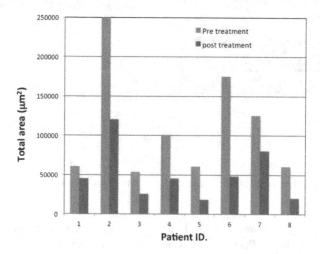

Fig. 3. Comparison between the areas of the low-reflection regions of the macula (liquid) pre- and post drug treatment in the case of pucker suffering patients

2.2 Structural Alterations of Cerebral Small Arteries in Patients with Essential Hypertension

Structural alterations of subcutaneous small resistance arteries, as indicated by an increased media to lumen ratio, are frequently present in hypertensive and/or diabetic patients [7], and may represent the earliest alteration observed. In addition, media to lumen ratio of small arteries evaluated by micromyography has a strong prognostic significance; however its extensive evaluation is limited by the invasiveness of the assessment, since a biopsy of the subcutaneous fat is needed [8].

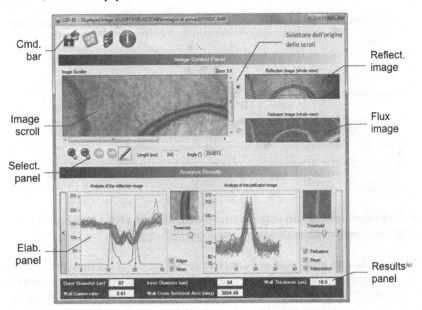

Fig. 4. Front panel of the software tool developed for the flow meter instrument

Non-invasive measurement of wall to lumen of retinal arterioles using scanning laser Doppler flowmetry (SLDF) has been recently introduced. However, this new technique was never compared with micromyographic measurements, generally considered the gold standard approach.

A Laboratory start-up, Nirox s.r.l., operated in conjunction with the Dept. of Medical and Surgical Sciences of our University, to develop a semi-automatic software tool (Fig. 4) combined to the use of a scanning Doppler Flowmeter (Heidelberg Instruments), which provides at the same time the information obtained by reflection images (the outer diameter of the vessel) and the information obtained by flow measurement (inner diameter), by means of which the media to wall ratio can be obtained [9].

With the aid of this tool, micromyographic measurements of subcutaneous small resistance artery structures were compared with the evaluation of wall to lumen ratio of retinal arteries by scanning laser Doppler flowmetry (SLDF).

The media to lumen ratio (M/L) of subcutaneous small arteries, and the wall to lumen ratio (W/L) of retinal arteries were found to be greater in hypertensives as compared to normotensives, while the internal diameter subcutaneous small arteries and the inner and outer diameter of retinal arteries were smaller in hypertensives as compared to normotensives. Moreover, blood flow in the retinal arteries is reduced in hypertensives as compared to normotensives. A close correlation was observed between M/L of subcutaneous small arteries and W/L of retinal arterioles: $r=0.77$, $R^2 = 0.59$, $p<0.001$ (Fig. 5).

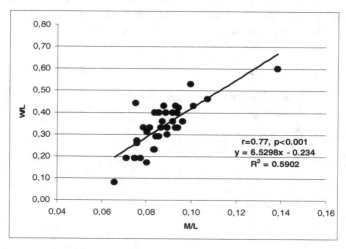

Fig. 5. Correlation between wall to lumen ratio (W/L) of retinal arterioles and media to lumen ratio (M/L) of subcutaneous small arteries: $r=0.77$, $p<0.001$, $R^2=0.59$

The above results show that non-invasive techniques such as flow measurements can be as effective as invasive techniques, and this would open a new scenario for mass screening of population in medical practices in search of early signal of hypertension-related damages. Up to now, instruments such as the one used are still too expensive for massive use by the basic physician. Research in the direction of low-cost optical tools (such as fundus cameras equipped with adaptive optics systems [10]) could represent a breakdown in the direction of the everyday use of this important diagnostic technique.

3   Application to Dentistry and to Orthodontics

2D vision is widely applied in dentistry, for the online documentation of surgical and orthodontic procedures, as well as for colour measurements with the aid of vision-based colour measuring instruments. 3D vision systems are becoming more and more popular in the last years, for two main applications. The first is the in-vivo 3D acquisition of single teeth or of entire oral cavities, by means of hand-held fringe-pattern or laser scanning miniaturized instruments, which scan the cavity, with full compensation for the patient's mouth movements during the scan [11]. The main goal is the 3D rendering of the tooth or the set of teeth, to create the exact replicas for implantology or orthodontic purposes.

The other main application of 3D vision is reverse engineering (RE) of oral plasters, for the CAD elaboration and Rapid Prototyping (RP) of the entire oral cavity, in support to the preparation of implants and of dentures. This application, although not eliminating the patient's discomfort when the plaster is created, has the advantage of a higher accuracy: an even higher advantage is, however, the cost saving, due to the fact that the measurement center can be the orthodontic laboratory, where the plasters are sent, without the need to equip every single dentist's practice with a 3D scanner.

One of the most important producers of 3D scanners (with its own brand or OEM) for orthodontist centres in Italy is Open Technologies s.r.l., a former start-up of the Laboratory of Optoelectronics, now part of a group of optoelectronic small industres located east of Brescia and representing what is called the optoeletronic pole of Brescia (Fig. 6).

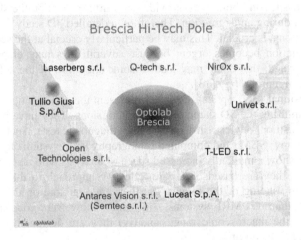

Fig. 6. The cluster of startups of the Laboratory of Optoelectronics of Brescia

Fig. 7. The Optical ReVeng Dental 3D optical scanner by OpenTecnhnologies s.r.l.

The instrument for 3D acquisition and Reverse Engineering is called Optical ReVeng Dental 3D (Fig. 7) and is a laptop-based 3D digitizer based on white-light fringe projection with two cameras [12]. The main characteristics of the system are: (i) a measurement accuracy of 5 µm, an acquisition time less than 180 s for the whole oral cavity (less than 30 s for a single unit), a repeatability of less than 2 µm, a richness of information of more than 3,000,000 triangles for the whole cavity, and the fact there is no need of opacification tools prior to acquisition. This makes the instrument the state of art in orthodontic technology in our country.

4   Vision in Forensic Sciences

The most important step both at post-mortem examination and in anthropological analysis is represented by the in-depth and accurate study of lesions. The analysis of lesions on soft tissues and bones can provide valid information on the shape and size of the tools used during an aggression. Therefore, a detailed 3D study on a tool-mark, be it on skin or bone, is crucial. This may be particularly crucial at the scene of crime, where documentation becomes urgent because several days may elapse before the autopsy is performed: this delay may lead to some alteration of the soft tissues and therefore of lesion morphology [13].

Three-dimensional (3D) optical digitizers represent the last frontier in attempting at accurately capturing the 3D shape of soft tissue lesions. They have been proposed as optimal acquisition devices instead of radio diagnostic tests, such as scanning electron microscopy (SEM) and computed tomography (CT) systems for a number of reasons: (i) they show a markedly increased speed of analysis with respect to 3D SEM and CT, because they are based on a pure-reflective approach to the measurement, and do not require slicing of the tissues, (ii) they are available on the market at by far lower costs than CT/MRI systems, (iii) they show very good measurement performances in the macroscopic range, and (iv) they are designed to perform the measurement in harsh conditions.

The objective of this study was to verify the feasibility of using an optical digitizer to perform the contactless 3D measurement of soft tissues, in order to assess a unique, precise, and reliable method of investigation useful in forensic pathology and anthropology. Moreover, this study aimed at stressing the usefulness of this technology in the forensic context in order to perform a reliable, fast, and complete recording of lesions and environment both in the analysis of the crime scene and at autopsy.

The instrument used to perform the acquisitions is the Vivid 910 digitizer (Konica Minolta, Inc), a market-available laser slit which scans the whole measurement area at variable speeds and resolutions, depending on the required measurement precision [14]. The measured data set represents the 3D shape of the surface of interest in the form of a point cloud. Suitable software is then used to align, edit, decimate and render the resulting triangle mesh. The resulting 3D model describes in a reliable, objective and quantitative way the surface features, and for this reason it is very suitable for studying the details of interest. The acquisition system is rugged, portable and can be easily positioned with respect to the surface, to optimize the adherence of captured data to the real lesion. A considerable number of tests have been performed on soft tissues: the interested reader can see the details in [15, 16]. Here, the test case of a woman (Fig. 8.a) who died after blunt force injury applied with a metal rod (Fig. 8.b) on the head is presented.

The optical digitizer was set up so that the measurement range was lower than 500 mm and the resolution was 0.3 mm. The measurement of the victim's face was carried out by acquiring and merging three views. The resulting point cloud was used as the skeleton on which a proper number of smaller views at a higher resolution (0.15 mm), taken in correspondence with important details of lacerated and contused wound, were added. In this way, a good trade-off between the quality of the measurement and the data amount was achieved. Then, the 3D mesh was created; this step was accomplished by using the Polyworks suite of programs (InnovMetric Inc, Ca).

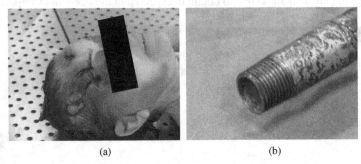

(a)   (b)

Fig. 8. The study case: (a) the head wounds and lacerations; (b) the metallic rod presumed to be the tool used during the aggression

The 3D model is shown in Fig. 9. The aim was to study the sensitivity of the optical measurements, i.e., to assess if it was possible to gauge the shape, the depth, and the contour of the skin wounds on the front head of the victim.

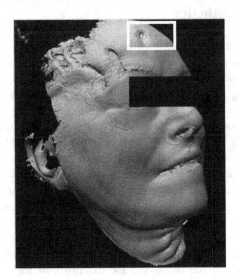

Fig. 9. 3D model of the victim face

Fig. 10 shows the triangle mesh corresponding to the lesion framed in Fig. 9. The model presents a high degree of adherence to the original shape and allows the extraction of the measurements, such as the depth, the width and the length of the wound. The metallic rod underwent the same analysis, in order to carry out a comparison between the lesions and the blunt tool general characteristics. The chance to operate in a 3D virtual space enabled to compare the rod model.

The two models were imported in the same workspace and were mutually oriented to verify the inclination and the direction of the blunt trauma. In Fig. 10.b, the comparison and the virtual matching of the lesion and the tool supposed to have been used during the aggression are shown. The evaluation of the area of the rod penetration into the lesion may also allow one to obtain the speed and the strength of impact.

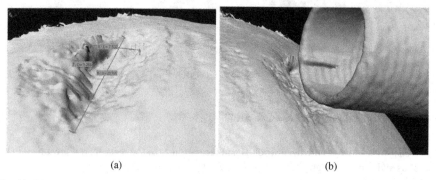

(a)   (b)

Fig. 10. 3D model of the lesion framed in Fig. 9. (a) depth, width and length measurements of the lesion; (b) use oft he 3D models to perform the metric matching between the lesion and the rod

5   Vision in Prosthetics

In maxillo-facial prosthetics, the use of optical, safe sensors combined with reverse engineering and rapid prototyping methods has gained increasing interest in the last years. Traditional reconstruction techniques are based on the following tasks: (i) use of impression making procedures, to obtain the negative patterns of the site of the deformity, (ii) plaster casting of negative patterns, to retrieve the positive defects, (iii) construction of wax positive replicas of the actual prosthesis, (iv) use of conventional flasking and investing procedures, to obtain the negative mould, and (v) casting of suitable materials into the negative mould, to obtain the prosthesis [17].

These procedures present a number of disadvantages. Firstly, impression making results in patient's discomfort and stress. The pressure that must be applied on the face to guarantee the required quality of the impression, inherently results in the deformation of soft tissues, and to the impossibility of acquiring the original face features and look. Secondly, the quality of the wax positive replica is dependent on the artistic skills of an experienced anaplasthologist. The performance of the process strongly depends on both the shape and the extension of the defect. Human error contribution, subjectivity in the reconstruction, low reproducibility of the process, and poor initial shape information often lead to serious unfitting of the final prosthesis, under both functional and aesthetical points of view. Thirdly, the mould production process is cumbersome and time consuming. The overall process is not adaptive, i.e., whenever the existing prosthesis is replaced, the overall process must be carried out from scratch.

3D vision techniques represent a real breakthrough in this context, transforming a purely manual, invasive technique into a semiautomatic and minimally invasive one. Our activity has been carried out in partnership with the specialists of the Removable Prosthodontics School of Dentistry, University of Brescia. Over the years we studied the process of fabrication of the final prosthetic element for a nose, an eye and an ear [18-20] The first is an example of a prosthesis designed in absence of a reference „healthy" organ to be used as a model, the second and third are examples of healthy organs being used „mirrored" to design the prosthetic unit. The first and the last of the three examples will be described here.

5.1   A Nose Prosthesis

In this case study, the patient suffered from a total loss of the nose, because of the excision of a tumour. Typology and dimension of the facial defect suggested the capture of the whole face, since the deformity was large and central with respect to the face. In addition it was necessary to acquire the lost part from healthy donors.

For the acquisition, our 3D acquisition system (again the Minolta Vivid 910) was mounted on a tripod, with the patient comfortably sitting on the dentistry chair. Different system setups were chosen, during each acquisition, to optimize the resolution in case of small details, without missing the entire face shape. The point clouds showed resolution ranging from 320 μm to 170 μm. The acquisition was performed also on the healthy nose donors.

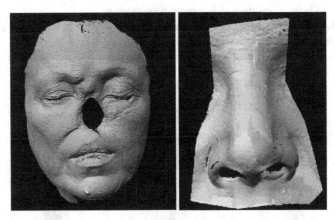

Fig. 11. Acquisition and alignment of the views. (a): aligned point clouds of the patient's face; (b) aligned point clouds of a donor nose

The acquired views were then aligned for both the face and the donor noses, as shown in Fig. 11. The maximum alignment error was 0.4 mm for the face and 0.7 mm for the nose. This was due to the influence of the donor movement and to the limited overlapping among the small nose views. The point cloud of each candidate nose was positioned on the point cloud of the face: this allowed both the operator and the patient to preview the final appearance of the face, to select the most aesthetic nose-face combination and to add modifications to optimise the shape, the position and the functional fitting of the prosthesis. Eventually, the selected nose was aligned to the face.

The „repaired" version of the nose/face point cloud and the original one were then used to create two triangle meshes, which are shown in Fig. 12: they represented the Sculptured model and the Reference model respectively.

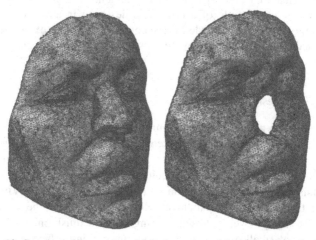

Fig. 12. Creation of the meshes. (a) Reference model; (b) Sculptured model

These models were then extruded, the former externally, the latter internally, and saved in two STL files, for their subsequent prototyping.

Both STL files were sent through the internet for RP machining. They were fabricated using the epoxy photo-polymerizing resin - 'Somos Watershed 11120' by the SLA 3500 Prototyping machine. The two physical models are shown in Fig. 13.

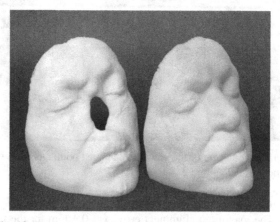

Fig. 13. Physical copies obtained by means of SLA machining

To fabricate the prosthetic element, the two physical models were physically overlapped to each other, and the wax was poured in the resulting cavity (Fig. 14.a). The wax pattern was then extracted from the mould and positioned on the prototype of the Reference model as shown in Fig. 14.b, to perform the try-in of the prosthesis and its refinement on this copy, without disturbing the patient.

The final prosthesis was obtained by conventional flasking and investing procedures. Fig. 15 shows the patient's aspect after the positioning of the prosthesis. The prosthesis was then manually refined on the patient's face, to match the skin colour and texture.

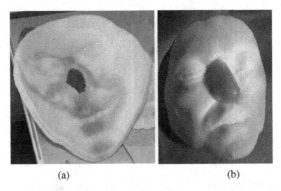

(a)   (b)

Fig. 14. Fabrication of the nose wax positive pattern and try-in of the wax nose pattern onto the physical copy of the Reference model

The patient comfort was optimal, since the acquisition step was quick, contactless and safe. The prosthesis try-in was not necessary, and the patient could be involved in the virtual sculpturing of the prosthesis (i.e., the choice of the template shape and its refinement). This aspect is of primary importance, especially in view of the subsequent replacements of the prostheses, which are necessary due to colour changes, aging, contamination and loss of fit.

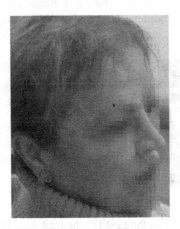

Fig. 15. Prosthesis of the nose

5.2   An Ear Prosthesis

As previously stated, this is an example where the make of the prosthesis was made easier by the presence of the healthy, corresponding organ. The patient's left ear was seriously damaged in consequence of a burn (Fig. 16.a). To fabricate the prosthetic element, the right ear, shown in Fig. 16.b, was used as the template.

The procedure was similar to that of the previous case, the main difference being the fact that the acquisition of the healthy ear replaced the acquisition of virtual donor organs.

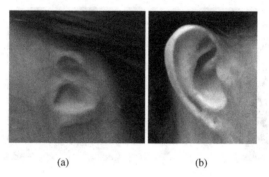

(a)   (b)

Fig. 16. The defect. (a) left, damaged ear; (b) right, safe ear

For both ears, several views were taken and aligned together, and the corresponding triangle meshes were obtained as shown in Fig. 17. The acquisition of the whole patient's face was also performed in three views, as shown in Fig. 18: this model was used as the skeleton, to accurately align the mirrored mesh of Fig. 17.b to the one in Fig. 17.c. Then the model of the ear was interactively aligned until the aesthetic appearance on the whole face was judged optimal. At this point, the skeleton was discarded. The two models were edited to fill residual holes and to reconstruct missing surface parts (mainly due to undercuts). Finally, they were finely connected in correspondence with their borders. The result of this step is shown in Fig. 19.

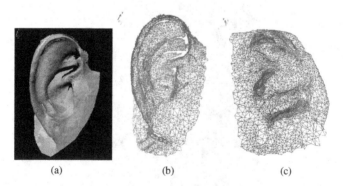

(a)   (b)   (c)

Fig. 17. Acquisition and creation of the model of the right ear. (a) alignment of the views; (b) mesh of the right ear; (c) mesh of the defect

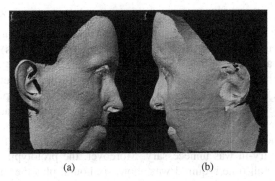

Fig. 18. Alignment of the views acquired in correspondence with the face. (a) right side; (b) left side

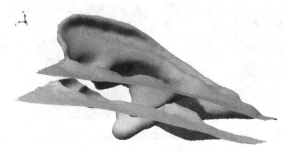

Fig. 19. Alignment of the models to obtain the final model of the ear prosthesis

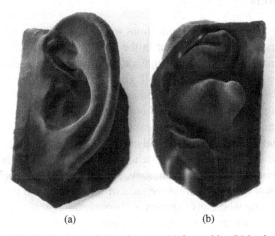

Fig. 20. The final prosthetic element. (a) front side; (b) back side

The mesh was topologically controlled to produce the physical copy. This has been fabricated again by means of rapid prototyping technology. The Connex 500 3D Printing System (Objet-Geometries Inc.) was used. This machine is capable of printing parts and assemblies made of multiple model materials, all in a single build. The materials used to fabricate the ear prosthetic element were the TangoBlackPlus

Shore A85 for the area corresponding to the auricle surface, and the TangoBlackPlus Shore A27 for the areas at the borders of the ear. The ear was obtained in about one hour; the process is very cheap (the cost is in the order of 70 Euros). Fig. 20 shows the front and the back sides of the final prosthesis.

Fig. 21 shows the patient's face after the application of the prosthetic element. It is worth noting that, in this figure, the prosthesis colour was not yet optimized. In fact, we wanted to check its functionality before optimizing it under the aesthetic point of view.

As for the previous example, also in this process the patient's comfort was optimal, and the prosthesis try-in was unnecessary. Moreover, the prototyping step was very cheap, and the overall time required was about six hours, plus the machining of the prosthesis.

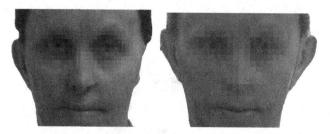

Fig. 21. Application of the prosthetic element to the patient's face

6   Conclusions

Vision is covering a wide span of applications in biomedicine as well as in a plethora of other disciplines. It is conceived that in the near future the availability of smaller and yet more intelligent cameras, integrated and distributed sensors, mobile applications, embedded 3D solutions, will give the pace for new exciting areas of biomedicine, including endoscopy, assisted and robotized surgery procedures, micro- and nanoscale imaging procedures. Cloud computing will probably be the technology of election for a number of image elaboration, storage and inter-laboratory sharing procedures. Start-ups emerging in this fields will probably flourish, given the fact that vision is a fantastic blend among hardware developments, skills in the search of new algorithms, innovation in telecom technology, software and man-machine interaction solutions. Biomedical imaging will set the pace of the evolution in the 3rd millennium world, and it is the area where investments will concentrate, given the social impact it has in the context of our present and future society.

References

1. Machine Vision Report (2012), http://www.emva.org
2. Website of the Laboratory of Optoelectronics, http://www.optolab-bs.it
3. Docchio, F., Sansoni, G.: University-industry synergies in photonics and optoelectronics: the case of Brescia. In: Proc. ICTON, July 1-5, p. 297 (2007)

4. Fujimoto, J.G.: Optical coherence tomography for ultrahigh resolution in vivo imaging. Nature Biotechnol. (October 31, 2003), doi:10.1038/nbt892

5. Prati, M., Donati, S., Tartaglia, V., Sansoni, G., Tironi, M., Chelazzi, P., Brancato, R., Azzolini, C.: Correlation Between Visual Acuity and Retinal Sensitivity Before and After Surgery for Macular Diseases. Invest. Ophthalmol. Vis. Sci. 49, E-Abstract 3203 (2008)

6. Donati, S., Sansoni, G., Tironi, M., Chelazzi, P., Brancato, R., Azzolini, C.: Evaluation of results of macular surgery: Role of microperimetry-related OCT imaging study. Abstract and Presentation, 8th. EURETINA congress, Vienna (2008)

7. Shiffrin, E.L., Hayoz, D.: How to access vascular remodelling in small and medium-sized muscular arteries in humans. J. Hypertens 15, 571–584 (1997)

8. Rizzoni, D., Porteri, E., Boari, G.E.M., De Ciuceis, C., Sleiman, I., Muiesan, M.L., Castellano, M., Mielini, M., Agabiti Rosei, E.: Prognostic significance of small artery structure in hypertension. Circulation 108, 2230–2235 (2003)

9. Rizzoni, D., Porteri, E., Duse, S., De Ciuceis, C., Rosei, C., La Boria, E., Semeraro, F., Costagliola, C., Sebastiani, A., Danzi, P., Tiberio, G.A., Giulini, S.M., Docchio, F., Sansoni, G., Sarkar, A., Agabiti Rosei, E.: Relationship between media-to-lumen ratio of subcutaneous small arteries and wall-to-lumen ratio of retinal arterioles evaluated noninvasively by scanning laser Doppler flowmetry. J. Hypertension 30, 1169–1175 (2012)

10. http://www.imagine-eyes.com

11. Ettl, S., Arold, O., Yang, Z., Häusler, G.: Flying triangulation – an optical 3D sensor for the motion-robust acquisition of complex objects. Appl. Opt. 51, 281–289 (2012)

12. http://www.scanner3d.it

13. Sansoni, G., Cattaneo, C., Trebeschi, M., Gibelli, D., Porta, D., Picozzi, M.: Feasibility of contactless 3D optical measurement for the analysis of bone and soft tissues lesions: new technologies and perspectives in forensic sciences. J. Forensic Sci. 54, 540–545 (2009)

14. Sansoni, G., Cattaneo, C., Trebeschi, M., Gibelli, D., Poppa, P., Porta, D., Maldarella, M., Picozzi, M.: Scene of crime analysis by a 3D optical digitizer: a useful perpective for forensic science. Am. J. Forensic Med. Pathol. 32, 280–286 (2011)

15. Cavagnini, G., Sansoni, G., Trebeschi, M.: Using 3D range cameras for crime scene documentation and legal medicine. In: Proc. SPIE Three-Dimensional Imaging Metrology, San Jose CA, vol. 7239, pp. 1–10 (2009)

16. Sansoni, G., Docchio, F., Trebeschi, M., Scalvenzi, M., Cavagnini, G., Cattaneo, C.: Application of three-dimensional optical acquisition to the documentation and the analysis of crime scenes and legal medicine inspection. In: Proc. of the 2nd IEEE International Workshop on Advances in Sensors and Interfaces, IWASI, Bari, pp. 217–226 (2007)

17. Beumer III, G., Curtis, T.A., Marunick, M.T.: Maxillofacial Rehabilitation - Prosthodontic and Surgical Considerations. Ishiyaku Euro America, St. Louis (1996)

18. Sansoni, G., Cavagnini, G., Docchio, F., Gastaldi, G.: Virtual and physical prototyping by means of a 3D optical digitizer: application to facial prosthetic reconstruction. Virtual and Physical Prototyping 4, 217–226 (2009)

19. Sansoni, G., Trebeschi, M., Cavagnini, G., Gastaldi, G.: 3D Imaging acquisition, modeling and prototyping for facial defects reconstruction. In: Proc. SPIE Three-Dimensional Imaging Metrology, San Jose, CA, vol. 7239, pp. 1–8 (2009)

20. Cavagnini, G., Sansoni, G., Vertuan, A., Docchio, F.: 3D optical Scanning: application to forensic medicine and to maxillofacial reconstruction. In: Proc. Int. Conference on 3D Body Scanning Technologies, Lugano, Switzerland, October 19-20, pp. 167–178 (2010)

Part I

Biomedical Electronics and Devices

Dry and Water-Based EEG Electrodes
in SSVEP-Based BCI Applications

Vojkan Mihajlović[1], Gary Garcia-Molina[1], and Jan Peuscher[2]

[1] Philips Research, High Tech Campus 34, Eindhoven, The Netherlands
[2] Twente Medical Systems International, Zutphenstraat 57, Oldenzaal, The Netherlands
{vojkan.mihajlovic,gary.garcia}@philips.com,
jan.peuscher@tmsi.com

Abstract. This paper evaluates whether water-based and dry contact electrode solutions can replace the gel ones in measuring electrical brain activity by the electroencephalogram (EEG). The quality of the signals measured by three setups (dry, water, and gel), each using 8 electrodes, is estimated for the case of a brain-computer interface (BCI) based on steady state visual evoked potential (SSVEP). Repetitive visual stimuli in the low (12 to 21Hz) and high (28 to 40Hz) frequency ranges were applied. Six people, that had different hair length and type, participated in the experiment. For people with shorter hair style the performance of water-based and dry electrodes comes close to the gel ones in the optimal setting. On average, the classification accuracy of 0.63 for dry and 0.88 for water-based electrodes is achieved, compared to the 0.96 obtained for gel electrodes. The theoretical maximum of the average information transfer rate across participants was 23bpm for dry, 38bpm for water-based and 67bpm for gel electrodes. Furthermore, the convenience level of all three setups was seen as comparable. These results demonstrate that, having optimized headset and electrode design, dry and water-based electrodes can replace gel ones in BCI applications where lower communication speed is acceptable.

Keywords: Dry electrodes, Water-based electrodes, EEG, Signal quality, Brain-computer interface, BCI, Steady state visual evoked potential, SSVEP.

1 Introduction

Brain computer interface (BCI) technology has not yet reached wider adoption except for the few cases where it is used by severely impaired patients (for a recent review see [1]). A number of research groups are trying to bring this technology to a more advanced level, mainly focusing on the most convenient of brain sensing solutions - the electroencephalogram (EEG) - which measures electrical activity of the brain.

Despite numerous advances, both in technological and ergonomic aspects, all three predominant noninvasive BCI modalities, namely steady state visual evoked potential (SSVEP), motor imagery, and P300 are still bound to laboratory settings and do not show clear signs of being ready for wider commercialization in coming years. Among the many problems that EEG-based BCI systems face, the most important ones include:

J. Gabriel et al. (Eds.): BIOSTEC 2012, CCIS 357, pp. 23–40, 2013.

1. Cumbersome and inconvenient procedures to prepare the user before BCI operation, low comfort during the BCI operation, and issues in detaching the system at the end of the usage.

2. Lower accuracy of the BCI command classification algorithms, especially when deployed outside lab conditions, leading to a lower information transfer rate (ITR).

3. Long time required for the user to adapt and learn to use the BCI, including the time required for the BCI to learn specific user parameters, i.e., long calibration procedure.

4. Unpleasant and intrusive interaction with a BCI system that results in users being aversive to the use of BCIs.

5. High number of users that cannot learn to use the BCI, i.e., a so-called BCI illiteracy.

While the first problem is common to all BCIs, except that the number and positioning of electrodes used in a particular system can differ, the latter ones have different impact depending on the BCI modality. In addressing these problems, we consider the SSVEP BCI as a promising solution because, when compared to other BCIs, it can provide high level of detection accuracy (i.e., high ITR), requires short calibration time, and has low BCI illiteracy [2, 3].

The steady state visual evoked potential refers to the response of the cerebral cortex to a repetitive visual stimulus (RVS) oscillating at a constant stimulation frequency. The SSVEP manifests as peaks at the stimulation frequency and/or harmonics in the power spectral density (PSD) of EEG signals [4]. Because of their proximity to the primary visual cortex, the occipital EEG sites exhibit a stronger SSVEP response. SSVEP based BCIs operate by presenting the user with a set of repetitive visual stimuli (RVSi). In most of current implementations, the RVSi are distinguished from each other by their stimulation frequency [5–8]. The SSVEP corresponding to the RVS receiving user's attention is more prominent and can be detected in the ongoing (i.e., background) EEG. Each RVS is associated with an action or a command which is executed by the BCI when the corresponding SSVEP response is detected.

The majority of current SSVEP-based BCIs use stimulation frequencies in the 4 to 30Hz frequency range [9]. RVS at these frequencies, as compared to higher frequencies, have several disadvantages that underpin the fourth problem defined previously: they are prone to visual fatigue which decreases the SSVEP strength, they entail a higher risk of photic induced epileptic seizure [10], and they overlap with the frequency bands of spontaneous brain activity. Higher stimulation frequencies are, thus, preferable for the sake of safety and comfort of the BCI user [3].

The major aspect addressed in this paper is the first problem of the inconvenient and uncomfortable preparation, usage, and detachment of a BCI. This problem stems from the fact that EEG recording procedures remained very similar to that of the early EEG days. The EEG is recorded using Ag/AgCl electrodes that are in contact with the scalp through electrolytic gel [11]. The electrolyte serves two purposes, i) it bridges the ionic current flow from the scalp and the electron flow in the Ag/AgCl electrode and ii) 'glues' the electrode to the scalp. To further improve signal quality, the scalp is frequently cleaned and, especially in clinical applications, skin on the scalp is abraded.

The abrasion process, as well as the usage of conductive gel (electrolyte) makes the whole EEG setup inconvenient for practical applications, especially in consumer settings. The application of electrolyte and the electrodes, even when typical EEG caps are used, requires expert assistance. The setup process is lengthy as it includes preparing the skin, applying the gel, positioning the electrodes (or the cap) and ensuring that the EEG signal quality level is acceptable. Additionally, the user (or expert) has to remove the electrolyte and clean the user's head afterwards, and also clean and dry the electrodes (and the cap) that were used. This also takes time and requires additional effort. This paper aims at addressing these issues, focusing on two alternative solutions, water-based and dry contact electrodes. It also addresses the issue of safe interaction in BCI applications using the high frequency RVSi.

The paper is organized as follows. An overview of the alternative approaches for EEG acquisition systems in BCI is given in the next section. Section 3 details the setups for measuring EEG, using dry, water-based, and gel electrodes, as well as the methods we used to evaluate the quality of the obtained signal in SSVEP BCI domain. Evaluation of the three setups with respect to signal quality is presented in Section 4. Section 5 addresses the potential practical application of water-based and dry electrodes in SSVEP BCIs, focusing on the impact of stimuli duration, and looking into convenience and comfort level of used setups. Results of the evaluation are discussed in Section 6. Section 7 concludes the paper.

2 EEG Acquisition Systems in BCI: Overview

Few research laboratories realized the problem of cumbersome EEG acquisitions systems and engaged in developing more convenient techniques for acquiring brain signals. The approaches range from developing hydrogel-based electrode [12] to numerous versions of dry electrodes. Dry electrode solutions include electrodes that are integrated into the wearable material (contactless electrodes) or affixed on top of the scalp (insulated electrodes) [13–18], electrodes that penetrate the outer layer of the skin [19–28], and dry contact electrodes that exhibit galvanic contact to the skin without the usage of additional electrolyte [29–32]. Contactless and insulated electrodes currently provide insufficient signal quality [33], while due to skin damage they can cause, users that use electrodes that penetrate the outer skin layers might be exposed to a higher risk of infection and skin irritation [34]. Given the recent developments and positive evaluation of dry contact electrode solutions [35–37], including the ones in the SSVEP-based BCI domain [38–40], we consider dry contact electrodes as the desired technology for convenient BCIs.

Addressing the SSVEP BCI field, in a recent publication [41], a successful BCI application of electrodes that use cotton soaked in water to replace the conductive gel is demonstrated. User investigation confirmed that so called 'water-based' electrodes are preferred over gel-based ones, and that no significant performance drop is reported.

Despite the advances in convenient (SSVEP) BCIs, to the best of our knowledge, none of the research publications have systematically characterized the performance of the new BCI acquisition systems in terms of signal quality in different electrodes, and the impact of signal quality, electrode selection, and classification algorithm parameters on the accuracy and ITR of the SSVEP BCI system. Furthermore, the impact of variety of users, having a different hair length and type on the signal quality is often omitted. In this paper we explore how headset setups using dry contact electrodes, water-based electrodes (replicated setup from [41]), and conductive gel electrodes compare to each other considering above mentioned aspects. We also evaluate the convenience levels for the end users, as well as the time required for preparing participants for the experiment.

We believe that robust BCI systems with dry and water-based electrode solutions would greatly simplify the usage, increase acceptance by users in certain clinical applications, and enable wider adoption of BCIs in consumer applications. Therefore, we emphasize the importance of a *practical EEG signal acquisition system* which does not require expert assistance, can be setup and removed by the user himself in a short period of time, and is designed to be ergonomic, convenient, unobtrusive, and comfortable during the measurement process.

3   Materials and Methods

In this section we present the brain signal acquisition technology consisting of an EEG signal amplifier and three different electrode setups, followed by presenting the details of the study design, data processing, and evaluation methods.

3.1   Amplifier Technology and Data Acquisition

The EEG data was recorded using the Twente Medical Systems International (TMSi) Porti system with 24 EEG channels. The Porti uses bipolar amplifier technology that amplifies the difference of the two inputs (so-called instrumentation amplifier technology) with a gain of 20 and includes a common mode rejection in the second stage of amplification. This technology prevents the issues caused by different gain in operation amplifiers and amplifies the input signal against the average reference of the incoming signals, i.e., the common mode signal. The common mode range is $-2V$ to $2V$, and the common mode rejection ratio is higher than 100db.

The Porti is used in a battery powered mode and it was connected to a PC via an optical cable. The highest possible sample rate of 2kHz was used (bit rate of 7.168Mbit/s). As the Porti system requirements include the usage of shielded cables for EEG electrodes and ground electrode with low impedance ($< 1k\Omega$), we used shielded cables for all electrodes and gel electrode as a ground for all three setups.

3.2   EEG Setups with Three Different Electrode Types

For all three setups we used 8 electrodes positioned at the occipital and parietal sites where the SSVEP exhibits the strongest response. The 8 electrode locations selected were O1, O2, Oz, PO3, PO4, POz, P1, and P2, according to the International 10-20 System. For all setups (except two configurations discussed in Section 5) conductive gel ground electrode was positioned at the participants right collar bone.

The dry electrode setup is constructed using 8 commercially available sintered Ag/AgCl ring electrodes (with 10mm outer and 5mm inner diameter) that have twelve

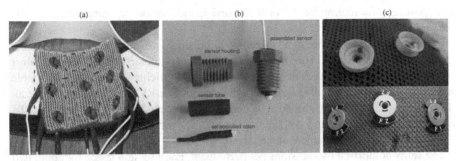

Fig. 1. EEG setups: a) Dry contact electrodes with pins integrated in the EEG headband; b) A water-based electrode and its components; c) Inner and outer side of the head cap with electrodes that need to be filled with conductive gel

2mm long rigid pins. These electrodes are attached to a soft textile patch which is connected to the elastic head band using six Velcro straps, as depicted in Figure 1a. Velcro straps are used to accommodate for head sizes and head shapes of different participants. The electrodes are connected using shielded cables to the TMSi Porti acquisition system. The setup is mounted on a participant's head similarly as a normal headband. To the best of our knowledge this is the first 'easy-to-mount' EEG setup with multitude of dry contact electrodes aimed at measuring SSVEP response.

For water-based setup we used electrodes that require tap water instead of electrolytic gel. First tests with such electrodeas are reported in [41]. Electrodes are made from a silver-chloride pallet and rolled up cotton, as shown in Figure 1b, and are connected to the Porti system via shielded cables. Commercially available EEG head cap with the screwing mechanism was used to position the water-based sensors on the head.

The setup for measuring brain signals using conductive gel was prepared using a standard 32 channel head cap (depicted in Figure 1c) for the usage with the TMSi Porti EEG acquisition system. Shielded cables were also used, as in the case of dry and water-based electrodes. Out of the 32 channels, only the 8 selected channels were filled with conductive gel (Signa gel from Parker Laboratories). In the preparation step we did not used any skin abrasion or clean up procedure in order to reproduce as closely as possible the setup in daily life applications.

3.3 Methods

This section describes the design of the experimental evaluation, the algorithms we used for handling artifacts, the estimation of the magnitude of the SSVEP response, and the manner in which we estimated the performance and comfort of different setups.

Study Design. Six participants (4 males and 2 females) aged 24, 26, 28, 29, 31, and 32, were recruited for the experiment. Participants were selected to cover different hair characteristics ranging from short, sparse, and thin hair, to long, dense, and thick hair. The participants were informed about the experiment and they all signed an informed consent before the start of the study. Special emphasis was put on verifying that the participants did not have any history of epileptogenic episodes or discomforts due to the exposure to oscillating light. For their participation in the experiment participants received a small reward.

Experiment was performed in a laboratory where only artificial light was present (the light screens were closed). The room was not specially shielded or controlled against environmental noise (to resemble real-life situation). Participants sat in a comfortable chair, at a distance of roughly 60cm from the screen. The RVS was rendered using an LED panel with 4 LEDs positioned at its corners, measuring 25cm on the diagonal. The LEDs were switched on and off simultaneously. The experimenter had an extensive experience in EEG measurement and analysis.

The study comprised two sessions per participant. Each session lasted for about two hours and consisted of a preparation segment, dry electrode evaluation segment, water-based electrode evaluation segment, gel electrode evaluation segment, and debriefing segment. The order of evaluation segments was chosen to enable testing of all three setups in a single session. Having to remove the gel after the usage of gel electrodes, or to dry the hair after the usage of water-based electrodes in a different order of segments, would make the study design more complex and lengthy. Also this stresses further the impracticality of especially gel solutions for daily applications.

During the preparation segment the procedure was explained to the participants and the three setups were shown to them. The participants then filled in the questionnaire expressing their perception of different electrode types and EEG acquisition systems that were used in the investigation. This procedure was only done during the first session.

Dry electrode evaluation segment consisted of positioning the textile patch with dry electrodes on participant's head. The experimenter visually inspected the EEG signal quality (high-pass filtered at 1Hz) and in case of no signal, larger impact of noise, and/or severe presence of artifacts in some of the channels, the headband was adjusted to improve the contact and achieve better EEG signal quality level. A chronometer was used to record the time required for this activity. Then the EEG signal was recorded while the participant was focusing on one of the 4 LEDs oscillating at 28, 32, 36, and 40Hz in one of the runs and 12, 15, 18, and 21Hz in the other run. The order of these runs was randomly selected for both sessions. This procedure was repeated for all 4 frequencies per run, each having segments of 5 seconds where LEDs were switched on, interspersed with segments of 4-6 seconds (randomized) where LEDs were switched off. The participants were instructed not to blink during the segments while the LEDs were on. The recorded EEG data, sampled at 2048Hz, for all frequencies were stored. At the end of this segment the headband with electrodes was removed.

For the water-based electrode evaluation segment, the experimenter soaked the water-based electrodes into the cup with tap water 5-10 minutes before the start of the setup. The setup consisted in positioning the EEG cap on a participants head and attaching the water-based electrodes to the EEG cap. In case the signal quality was not good enough according to the experimenter, more water was added to some of the electrodes and/or the electrode positioning was adjusted. The time required to perform this procedure was recorded (excluding the time required to soak the electrode in water before the setup procedure). Then the EEG signal was recorded in two runs, the same way as explained in the previous paragraph for the dry electrodes. After the recording the head cap and the electrodes were removed.

For the gel electrode evaluation segment, the experimenter positioned the EEG cap on a participants head and filled the holes (electrodes) in the cap with conductive gel. The experimenter controlled the EEG signal quality and if needed added more gel to improve the contact and achieve desired EEG signal quality level. This activity was also timed. Then the EEG signal was recorded in the two runs, the same way as for water-based and dry electrodes. After the recording, the EEG cap was removed from the participants head. On completion of this session, a hair wash coupon was given to the participant.

In the debriefing segment (only at the end of the second session) participants had to fill in a questionnaire about their experience with the different electrode types and mounting systems. They were also encouraged to give general comments on the setups and the study design.

Signal Analysis. In the SSVEP BCI framework, the goal of signal processing methods is to detect the presence of an SSVEP at a given stimulation frequency in the EEG. In general, the problem consists of deciding if within a certain time window, the attention of the subject on an RVS has been sufficient to elicit an SSVEP response. The main challenge is to avoid the impact of various artifacts and noise (including background EEG) on the SSVEP, as well as the selection of the best components (e.g., electrodes, temporal segments) that contribute the most in the SSVEP response. These two aspect correspond to artifact handling methods and algorithms for optimal SSVEP detection.

To minimize the impact of severe artifacts, expected when using dry and water-based electrodes, we employed an algorithm for rejecting epochs with artifacts. We selected the epoch duration of 1s with 75% overlap. The algorithm excluded the epochs where the absolute amplitude peak inside the epoch was larger than the empirically selected threshold. The thresholds were estimated based on the standard deviation within the recorded segment. We used a numeric value that is 5 times larger than the standard deviation of the signal in each electrode. Such threshold was used for all three electrode types. In addition, the standard deviation of the recording was used in estimating the level of noise in a particular channel (see Section 4).

The strength of SSVEP can be estimated using various methods, ranging from univariate based power spectral density (PSD) estimation to the use of multivariate spatial filtering [9]. Since our intention was to compare the SSVEP strength measured with different electrode setups and to infer the difference in performance, we employed a PSD estimation method using the Welch algorithm [42], which also provided us with easily traceable and interpretable results. The PSD was estimated on one-second long epochs with 75% overlap (that do not contain artifacts). For selecting the best channels, the absolute PSD value across the 1 to 40Hz frequency spectra was used to estimate the presence of noise in the EEG (see Section 4).

Evaluation Protocol. BCI performance is usually assessed in terms of classification accuracy, classification speed, and the number of available choices. In SSVEP-based BCIs, the classification accuracy is primarily influenced by the strength of the SSVEP response, the signal-to-noise ratio (SNR), and the differences in the properties of the stimuli. That is why we focus on reporting the accuracy of three setups. As the

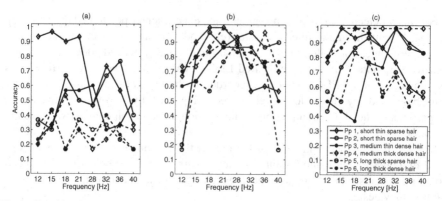

Fig. 2. Classification accuracy for three different setups with 6 participants (Pp 1 to Pp 6) when using PSD at the occipital sites (i.e., O1, O2, and Oz): a) Dry contact electrodes; b) Water-based electrodes c) Conductive gel electrodes

classification speed depends on the time it takes for the SSVEP to be of sufficient strength, we also report the information transfer rate (ITR). ITR is estimated using the approach detailed in [43].

In addition to the bit rate, it is also important to consider the safety and comfort of SSVEP-based BCIs. That is why participants had to rate the comfort level of each of the setups, and provide additional information on whether and under what circumstances would they use a particular setup.

4   Signal Quality and Performance

This section discusses the performance of dry, water-based, and gel-based setups considering the presence of noise in the EEG signal, usage of harmonics, and the optimal selection of electrodes.

4.1   Baseline Performance and the Presence of Noise

Accuracy When Using Occipital Electrodes. We expected that the SSVEP response is strongest in the occipital sites and therefore as a baseline measurement we selected three electrodes at occipital sites, namely, O1, O2, and Oz. Figure 2 depicts the classification accuracy (on 5s-long segments) when maximum PSD in these electrodes is used for each of the participants and for each of the setups. The accuracy for dry electrodes (Figure 2a) is rather low, ranging from the chance level, i.e., 25% for participants with long and/or thick hair (Participant 4, 5, and 6) to more than 50% for participants with shorter and/or tinner hair (Participant 1, 2, and 3). These baseline results demonstrate at the first stage of our analysis the problem of measuring the SSVEP response, and EEG in general, using dry electrodes with people with long and thick hair.

The detection of 12Hz response is challenging for water-based and gel electrodes, as illustrated in Figure 2b and Figure 2c. This demonstrates the problem of distinguishing the changes in the alpha power domain due to the overlap with the dominant alpha frequency. Similar issues can be seen with the 40Hz response, which is mainly due to the decrease in the SSVEP strength at the very high frequency.

As expected, the overall performance of the water-based electrodes was much higher than for the dry electrodes. The surprising result was that the overall accuracy obtained with gel electrodes was comparable to the one obtained with water-based ones. Although this coincides with the results obtained in [41], as we did not used any advanced algorithm for SSVEP response estimation, these results were unexpected. To better understand these results we analyzed the impact of noise on the signal, and the signal quality in different electrodes.

Spectral Content across Electrode Types and Positions. By comparing the raw signal and the power spectra obtained within different EEG channels we observed the following:

1. The noise component in the EEG signal, being environmental or physiological, can be observed for all setups.
2. The severity of noise contribution in the signal and the number of EEG channels contaminated by the noise is higher for the dry setup than the water-based setup, and for the water-based than the gel setup.
3. The impact of noise per electrode can vary throughout a single recording session and it differs for different recording sessions.
4. In most cases, the higher the level of noise in an EEG channel, the lower the SSVEP response.

The first observation can be explained by the environment which was not specially shielded for such kind of experiment and could have been contaminated by electromagnetic waves. Also, motion artifacts stemming from muscle tension and head and body movements were present, due to the lengthy recording procedure. The second observation was expected due to the type of the skin-electrode contact. The third one was not expected for gel electrodes, although it can be partially explained by the fast preparation procedure that did not include skin cleaning and de-greasing before the measurement. It was expected for water-based and especially dry electrodes. Finally, the last observation was a learning point for us and we wanted to incorporate this fact in devising the electrode selection algorithm that will improve the accuracy levels obtained with different setups, as explained below.

4.2 The Impact of Noise Estimation on the Electrode Selection

To estimate the noise level in the signal, we used two simple approaches. The first one was based on the standard deviation (STD) in each channel, i.e., we assumed that the lower the STD in the channel, the lower the noise. The second one used the amount of white noise in each channel, estimated in the frequency range of 1Hz to 40Hz. In both cases the estimations were performed using the EEG segments where stimulation was not presented. To keep the results comparable to the baseline ones in terms of the number of electrodes, we used the 3 electrodes with the lowest noise. The comparison of the achieved accuracy is depicted in Figure 3. The figure clearly illustrates the benefits

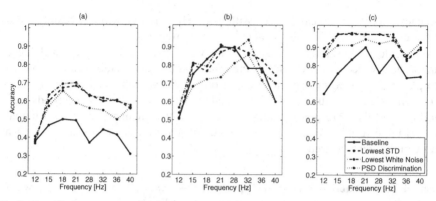

Fig. 3. Classification accuracy for three setups with the optimal selection of 3 electrodes for PSD computation: a) Dry contact electrodes; b) Water-based electrodes c) Conductive gel electrodes

of using the channels with the lowest noise for dry and gel electrodes, irrespective of the approach used for noise estimation. The effect for water-based electrodes is not so pronounced.

4.3 SSVEP Discrimination Power and Electrode Selection

The second feature that we explored for optimal electrode selection was the discrimination level of the PSD of the RVS during the stimuli period versus PSD of the RVS during the non-stimuli period. For each stimuli period we selected the three electrodes with the higher discrimination power for all 4 stimuli. The extent to which this approach compares to the baseline and noise estimation approaches is also depicted in Figure 3. Although it outperformed baseline results for dry and gel electrodes, the accuracy was lower than for the noise estimation approaches across all three setups and even lower than the baseline run for the water-based setup (see Figure 3b). Consequently, we infer that for the optimal electrode selection it is of high importance to select the EEG channels that have the lowest impact of noise in order to make a good estimation of the SSVEP power. Thus, for the rest of the paper we use the optimal selection of electrodes with the lowest white noise component.

4.4 Usage of Harmonics and Different Electrode Number

Figure 4a illustrates the effect of using first harmonics in the noise estimation as well as PSD estimation on classification accuracy. For dry electrodes, significant increase was observed only for the high frequencies, while the improvements of accuracy are significant for both low and high frequencies of water-based and gel setups. With the use of harmonics, the mean accuracy across subjects can be increased to more than 60% for dry setup (except for 12Hz stimuli frequency), to more than 70% accuracy for water-based electrodes, and to more than 88% accuracy for gel electrodes.

So far in the analysis we have used 3 electrodes as we reckoned that this number would be sufficient to achieve good SSVEP detection performance. To test this hypothesis and to investigate the impact of the number of electrodes used in the analysis, we

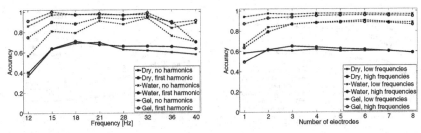

Fig. 4. Classification accuracy across three different setups when: a) using first harmonics; b) different number of electrodes is used

applied PSD-based algorithm (that uses harmonics) to the cases of one to eight electrodes. The results are illustrated in Figure 4b. The figure shows that on average the accuracy is stable across all setups, if 3 to 6 electrodes were used. Also the accuracy level when using 2 electrodes is not much lower than when using three or more.

5   Practical Application

This section illustrates the potential of applying the presented setups in the real life situations. The effect of stimuli duration on the classification rate is addressed in terms of finding the highest ITR. Then, the impact of replacing the conductive gel ground electrode with the dry and water-based one is demonstrated. The section finishes with the discussion on user comfort, convenience, and time required to prepare different setups for practical use.

5.1   Stimuli Duration and ITR

The increase of average accuracy across three setups when using different stimuli duration times is illustrated in Figure 5a. In contrast to our expectation, the decrease of accuracy was minor when shortening the stimuli period from 5s to 3s and not so steep from 3s to 0.75s. This resulted in very high theoretical bit rates (shown in Figure 5b), going up to 26bpm (1s stimuli duration) for low and 21bpm (0.875s stimuli duration) for high frequencies with dry setup, 41bpm (1.5s stimuli duration) for low and 40bpm (0.875s stimuli duration) for high frequencies with water-based setup, and 69bpm (1.125s stimuli duration) for low and 65bpm (1.125s stimuli duration) for high frequencies with gel setup.

Although this transfer rates probably cannot be achieved in practice due to transition effects and the time required for a person to refocus from one stimuli to the other, this result indicates that even with the technology such as dry or water-based electrodes, quite good communication speed can be reached, i.e., 23bpm for dry and 38bpm for water-based electrodes, when averaged over low and high frequency bands. Furthermore, the decrease of the ITR in high frequency range is only minor compared to the low frequency range ITR.

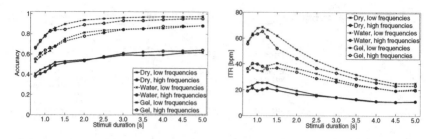

Fig. 5. The impact of stimuli duration ranging from 0.75s to 5s on: a) average classification accuracy; b) ITRs

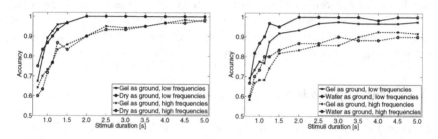

Fig. 6. Comparison of classification accuracy levels with: a) dry contact electrodes when using gel and dry electrode as ground; b) water-based electrodes when using gel and water-based electrode as ground

5.2   Using Dry and Water-Based Electrode as a Ground

To test whether we can design a complete EEG acquisition system using dry or water-based electrodes we performed additional experiments with Participant 1 where we replaced the conductive gel ground electrode with the dry or water-based one, respectively. The comparison of classification accuracy using different stimuli duration for dry electrode setup is shown in Figure 6a. The figure illustrates that the accuracy when using gel and dry electrodes as a ground is almost the same for both, low and high frequencies. Moreover, for this participant (short thin hair), the accuracy is higher than 90% leading to the maximum theoretical bitrate of 86bpm (1s stimuli duration) for low and 65bpm for high frequencies (1.25s stimuli duration). Similarly, Figure 6b illustrates the comparison of water-based setups when using gel and water-based electrode as a ground. In this case, the usage of water-based ground electrode results in improved accuracy, leading to bitrates of 88bpm (1.25s stimuli duration) for low and 61bpm (1.125s stimuli duration) for high frequencies.

In conclusion, although we cannot infer statistical significance of the presented results, we can hypothesize that using acquisition setup that is completely based on dry or water electrodes would not significantly decrease the achieved classification accuracy over the setup that uses conductive gel as a ground.

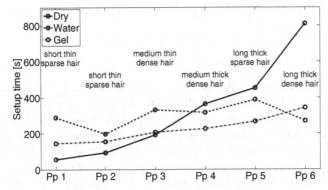

Fig. 7. Time (averaged over two sessions) required for the experimenter to prepare the participant for the recording

5.3 Convenience and Setup Time

Before and after the usage of the headsets, participants reported their anticipated and experienced convenience level (see evaluation protocol detailed in Section 3). Although participants expected that the dry electrode setup will be the least convenient one with the median score of 3.5 on the scale of 1 (unconvenient) to 10 (convenient), compared to 4.5 and 6 for water-based and gel ones respectively, this was not reflected in the rating after the usage of the electrodes. The final median score was 5 for dry, 6 for water-based, and 5 for gel electrodes. These results also indicate that the water-based solution would be preferred over the gel one. Note that the users were asked to rate not only the perceived comfort, but also the practicality of the headsets and the effort required to use them.

Based on the comments and the discussion with participants we learned that the participants with shorter hair had the preference for using the dry electrodes, over water-based and gel ones, while for the participants with the longer hair water-based electrodes were preferred as the procedure for positioning the dry headset was too long (see also Figure 7) and/or the headband was too tight. Most participants did not like the feeling of gel in their hair and the fact that they had to wash the hair after the usage of gel setup.

Direct comparison of the preparation time, depicted in Figure 7, shows that water-based electrodes require more preparation time than gel ones, while dry electrodes require less time than the other two for people with shorter and thiner hair style and more time for people with longer and/or thick hair style. The former can be partially explained by the cumbersome 'screwing' procedure required for placing the water-based electrodes in the EEG headset holes, which had to be redone in case the signal was not good enough. For the gel setup the procedure required adding more gel between the scalp and the electrode until the signal was at the acceptable level, which was much faster. The latter can be explained by the design of the dry electrodes themselves. The length and the rigidness of the pins was a major problem for obtaining a good signal for participants with longer hair. The amount of hair under the electrode was in many cases preventing the pins from getting in contact with the skin on the scalp. Finally, quality of the signal obtained was much lower compared to the participants with shorter and thin hair types (see, e.g., Figure 2)

6 Discussion

The question we aim to answer in this paper is whether and to what extent can dry contact and water-based electrodes replace the conductive gel ones for EEG measurements. As a means to test the performance we selected the SSVEP, since this is a well understood phenomena and as it is one of the most popular neural sources in BCI applications. The analysis has revealed that both electrode types can replace the gel-based configuration, but at the expense of performance. The mean accuracy rate drops by $10 - 25\%$ when using water-based electrodes and by $35 - 45\%$ when using dry electrodes. The information transfer rate in the optimal case is one half for water-based and one third for dry electrodes, compared to gel-based ones.

The baseline evaluation where we used only the occipital sites (i.e., O1, O2, and Oz) demonstrated low accuracy levels for dry electrodes, and lower than expected values for gel electrodes which were comparable to the water-based ones (see Figure 2). Our hypothesis that the selection of 3 electrodes that have the lowest contamination by noise would improve the accuracy for dry electrodes, showed to be true, as it can be seen in Figure 3. The same was observed when using gel electrodes. However, contrary to our expectations this was not the case for water-based electrodes. Further investigation is required to fully understand this effect. Nevertheless, we can infer that the step of selecting the electrodes with the 'cleanest EEG signal' is crucial in achieving higher performance in the SSVEP classification task.

Although we expected that using the discriminative spectral power around the peak of the RVS frequency would be also beneficial for selecting the optimal set of electrodes, this was proven wrong. The accuracy improved for dry and gel electrodes compared to the baseline, but for water-based ones this was not the case. The accuracy for all three setups was lower then when using 3 electrodes with the lowest noise contamination (see Figure 3). It is worth noting that we used two simple algorithms that compute the presence of white noise and the standard deviation in the EEG channels which resulted in similar performance. We argue that using algorithms that have better characterization of environmental or motion artifacts, tailored to the used electrode and amplifier technology, could further improve the performance.

The utilization of first harmonics increased the accuracy in all cases, except for the dry electrodes in the low frequencies, as shown in Figure 4. The latter and the small increase in the accuracy levels for dry electrodes overall, still remain an effect that requires further investigation. However, the performance of different setups when using different number of optimal electrode positions confirmed that the maximum accuracy is reached when using $3 - 6$ electrodes. Furthermore, slight drop in performance when using two electrodes suggests that a setup with 2 dry (or water-based) electrodes with a low noise contamination might be sufficient for SSVEP classification, as also shown in [40, 44].

Maximum accuracy across different setups is achieved when using 4 electrodes with the lowest presence of white noise and when the harmonics were used. The performance per participant in this optimal case is depicted in Figure 8. The figure illustrates

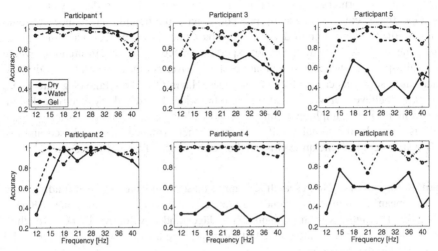

Fig. 8. Comparison of optimal classification accuracy across participants

that the comparable accuracy to the gel solution can be achieved with dry and water-based electrode solutions for users with short thin hair (Participants 1 and 2). However, for participants with medium and long hair, the performance of water-based electrodes slightly decreases, and exhibits a significant drop of accuracy for dry electrodes. The results are in favor of our hypothesis that better design and integration of dry and water-based electrodes in a headset solution is a prerequisite for good performance in a wide range of users, having different hair type and length.

Looking into practical aspects of the proposed alternatives to gel electrodes that are discussed in Section 5, we can infer that users do not perceive dry and water-based electrodes as less convenient than the gel ones, especially having in mind that they do not require gel removal after the usage. Given the potential short preparation time, i.e., having optimized electrode and headset design such as the one for dry electrodes presented in [45], and having the possibility to use only dry or only water-based electrodes without additional loss in signal quality as shown in Figure 6, practical EEG acquisition systems are achievable. Such systems can provide ITRs of 40bpm for water-based and of 20bpm for dry electrodes, and with the optimized design and application of better signal processing and classification algorithms, potentially even higher transfer rates for SSVEP BCI applications.

7 Conclusions

This paper demonstrates that EEG systems that use water-based electrodes and dry contact electrodes can be a viable alternative to traditionally used conductive gel systems for certain application areas. We have shown that an EEG system that uses dry contact electrodes and water-based electrodes can be applied to the SSVEP BCI. The mean information transfer rate in the optimal case, estimated on 6 participants, was 38bpm for water-based and 23bpm for dry electrodes, compared to 67bpm for gel electrodes.

Although the performance was lower than for conductive gel electrodes, dry and water-based electrodes can ease the setup procedure, have the potential to reduce the setup time with the proper design of dry and water-based headset and electrodes, and can facilitate the usage of the system without expert assistance. Therefore, in situations where practical aspects of the EEG system are preferred over communication speed, dry and water-based solutions can replace the gel ones. Having the advantage of greater practicality, we believe that with the advances in the water-based and dry electrode design, improvements in the amplifier technology, and with the usage of more advanced signal analysis methods, the signal quality and overall performance of water-based and dry EEG systems could come quite close to the performance of the gel-based ones.

Acknowledgements. The research leading to these results has received funding from the European Community Seventh Framework Programme under grant BRAIN, n°224156. The authors wish to thank Mark Jäger and Tsvetomira Tsoneva for their suggestions on improving the content of the paper. We also thank our colleagues who participated in the experiment.

References

1. Kübler, A., Birbaumer, N.: Brain-Computer Interfaces and Communication in Paralysis: Extinction of Goal Directed Thinking in Completerly Paralysed Patients. Clinical Neurophysiology 119, 2658–2666 (2008)
2. Cheng, M., Gao, X., Gao, S.: Design and Implementation of a Brain-Ccomputer Interface with High Transfer Rates. IEEE Transactions on Biomedical Engineering 49, 1181–1186 (2002)
3. Garcia-Molina, G., Zhu, D., Abtahi, S.: Phase Detection in a Visual-Evoked-Potential Based Brain Computer Interface. In: Proceedings of the 18th European Signal Processing Conference, EUSIPCO, pp. 949–953 (2010)
4. Regan, D. (ed.): Human Brain Electrophysiology: Evoked Potentials and Evoked Magnetic Fields in Science and Medicine, 1st edn. Elsevier (1989)
5. Gao, X., Xu, D., Cheng, M., Gao, S.: A BCI-Based Environmental Controller for the Motion-Disabled. IEEE Transactions on Neural Systems and Rehabilitation Engineering 11, 137–140 (2003)
6. Lalor, E.C., Kelly, S.P., Finucane, C., Burke, R., Smith, R., Reilly, R.B., McDarby, G.: Steady-State VEP-Based Brain-Computer Interface Control in an Immersive 3D Gaming Environment. Eurasip Journal on Applied Signal Processing 19, 3156–3164 (2005)
7. Friman, O., Lüth, T., Volosyak, I., Gräser, A.: Spelling with Steady-State Visual Evoked Potentials. In: Proceedings of the 3rd International IEEE EMBS Conference on Neural Engineering, pp. 354–357 (2007)
8. Garcia-Molina, G., Mihajlović, V.: Spatial Filters to Detect Steady-State Visual Evoked Potentials Elicited by High Frequency Stimulation: BCI Application. Biomedical Engineering 55, 173–182 (2010)
9. Zhu, D., Bieger, J., Garcia-Molina, G., Aarts, R.M.: A Survey of Stimulation Methods Used in SSVEP-Based BCIs. Computational Intelligence and Neuroscience (2010)
10. Fisher, R.S., Harding, G., Erba, G., Barkley, G.L., Wilkins, A.: Photic- and Pattern-Induced Seizures: a Review for the Epilepsy Foundation of America Working Group. Epilepsia 46, 1426–1441 (2005)

11. Webster, J.G. (ed.): Medical Instrumentation: Application and Design, 3rd edn. Wiley (1997)
12. Alba, N.A., Sclabassi, R.J., Sun, M., Cui, X.T.: Novel Hidrogel-Based Preparation-Free EEG Electrode. IEEE Transactions on Neural Systems and Rehabilitation Engineering 18, 415–423 (2010)
13. Alizadeh-Taheri, B., Smith, R.L., Knoght, R.T.: An Active, Microfabricated, Scalp Electrode Array for EEG Recording. Senosors and Actuators A, 606–611 (1996)
14. Harland, C.J., Clark, T.D., Prance, R.J.: Remote Detection of Human Electroencephalograms using Ultrahigh Input Impedance Electric Potential Sensors. Applied Physics Letters 81, 3284–3286 (2002)
15. von Ellenreider, N., Spinelli, E., Muravchik, C.H.: Capacitive Electrodes in Electroencephalography. In: Proceedings of the 28th IEEE EMBS Annual International Conference, pp. 1126–1129 (2006)
16. Sullivan, T.J., Deiss, S.R., Cauwenbergs, G.: A Low-Noise, Non-Contact EEG/ECG Sensor. In: Proceedings of the IEEE Biomedical Circuits and Systems Conference, BIOCAS, pp. 154–157 (2007)
17. Fonseca, C., Cunha, J.P.S., Martins, R.E., Ferreira, V.M., de Sá, J.P.M., Barbose, M.A., da Silva, A.M.: A Novel Dry Active Electrode for EEG recording. IEEE Transactions on Biomedical Engineering 54, 162–165 (2007)
18. Chi, Y.M., Deiss, S.R., Cauwenberghs, G.: Non-Contact Low Power EEG/ECG Electrode for High Density Wearable Biopotential Sensor Networks. In: Proceedings of the Sixth International Workshop on Wearable and Implantable Body Sensor Networks, BSN, pp. 246–250 (2009)
19. Ruffini, G., Dunne, S., Farrés, E., Marco-Pallarés, J., Ray, C., Mendoza, E., Silva, R., Grau, C.: A Dry Electrophysiology Electrode Using CNT Arrays. Sensors and Actuators A: Physical 132, 34–41 (2006)
20. Ruffini, G., Dunne, S., Farrés, E., Marco-Pallarés, J., Ray, C., Mendoza, E., Silva, R., Grau, C.: First Human Trials of a Dry Electrophysiology Sensor Using Carbon Nanotube Array Interface. Sensors and Actuators A: Physical 144, 275–279 (2008)
21. Griss, P., Tolvanen-Laakso, H.K., Stemme, P.M.G.: Characterization of Micromachined Spiked Biopotential Electrodes. IEEE Transactions on Biomedical Engineering 49, 597–604 (2002)
22. Gramatica, F., Carabalona, R., Casella, M., Cepek, C., Fabrizio, E.D., Rienzo, M.D., Gavioli, L., Matteucci, M., Rizzo, F., Sancrotti, M.: Micropatterned Non-Invasive Dry Electrodes for Brain-Computer Interface. In: Proceedings of the 3rd IEEE-EMBS International Summer School and Symposium on Medical Devices and Biosensors, pp. 69–72 (2006)
23. Chiou, J.C., Ko, L.W., Lin, C.T., Hong, C.T., Jung, T.P., Liang, S.F., Jeng, J.L.: Using Novel MEMS EEG Sensors in Detecting Drowsiness Application. In: Proceedings of the IEEE Biomedical Circuits and Systems Conference, BioCAS, pp. 33–36 (2006)
24. Matteucci, M., Carabalona, R., Casella, M., Fabrizio, E.D., Gramatica, F., Rienzo, M.D., Snidero, E., Gavioli, L., Sancrotti, M.: Micropatterned Dry Electrodes for Brain–Computer Interface. Microelectronic Engineering 84, 1737–1740 (2007)
25. Lin, B.C.T., Ko, L.W., Chiou, J.C., Duann, J.R., Huang, R.S., Liang, S.F., Chiu, T.W., Jung, T.P.: Noninvasive Neural Prostheses Using Mobile and Wireless EEG. Proceedings of the IEEE 96, 1167–1183 (2008)
26. Chang, C.W., Chiou, J.C.: Surface-Mounted Dry Electrode and Analog-Front-End Systems for Physiological Signal Measurements. In: Proceedings of the IEEE/NIH Life Science Systems and Applications Workshop, LiSSA, pp. 108–111 (2009)
27. Ng, W., Seet, H., Lee, K., Ning, N., Tai, W., Sutedja, M., Fuh, J., Li, X.: Micro-Spike EEG Electrode and the Vacuum-Casting Technology for Mass Production. Journal of Materials Processing Technology 209, 4434–4438 (2009)

28. Dias, N., Carmo, J., da Silva, A.F., Mendes, P.M., Correia, J.: New Dry Electrodes Based on Iridium Oxide (IrO) for Non-Invasive Biopotential Recordings and Stimulation. Sensors and Actuators A: Physical 164, 28–34 (2010)
29. Taheri, B.A., Knight, R.T., Smith, R.L.: A Dry Electrode for EEG Recording. Electroencephalography and Clinical Neurophysiology 90, 376–383 (1994)
30. Matthews, R., McDonald, N.J., Hervieux, P., Turner, P.J., Steindorf, M.A.: A Wearable Physiological Sensor Suite for Unobtrusive Monitoring of Physiological and Cognitive State. In: Proceedings of the 29th IEEE EMBS Annual International Conference, pp. 5276–5281 (2007)
31. Gargiulo, G., Calvo, R.A., Bifulco, P., Cesarelli, M., Jin, C., Mohamed, A., van Schaik, A.: A New EEG Recording System for Passive Dry Electrodes. Clinical Neurophysiology 121, 686–693 (2010)
32. Grozea, C., Voinescu, C.D., Fazli, S.: Bristle-Sensors–Low-Cost Flexible Passive Dry EEG Electrodes for Neurofeedback and BCI Applications. Journal of Neural Engineering 8 (2011)
33. Chi, Y.M., Jung, T.P., Cauwenberghs, G.: Dry-contact and Non-contact Biopotential Electrodes. IEEE Reviews in Biomedical Engineering 3, 106–119 (2010)
34. Ferree, T.C., Luu, P., Russell, G.S., Tucker, D.M.: Scalp Electrode Impedance, Infection Risk, and EEG Data Quality. Clinical Neurophysiology 112, 536–544 (2001)
35. Chi, Y.M., Wang, Y.T., Wang, Y., Maier, C., Jung, T.P., Cauwenberghs, G.: Dry and Non-contact EEG Sensors for Mobile Brain-Computer Interfaces. IEEE Transactions on Neural Systems and Rehabilitation Engineering PP, 1–7 (2011)
36. Kitoko, V., Nguyen, T.N., Nguyen, J.S., Tran, Y., Nguyen, H.T.: Performance of Dry Electrode with Bristle in Recording EEG Rhythms Across Brain State Changes. In: 33rd Annual International Conference of the IEEE EMBS, pp. 59–62 (2011)
37. Zander, T.O., Lehne, M., Ihme, K., Jatzev, S., Correia, J., Kothe, C., Picht, B., Nijboer, F.: A Dry EEG-System for Scientific Research and Brain–Computer Interfaces. Frontiers in Neuroscience 5 (2011)
38. Popescu, F., Fazli, S., Badower, Y., Blankertz, B., Müller, K.R.: Single Trial Classification of Motor Imagination Using 6 Dry EEG Electrodes. PloS One 2, e637 (2007)
39. Sellers, E.W., Turner, P., Sarnacki, W.A., McManus, T., Vaughan, T.M., Matthews, R.: A Novel Dry Electrode for Brain-Computer Interface. In: Jacko, J.A. (ed.) Human-Computer Interaction, Part II HCII 2009. LNCS, vol. 5611, pp. 623–631. Springer, Heidelberg (2009)
40. Luo, A., Sullivan, T.J.: A User-Friendly SSVEP-Based Brain–Computer Interface using a Time-Domain Classifier. Journal of Neural Engineering 7, 1–10 (2010)
41. Volosyak, I., Valbuena, D., Malechka, T., Peuscher, J., Gräser, A.: Brain-Computer Interface using Water-Based Electrodes. Journal of Neural Engineeting 7 (2010)
42. Welch, P.D.: The Use of Fast Fourier Transform for the Estimation of Power Spectra: A Method Based on Time Averaging over Short, Modified Periodograms. IEEE Transactions of Audio and Electroacoustics 15, 70–73 (1967)
43. Dornhege, G., del R. Millán, J., Hinterberger, T., McFarland, D.J., Müller, K.R. (eds.): Medical Instrumentation: Application and Design, 1st edn. The MIT Press (2007)
44. Garcia-Molina, G., Zhu, D.: Optimal Spatial Filtering for the Steady State Visual Evoked Potential: BCI application. In: 5th International IEEE EMBS Neural Engineering Conference (2011)
45. Mihajlović, V., Jäger, M., Asvadi, S., Asjes, R.: Flexible Dry Electrodes for Usage in Daily Life Situations (2012) (submitted for publication)

Nitrite Biosensing Using Cytochrome C Nitrite Reductase: Towards a Disposable Strip Electrode

Cátia Correia[1], Marcelo Rodrigues[1], Célia M. Silveira[1], José J.G. Moura[1], Estibaliz Ochoteco[2], Elena Jubete[2], and M. Gabriela Almeida[1,3,*]

[1] REQUIMTE—Dept. de Química, Faculdade de Ciências e Tecnologia, Universidade Nova de Lisboa, 2829-516 Monte Caparica, Portugal
`cs.correia@campus.fct.unl.pt, marcelofrancisco19@hotmail.com,`
`{c.silveira,jose.moura}@fct.ul.pt`
[2] CIDETEC-IK4- Sensors and Photonics Unit, Parque Tecnológico de San Sebastián, Pº Miramon, 196, 20009 Donostia – San Sebastián, Spain
`{eochoteco,ejubete}@cidetec.es`
[3] Escola Superior de Saúde Egas Moniz, Monte de Caparica, 2829-511 Caparica, Portugal
`mg.almeida@fct.unl.pt`

Abstract. This paper presents the results of a primary study that aims to produce miniaturized biosensing devices for nitrite analysis in clinical samples. Following our previous works regarding the development of amperometric nitrite biosensors using the nitrite reducing enzyme (ccNiR) from *Desulfovibrio desulfuricans* ATCC 27774, here we aimed at reducing the size of the experimental set-up according to the specific needs of biomedical applications. For this, thick-film strip electrodes made of carbon conductive inks deposited on plastic supports were modified with the ccNiR enzyme, previously mixed with the conductive graphite ink. Firstly, though, the electrode preparation was optimized (enzyme amount, organic solvent and curing temperature). Then, the biocompatibility of ccNiR with these harsh treatments and the analytical performance of the modified electrodes were evaluated by cyclic voltammetry. Finally, the carbon paste screen-printed electrodes were coated with the *ccNiR/carbon ink* composite, displaying a good sensitivity (5.3×10^{-7} $A.uM^{-1}.cm^{-2}$) within the linear range of 0.001 - 1.5 mM.

Keywords: Nitrite, Cytochrome c nitrite reductase, Electrochemical biosensors, Screen printed electrodes.

1 Introduction

The detection of nitrites in physiological fluids such as plasma and urine is commonly used for clinical diagnosis and has gained an increasing importance in biomedical research. In fact, the nitrate-nitrite-NO pathway is emerging as an important mediator of blood flow regulation, cell signaling, energetics and tissue responses to hypoxia [1-3]. Most of the strategies used for analytical determination of NO_3^- and NO converge to the quantification of NO_2^-. However, the classical protocols for nitrite assessment lack

---

* Corresponding author.

J. Gabriel et al. (Eds.): BIOSTEC 2012, CCIS 357, pp. 41–50, 2013.
© Springer-Verlag Berlin Heidelberg 2013

the sensitivity and selectivity needed for the analysis of physiological samples [4-7]. For example, urine test strips are routinely used for screening nitrites in patients with infection, but results are just qualitative as they are obtained by visual comparison to a color chart. Plasma analysis is much less frequent, owing to limitations of the analytical methods, including blood sampling and processing [5,8]. As a consequence, there is a growing demand for improved analytical tools, increasingly sensitive, reliable and, preferentially, easy-to-use and inexpensive.

An alternative approach relies on the construction of biosensing devices using stable enzymes with high catalytic activity and specificity for nitrite. Due to its high selectivity, turnover and stability, the multihemic cytochrome c nitrite reductase (ccNiR) from the sulphate reducing bacterium *Desulfovibrio desulfuricans* ATCC 27774, which performs the six electron reduction of nitrite to ammonia (eq. 1) [9], has proven to be a promising candidate for the development of an electrochemical nitrite biosensor [10-15].

$$NO_2^- + 8 H^+ + 6 e^- \rightarrow NH_4^+ + 2 H_2O \tag{1}$$

Miniaturization is critical for both health care and physiological studies. The screen-printing technology has been widely used for the large-scale fabrication of disposable biosensors. Besides the portable dimensions, screen-printed electrodes (SPEs) are low-cost and versatile in terms of formats and materials [16].

In this study, the working electrodes were modified with a layer of a carbon based conductive ink previously diluted in propanone or methyl ethyl ketone (MEK) and mixed in different proportions with ccNiR. The enzyme activity after immobilization in this harsh environment (solvents exposure and heat dry) was evaluated by cyclic voltammetry and has proved to be highly satisfactory. The electrode preparation was further optimized and applied on thick-film strip electrodes which were fabricated beforehand by printing a similar conductive carbon paste on plastic supports.

2   Materials and Methods

2.1   Reagents

Acetone (propanone; 99%; b.p. 56°C) and propanone (methylethylketone, MEK; 99%, b.p. 79°C) were from Pronalab. The remaining chemicals were analytical grade and were used without further purification. Solutions were prepared with deionized (DI) water (18 MΩcm) from a Millipore MilliQ purification system.

The graphite conductive ink was obtained from Acheson. Alumina slurries (0.05 and 1.0 µm) were from Buehler.

ccNiR was purified from *Desulfovibrio desulfuricans* ATCC 27774 cells grown in nitrate, as previously described by Almeida and co-workers [9].

2.2   Electrochemical Measurements

For the optimization studies, a conventional three-electrode electrochemical cell was used, with an Ag/AgCl reference electrode, a Pt counter electrode (both from

Radiometer) and a home-made working electrode made of pyrolytic graphite disks (4 mm diameter) and modified with the enzyme/ink layer.

The characterization of the optimized electrode was performed after replacing the previous system by carbon paste screen-printed electrodes (CPSPEs) with a three electrode configuration (Fig. 1), including an Ag/AgCl pseudo-reference, a graphite paste counter electrode and a graphite paste working electrode (3.1 mm diameter). The CPSPEs were fabricated at CIDETEC facilities, as described by Ochoteco and co-workers [17].

Fig. 1. A screen-printed three-electrode system. (1) working electrode; (2) reference electrode; (3) counter-electrode.

The one-compartment electrochemical cell containing 0.1 M KCl in 0.05 M Tris-HCl buffer, pH 7.6 as supporting electrolyte, was thoroughly purged with Argon before each experiment. Measurements were performed with a potentiostat Autolab PSTAT 12 (Eco-Chemie) monitored by the control and data acquisition software GPES 4.9. The cyclic voltammograms (CV) were plotted at room temperature ($22 \pm 2°C$), with a scan rate of 20 mV/s, in the potential window [0.0; -0.8] V (*vs* reference system). To evaluate the biosensors response to the analyte (0.001 - 7 mM), the cell was successively spiked with standard solutions of nitrite. After each addition, the electrochemical cell was deoxygenated and the CV was registered. The catalytic currents (I_{cat}) were measured at the inversion potential (-0.8V); all values were subtracted from the non-catalytic current recorded in the absence of nitrite (I_c). Each experiment was replicated at least two times. For amperometric measurements, the working potential was settled at -0.5 or -0.7 V, with a speed rotation (ω) of 0 or 1000 rpm.

2.3   Bioelectrodes Preparation

Prior to coating, the pyrolytic graphite electrodes (PGEs) were polished with alumina slurry in cloth pads. Then, the electrodes were thoroughly washed with DI water and ethanol and ultrasonicated in water for 5 min. The electrodes surface was further washed with DI water and dried with compressed air.

CPSPEs were used as produced, with no pre-activation. The conductive carbon inks were previously diluted with an organic solvent (acetone or MEK) in a 1:1 ratio and homogenised with the help of an ultrasound bath. The ink suspensions were then mixed with ccNiR in different proportions (4:1, 2:1, 1:1 and 1:2 ink/enzyme). Finally, a 5 μL drop was placed on the surface of the working electrodes which were cured for 20 min. inside an oven set at 40°C or 60°C. Control experiments were carried out with no curing treatment and/or no carbon ink; in such cases, the ccNiR layer was dried at room temperature.

3   Results

3.1   Effect of the Carbon Ink on the Bioelectrode Response

Preliminary experiments were carried out in order to check if the chemicals (organic solvents) and the thermal treatment (curing) required for printing the working electrode component in CPSPEs were compatible with nitrite reductase activity. In this regard, three different PG electrodes were modified with i) ccNiR only, ii) ccNiR mixed with carbon ink diluted in acetone and iii) the same as (ii) but with an extra curing step at 60°C.

Non-catalytic Response. Regardless of the composition of the electrode coating, in the absence of the enzyme substrate the CVs displayed a small and broad cathodic wave at *ca.* –0.380 V *vs* Ag/AgCl, as illustrated in Fig. 2A.

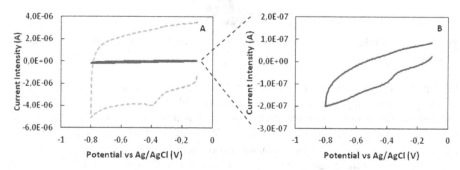

Fig. 2. CVs of ccNiR coated PG electrodes, traced at 20 mV/s, in 0.1 M KCl, 0.05 M Tris-HCl, pH 7.5. (A) (- - -) enzyme mixed with carbon ink (diluted in acetone); (——) enzyme casted directly onto a bare electrode, as zoomed in (B). The modified electrodes were dried at room temperature (≈ 22°C).

This non-catalytic signal can be assigned to the redox co-factors of ccNiR, as previously seen in bare PGEs and carbon nanotubes modified electrodes [14,18]. Due to the consequent enlargement of the electroactive area, in the presence of the carbon ink both Faradaic and capacitive currents increased over 20 times when compared to ones obtained with bare electrodes (Fig. 2B).

Response to Nitrite - Temperature and Solvent Effects. In the presence of the enzyme substrate, the CVs exhibited a sigmoid shape (Fig. 3) reflecting the direct charge transfer between the electrode surface and ccNiR and the subsequent catalytic reduction of nitrite into ammonium.

In general, the plots catalytic current (I_{cat}) *vs* nitrite concentration (Fig. 4) could be fitted to a hyperbolic equation denoting a Michaelis-Menten profile (eq. 2) i.e., at high nitrite concentrations the current response deviated from linearity and approached a constant value, reflecting the maximum rate of the enzymatic reaction under nitrite saturating conditions (I_{max}). Interestingly, this value has increased about three times in the presence of the conductive ink, which should be related with the surface area increase (Fig. 4).

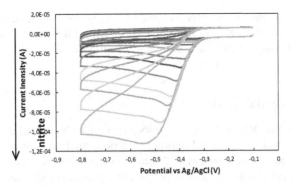

Fig. 3. CVs of a ccNiR/carbon ink/acetone coated PG electrode in the presence of increasing nitrite concentrations (0 - 11 mM). Electrolyte: 0.1 M KCl in 0.05 M Tris-HCl buffer, pH 7.5; scan rate, 20 mV/s.

$$I_{cat} = I_{max} \times [NO_2^-] / (K_m^{app} + [NO_2^-]) \qquad (2)$$

Data fitting to the Michaelis-Menten kinetic model using the software GraphPad Prism 4.0 indicated extremely high apparent Michaelis-Menten constants (K_m^{app}); typical values are close to 0.9 ± 0.1 mM, which is about 130 times higher than the K_m previously determined by protein film voltammetry [19]. This means that the diffusion of nitrite within the carbon ink is a very slow process, thereby controlling the biosensor response, as further corroborated by the wide linear range of the calibration curves (see section 3.3).

Overall, results suggest that the carbon ink/acetone composite had no critical effect on the catalytic activity.

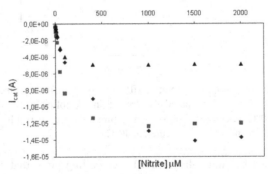

Fig. 4. Electrocatalytic response to nitrite of ccNiR (3.3 µg) modified PG electrodes: (▲) with no carbon conductive ink and thermal treatment (sensitivity: 3.2×10^{-7} A.µM^{-1}.cm^{-2}); (■) pre-mixed with the carbon conductive ink diluted in acetone (sensitivity: 7.3×10^{-7} A.µM^{-1}.cm^{-2}); (◆) pre-mixed with the carbon conductive ink diluted in acetone and cured at 60°C (sensitivity: 3.3×10^{-7} A.µM^{-1}.cm^{-2})

The analytical performance of each bioelectrode was then characterized taking into account the catalytic efficiency (I_{max}/I_c), sensitivity of detection (slope of the linear range), correlation coefficient (r^2) and quantification range. When comparing the response profiles obtained with or without electrode curing (both in the presence of conductive ink), one can see that the thermal treatment does not have a strong influence on I_{max} (Fig. 4). On the other hand, the sensitivity of the sensor, as given by the slope of the linear range of the plot, decreased about 55%. This indicates that partial protein denaturation has occurred.

3.2   Electrode Optimization

Enzyme/Carbon Ink Ratio. Different proportions of enzyme and carbon ink suspended in acetone were early tested in order to choose the best composition. The one using the highest amount of protein (1:2 ratio, corresponding to 3.3 µg ccNiR) displayed the best results (not shown) without relevant loss of activity, and was selected for further studies.

Organic Solvent and Curing Temperature. Normally, inks for screen printing contain organic solvents that are later evaporated by heating. If other ingredients like ccNiR need to be included, it is highly recommended to lower the viscosity of the paint in order to facilitate the mixing process. For this reason, prior to enzyme incorporation, the carbon ink used in this work was diluted with two different organic solvents - MEK or acetone. It is worth noting that acetone is less commonly used for

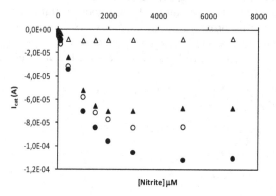

Fig. 5. Electrocatalytic response to nitrite of ccNiR (3.3 µg) modified PG electrodes, pre-mixed with the carbon conductive ink, prepared and treated in the following ways: (△) diluted in MEK and cured at 40°C; (▲) diluted in MEK and cured at 60°C; (○) diluted in acetone and cured at 60°C. (●) diluted in acetone and cured at 40°C.

inks dilution than MEK, although it has a lower boiling point that could permit the use of lower curing temperatures. Actually, the response to nitrite was much higher when this solvent was used instead of MEK (Fig. 5) whereas the linear range was wider. Therefore, acetone has proved to be less harmful to the protein.

In order to evaporate residual organic solvents, most CPSPEs have to be dried thermally. Although a temperature of 60°C is normally selected for the curing process of those used in the present work, due to the presence of the biocatalyst, we have also tested the lowest permitted heating temperature, i.e., 40°C. Interestingly, the differences on nitrite reducing activity were generally small, except when MEK was used for ink dilution, which generated much lower catalytic currents. Most likely, this solvent did not evaporate completely at 40°C, enhancing its detrimental effect on the enzyme activity.

In accordance to the results obtained in this combined study, we have selected acetone for ink solubilization and a curing temperature of 40°C.

3.3   Analytical Parameters

The analytical performance of the bioelectrodes was further evaluated by chronoamperometry, poising the working electrodes at -0.5 or 0.7 V either (Fig. 6) or not (not shown) with self-rotation. Results were compared with the ones obtained by cyclic voltammetry (Table 1).

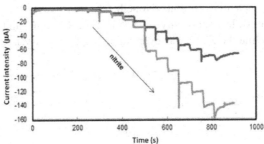

Fig. 6. Amperometric measurements of a ccNiR/carbon ink (acetone, 60°C)/PG electrode poised at different working potentials: (—) -0.5 V; (—) -0.7 V (*vs* Ag/AgCl), in the presence of increasing concentrations of nitrite. ω = 1000 rpm

Table 1. Analytical parameters of optimal ccNiR/carbon paste/PG electrodes (3.3 μg enzyme in an acetone:ink mix; curing temperature, 40°C), as obtained by chronoamperometry and cyclic voltammetry assays

Technique	Potential (V)	Electrode Rotation	Sensitivity (A.μM^{-1}.cm^{-2})	r^2	Linear range (μM)	Catalytic efficiency
	-0.5	No	$5x10^{-7} \pm 3x10^{-7}$	0.995	0.666 - 132	27 ± 26
Chrono-amperometry	-0.7	No	$5x10^{-7} \pm 3x10^{-7}$	0.995	0.167 - 132	48 ± 12
	-0.5	Yes	$2.6x10^{-6} \pm 6x10^{-7}$	0.991	0.666 - 13.2	54 ± 24
	-0.7	Yes	$2.0x10^{-6} \pm 3x10^{-7}$	0.996	0.666 - 264	56 ± 23
CV	-0.8	No	$2.3x10^{-7} \pm 2x10^{-8}$	0.995	$6.64 - 1.97x10^{3}$	26 ± 11

The amperometric response upon nitrite addition was always prompt, indicative of a fast heterogeneous electron transfer. Although not being a practical option in real situations, the electrode rotation delivers a much better performance (Table 1). In fact, because there are no mass transport limitations, both the sensitivity and catalytic efficiency of the biosensor were higher in these conditions, no matter the operating voltage. On the other hand, the linear range was much wider (almost three orders of magnitude) if measurements were made through cyclic voltammetry. In these conditions, the current response is diffusion controlled, which slows down the overall process, thereby decreasing the sensitivity and broadening the quantification range. For this reason, subsequent analyses were preferably made by cyclic voltammetry. Nevertheless, if very low detection limits are required (low nano range), chronoamperometry measurements should be carried out at the lowest potential (-0.7 V *vs* Ag/AgCl), using non-rotating electrodes.

3.4   Application of Carbon Paste Screen Printed Electrodes (CPSPEs)

Following the optimization of the ccNiR containing conductive paints, the enzyme/carbon inks were deposited on CPSPEs. The CVs displayed higher background currents (Fig. 7A), which is most likely related to the roughness of the SPEs surfaces, generating higher capacitive currents. Nonetheless, the analytical parameters (Fig. 7B) were quite similar, with a sensitivity of 5.3×10^{-7} $A.\mu M^{-1}.cm^{-2}$ and a linear range of 0.001 - 1.5 mM, while the K_M^{app} was even higher (2.2 ± 0.4 mM).

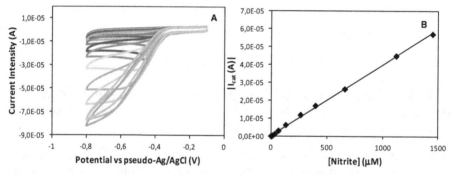

Fig. 7. (A) CVs of a CPSPE coated with a ccNiR/carbon ink/acetone composite, in the presence of increasing nitrite concentrations (0 - 6 mM) and (B) corresponding calibration curve. Electrolyte: 0.1 M KCl in 0.05 M Tris-HCl buffer, pH 7.5; scan rate, 20 mV/s.

However, despite the successful incorporation of ccNiR within the thermally cured conductive ink and the good sensitivity of detection displayed by the enzyme coated CPSPEs, the low limit of the linear range is higher than the ones provided by the previous biosensor configurations [10-15] and should be further improved if the main target is plasma and blood analysis.

3.5   Interferences

The chemistries of nitrate and nitrite, which are present in mammalian physiological systems either from dietary provision or endogenous formation, are nearly indissociable [2,3]. Aiming at evaluating the potential interfering effect of nitrate on the analytical assay, the biosensor response to nitrite (666 µM)[1] was compared with the one recorded in the presence of an equal amount of nitrate; though, the

---

[1] This concentration fits in the first half of the calibration curve.

interference level (+3%) was not significant. In contrast, the interference caused by hydroxylamine, an alternative ccNiR substrate, was about 10%; although much lower than the values obtained with some other electrode configurations [10-15], it can not be neglected. Though, the risks of having both species in the same physiological sample are not great.

4   Conclusions

This R&D project was designed to address a critical and growing need for real-time monitoring of nitrites and to provide better analytical tools for its clinical diagnosis. Our previous results have demonstrated the feasibility of using ccNiR in the construction of bioelectrodes for a selective nitrite analysis [10-15]. Herein we have shown the biocompatibility of the painting materials and the electrode curing procedure with ccNiR activity. The success of this work opens up the possibility of including the enzyme directly in the printing paste used for the fabrication of thick-film electrodes, facilitating the mass production of easy-to-use nitrite biosensors. If coupled to a portable potentiostat, these enzyme containing disposable electrode strips will turn a long and elaborated laboratory protocol into a simple task, quickly executed onsite.

Acknowledgements. The authors thank the financial support from Associated Laboratory REQUIMTE.

References

1. Bryan, N.S., Fernandez, B.O., Bauer, S.M., Garcia-Saura, M.F., Milsom, A.B., Rassaf, T., Maloney, R.E., Bharti, A., Rodriguez, J., Feelisch, M.: Nitrite is a Signaling Molecule and Regulator of Gene Expression in Mammalian Tissues. Nat. Chem. Biol. 1, 290–297 (2005)
2. Hord, N.G., Tang, Y., Bryan, N.S.: Food sources of nitrates and nitrites: The Physiologic Context for Potential Health Benefits. Am. J. Clin. Nutr. 90, 1–10 (2009)
3. Lundberg, J.O., Gladwin, M.T., Ahluwalia, A., Benjamin, N., Bryan, N.S., Butler, A., Cabrales, P., Fago, A., Feelisch, M., Ford, P.C., Freeman, B.A., Frenneaux, M., Friedman, J., Kelm, M., Kevil, C.G., Kim-Shapiro, D.B., Kozlov, A.V., Lancaster Jr., J.R., Lefer, D.J., McColl, K., McCurry, K., Patel, R.P., Petersson, J., Rassaf, T., Reutov, V.P., Richter-Addo, G.B., Schechter, A., Shiva, S., Tsuchiya, K., van Faassen, E.E., Webb, A.J., Zuckerbraun, B.S., Zweier, J.L., Weitzberg, E.: Nitrate and Nitrite in Biology, Nutrition and Therapeutics. Nature 12, 865–869 (2009)
4. Almeida, M.G., Serra, A.S., Silveira, C., Moura, J.J.G.: Nitrite Biosensing Via Selective Enzymes – a Long But Promising Route. Sensors 10(12), 11530–11555 (2010), doi:10.3390/s101211530
5. Ellis, G., Adatia, I., Yazdanpanah, M., Makela, S.K.: Nitrite and Nitrate Analyses: a Clinical Biochemistry Perspective. Clin. Biochem. 31(1998), 195–220 (1998)
6. Moorcroft, M.J., Davis, J., Compton, R.G.: Detection and determination of nitrate and nitrite: a review. Talanta 54(5), 785–803 (2001)
7. Dutt, J., Davis, J.: Current strategies in nitrite detection and their application to field analysis. J. Environ. Monit. 4, 465–471 (2002)

8. Dejam, A., Hunter, C.J., Schechter, A.N., Gladwin, M.T.: Emerging role of nitrite in human biology. Blood Cells Mol. Dis. 32, 423–429 (2004)

9. Almeida, M.G., Macieira, S., Gonçalves, L.L., Huber, R., Cunha, C.A., Romão, M.J., Costa, C., Lampreia, J., Moura, J.J.G., Moura, I.: The Isolation and Characterization of Cytochrome c Nitrite Reductase Subunits (NrfA and NrfH) from Desulfovibrio desulfuricans ATCC 27774. Re-evaluation of the spectroscopic data and redox properties. Eur. J. Biochem. 270, 3904–3915 (2003)

10. da Silva, S., Cosnier, S., Almeida, M.G., Moura, J.J.G.: An Efficient Poly(pyrrole)-Nitrite Reductase Biosensor for the Mediated Detection of Nitrite. Electrochem. Comm. 6, 404–408 (2004)

11. Almeida, M.G., Silveira, C.M., Moura, J.J.G.: Biosensing Nitrite Using the System Nitrite Redutase/Nafion/Methyl Viologen - A Voltammetric Study. Biosens. Bioelectron. 22, 2485–2492 (2007)

12. Chen, H., Mousty, C., Cosnier, S., Silveira, C., Moura, J.J.G., Almeida, M.G.: Highly Sensitive ccNiR Biosensor Electrical-Wired by [ZnCrAQS$_2$] for Nitrite Determination. Electrochem. Comm. 9, 2240–2245 (2007)

13. Silveira, C.M., Gomes, S.P., Araújo, A.N., Couto, C.M.C.M., Montenegro, M.C.B.S.M., Silva, R., Viana, A.S., Todorovic, S., Moura, J.J.G., Almeida, M.G.: An Efficient Mediatorless Biosensor for Nitrite Determination. Biosens. Bioelectron. 25, 2026–2032 (2010)

14. Silveira, C., Baur, J., Cosnier, S., Moura, J.J.G., Holzinger, M., Cosnier, S., Almeida, M.G.: Enhanced direct electron transfer of a multihemic nitrite reductase on SWCNT modified electrodes. Electroanalysis 22, 2973–2978 (2010)

15. Zhang, Z., Xia, S., Leonard, D., Jaffrezic-Renault, N., Zhang, J., Bessueille, F., Goepfert, Y.S., Wang, X., Chen, L., Zhu, Z., Zhao, J., Almeida, M.G., Silveira, C.M.: A novel nitrite biosensor based on conductometric electrode modified with cytochrome c nitrite reductase composite membrane. Biosens. Bioelectron. 24, 1574–1579 (2009)

16. Jubete, E., Loaiza, O.A., Ochoteco, E., Pomposo, J.A., Grande, H., Rodrıguez, J.: Nanotechnology: A Tool for Improved Performance on Electrochemical Screen-Printed (Bio)Sensors. J. Sensors, Article ID 842575, 13 pages (2009), doi:10.1155/2009/842575

17. Ochoteco, E., Jubete, E., Pomposo, J.A., Grande, H., et al.: Patent Application Number E200930539 (2009)

18. Silveira, C.M., Besson, S., Moura, I., Moura, J.J.G., Almeida, M.G.: Measuring the Cytochrome c Nitrite Reductase Activity – Practical considerations on the enzyme assays. Bioinorg. Chem. Applic. 2010, Article ID 634597, 8 pages (2010), doi:10.1155/2010/634597

19. Almeida, M.G., Silveira, C.M., Guigliarelli, B., Bertrand, P., Moura, J.J.G., Moura, I., Leger, C.: A Needle in a Haystack: the Active Site of the Membrane-bound Complex Cytochrome c Nitrite Reductase. FEBS Lett. 581, 284 (2007)

A Real-Time and Portable Bionic Eye Simulator

Horace Josh, Benedict Yong, and Lindsay Kleeman

Department of Electrical and Computer Systems Engineering and Monash Vision Group,
Monash University, Wellington Road, Clayton, Australia
{horace.josh,benedict.yong,lindsay.kleeman}@monash.edu.au

Abstract. The Monash Vision Group is developing a bionic eye based on an implantable cortical visual prosthesis. The visual prosthesis aims to restore vision to blind people by electrical stimulation of the visual cortex of the brain. Due to the expected naivety of early prostheses, there is need for the development of innovative pre-processing of scene information in order to provide the most intuitive representation to user. However, in order to explore solutions to this need, prior to availability of functional implants, a simulator system is required. In this paper, we present a portable, real-time simulator and psychophysical evaluation platform that we have developed called the 'HatPack'. It makes use of current neurophysiological models of visuotopy and overcomes limitations of existing systems. Using the HatPack, which is compiled into a neat, wearable package, we have conducted preliminary psychophysics testing, which has shown the significance of available greyscale intensity levels and frame rates. A learning effect associated with repeated trials was also made evident.

Keywords: Visual prosthesis, Portable simulator, Visual cortex, Visuotopic mapping, Phosphene, Bionic vision.

1 Introduction

A study conducted in 1968 showed that electrical stimulation of the visual cortex of a human brain resulted in the elicitation of bright spots of light, called 'phosphenes', in the visual field of the subject [4]. Supporting results were also found in [1,8,9]. Further studies [14,24] have shown that it is also possible to generate phosphenes via electrical stimulation of the retina and optic nerve. These early studies provided a basis for widespread research into the development of functional visual prostheses.

A visual prosthesis, also often referred to as a 'bionic eye', is an implantable biomedical device that aims to restore vision to the blind. The core component of these devices is an array of electrodes, driven by specialised electronics. The electrodes inject electrical current into a particular section of the patient's visual pathway in order to generate an 'image' in the visual field.

The term visual pathway refers to the path that signals take from the retina in the eye where they are generated to the primary visual cortex at the back of the brain. Light that is incident on photoreceptors in the retina, a layer of cells at the back of the eye, results in the generation of signals. These signals are passed through the optic nerve and Lateral Geniculate Nucleus (LGN) before arriving at the primary visual cortex (V1), which

J. Gabriel et al. (Eds.): BIOSTEC 2012, CCIS 357, pp. 51–67, 2013.

is at the back of the brain. From V1, signals diverge to subsequent levels of visual cortex where higher level processing takes place. In a blind individual, parts of the visual pathway may not function. Therefore, visual signals do not reach the visual cortex. A successful prosthesis would bypass these inoperative sections in order to deliver signals to V1.

The Australian Research Council funded a new collaborative research initiative in 2009 to develop a functional visual prosthesis. One of the two proposals accepted for this initiative was by a Monash University led team of researchers, now known as the Monash Vision Group (MVG) [16]. Established in 2010, the MVG aims to develop a visual prosthesis (Monash Bionic Eye) centred on a cortical implant, making use of approximately 600 electrodes.

As research grows in this new area of bionics, there is a great need for simulation or visualisation of the possible results of such an implant. Bionic eye simulators serve as good platforms for researchers to investigate the effectiveness of implemented algorithms, tune parameters, and realise the importance of certain parameters prior to actual clinical trials. The simulators would be used most in psychophysical trials - trials involving normally sighted individuals attempting to complete tasks with the limited vision provided by a simulator. However, the simulators would also be of use to the general public for educational purposes and to handle the expectations of families and friends of potential patients. Input to the system is in the form of an image or image stream. This image data goes through processing that transforms it into a representation that attempts to mimic the elicitation of phosphenes through electrode stimulation. The processed image data is then stored and/or displayed on a screen for viewing by the user.

Many visual prosthesis simulators have already been developed and some of the more recent work is found in [6,12,22,23,26]. Nevertheless, there are some significant limitations that arise in their implementations. The majority of these simulators perform their image processing on a computer using image processing libraries and so are often limited to use within an area close to a stationary computer. Depending on the complexity of processing and the available processing power of the equipment in use, these systems may sometimes suffer from latency and frame rate issues. In the case of simulators for cortical visual prostheses, visuotopic mapping the mapping of electrode placement on the visual cortex to elicitation of phosphenes in the visual field, has often been overlooked or used simplified models.

Our system aims to address the shortcomings of currently implemented systems. In comparison to other cortical FPGA based systems [12,22], the HatPack is very mobile and has been used to do untethered preliminary psychophysics testing. It is based on a Field Programmable Gate Array (FPGA) system implementation. FPGAs are microchips that offer extremely dense amounts of electronically reconfigurable logic gates. FPGA systems offer the advantages of low latency, highly parallel implementation and the ability to integrate with large numbers of external devices through the high availability of peripheral interface pins. Figure 1 shows the main components of our simulator system. A CMOS camera captures a stream of image data, which is then processed on an FPGA development board and finally displayed on a head-mounted display and optionally on an external monitor as well. An infra-red remote control interface is used to

Fig. 1. Main components of our simulator system: A) CMOS Camera B) FPGA Development Board C) IR Remote D) Head-Mounted Display E) External Monitor

enable/disable the various functions. A more detailed description of the system components is provided in Section 2.

2 System Setup

As shown in Figure 1, the HatPack is comprised of the following main components: a camera for acquiring images, an FPGA development board for performing all image processing functions and visuotopic mapping, a head-mounted display as well as optional external monitor for display of the resulting image stream, and finally an infra-red remote control for toggling of functions.

The camera that we have chosen to use is a low cost CMOS camera (Sparkfun Electronics CM-26N/P) which has an analogue signal output. It captures images at a resolution of 640×480 pixels, at a frame rate of 59.94Hz and has a viewing angle of $70°$. Reasons for choosing this particular camera include low cost, small physical size, switchable PAL/NTSC output, and the simplicity of a three wire power/signal connection which also allows for longer cable lengths.

At the centre of our system, we have a Terasic DE2-115 FPGA development board, which is based on an Altera Cyclone IV EP4CE115F29C7 FPGA chip. We chose this development board for its low cost, lower power consumption, high logic element and on-chip memory count, wide range of available peripheral devices and I/O pins, and our familiarity with its design and operation.

An infra-red remote, that comes standard with the DE2-115, was utilised for capturing user input. It provides a simple and easy way of toggling and controlling all implemented functions.

indent For display of the final output, we have chosen a head-mounted display (HMD) unit (Vuzix iWear VR920), sometimes referred to as virtual reality goggles. This HMD offers a 640×480 pixel display resolution with a viewing angle of $32°$. The VR920 was chosen for its low cost, compatible resolution, lightweight design, and its ability to take an analogue VGA signal as its input. Since our system outputs video via a VGA port, we were able to use a simple passive splitter cable to provide dual output (HMD as well as an external monitor).

For the HatPack to be mobile, all hardware needed to be integrated into a neat, wearable package. We achieved the result shown in Figure 2. The majority of components

Fig. 2. Integrated system

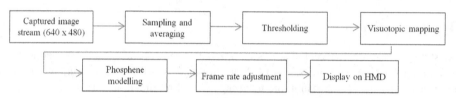

Fig. 3. Flowchart of main functions of the system

are fastened inside a hard plastic laptop casing, which is then placed in a neoprene laptop bag with cables running to the camera and HMD that the user is wearing. A 12V
rechargeable lithium-ion battery pack is used to power the system.

3   System Implementation

The flowchart shown in Figure 3 outlines the implementation of the main functions of
the HatPack. A high resolution image stream (640 × 480 pixels) is captured by the
CMOS camera, which is delivered to the DE2-115 development board via a standard
NTSC analogue connection. After decoding of the NTSC signal is complete, the pixels
are sampled and averaged. The sampled data is thresholded in order to simulate possible limitations of electrode stimulation. A pre-generated visuotopic mapping lookup
table is then used to determine the placement of the phosphenes on the output display.
A discrete Gaussian falloff profile is used to simulate the physiological phenomena of
a phosphene dot in the visual field. Before output on the screen, the frame rate of the
system can be set in real-time in order to simulate varying stimulation frequencies of
electrodes. A more detailed explanation of these main system features is given in Subsections 3.1, 3.2, 3.3, 3.4, 3.5, and 3.6.

Furthermore, features such as edge detection, histogram assisted threshold selection,
and dead electrode simulation, have been implemented in order to allow for evaluation
of the effects of such image processing techniques on the perception of the provided
low resolution data (Subsection 3.7).

All processing performed on the image stream from the camera is implemented using Verilog hardware description language. Unlike conventional code that is written for execution on a processor that runs at a specific clock speed, Verilog describes the way logic gates are to be arranged and connected and so is compiled into a synthesisable logic solution that can be either synchronous (operate with reference to a clock), asynchronous (without reference to a clock) or a mixture of the two. A Verilog solution was chosen due to the ability to create functions that can run in parallel, resulting in a low latency real-time system.

3.1 Visuotopic Mapping

Early physiological research [21,25] proved that 'points' in the visual field correspond to specific locations on the visual cortex, inferring a 'map' or transfer function between visual field points and the visual cortex. Furthermore, that map is mostly continuous in that neighbouring points in the visual field correspond with neighbouring points on the visual cortex. The map or transfer function which describes the translation of points between the visual cortex to its corresponding points on the visual field is known as the visuotopic map.

Due to the physiological non-linear properties of the visual cortex, the visuotopic map is also non-linear and 'distorted'. In humans, the phenomenon known as cortical magnification describes how a small region at the centre of the visual field, known as the fovea, corresponds with a much larger area of the visual cortex [11,13]. Early work by Schwartz [21] indicated an approximation to the mapping by a 'log-polar' representation, where linear points on the visual cortex correspond to eccentrically logarithmic and angularly linear points in the visual field. The foveal region is represented this way as a dense packing of points in the centre of the visual field which corresponds to a disproportionately larger region on the visual cortex. Also important to note is that the visual cortex is spread over both halves of the brain with the left visual cortex corresponding with the right visual hemifield and vice versa, due to cross-over of the optic nerves [3].

Mathematical models that came from this include the Monopole model (defined from the 'log-polar' observations) [17,19,21], the Wedge-Dipole model (adds a second parameter to Monopole model to account for curvature in the periphery region of the visual cortex) [2,17] and more recently the Double-Sech model (adds a shear function to the Wedge-Dipole model to account for changing local isotrophy as well as increasing accuracy of mapping at higher levels of visual cortex V2, V3) [18,19].

As the implant is anticipated to consist of a linear array of electrodes, the resulting phosphene pattern would not be linear but rather follow this log-polar mapping. It would be useful and more accurate to model the output visualisation based off a mathematical model of the visuotopic mapping. Since the implant is expected to be placed in the primary visual cortex V1 and closer to the foveal side of the visual cortex, the Monopole model was chosen to model the output visualisation as it was mathematically simpler and still provides reasonable accuracy.

The Monopole equation (1) describes the left visual cortex 'w' as a complex function of the right visual hemifield 'Z_w'. '$\mathbb{C}$' is the set of complex numbers, and 'k' is a dilation factor constant.

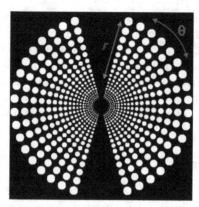

Fig. 4. Resultant visual field of implemented visuotopic map

$$w = k \, log \, (Z_w + a) \in \mathbb{C} \tag{1}$$

Visual field Z_w can be represented as a complex exponential where r represents the eccentricity and θ represents polar angle.

$$Z_w = re^{i\theta} \in \mathbb{C} \tag{2}$$

Rearranging the Monopole equation describes visual field Z_w as a function of visual cortex w.

$$Z_w = e^{\left(\frac{w}{k}\right)} - a \in \mathbb{C} \tag{3}$$

The electrode array of the implant was assumed to be a linear array placed on the visual cortex closer to the foveal region. The visuotopic map was created using MATLAB and ported over to the FPGA for use as a large lookup table. Approximate values were used for the Monopole equation parameters, which are reasonably consistent with the various values used in the literature: $k=15$, $a=0.7$ [12,17,18]. The exact dimensions and intended locations of the implant are still not known, the eccentricity and polar angle were limited to an 18×18 linear array on the visual cortex that cover the following values on: $r=[10,40]$, $\theta=[-0.8\left(\frac{\pi}{2}\right), 0.8\left(\frac{\pi}{2}\right)]$. This only represents the left visual cortex, corresponding with the right visual hemifield. The 18×18 array was duplicated for the right visual cortex, creating another array on the left visual hemifield. This produces a total electrode count of 648. These assumptions were taken to make better use of the limited screen resolution of the head-mounted display while remaining realistic to the 'log-polar' mapping of the visual cortex. However, new maps can be simply regenerated on MATLAB to accommodate any changes to this and implemented into our system. The resultant visual field of our implemented map is shown in Figure 4.

3.2 Averaging Sampler

Figure 5 outlines our averaging sampler implementation. After NTSC decoding, the image stream from the camera is made available one pixel at a time in a sequential fashion. As each pixel arrives at the sampling section of the system, its X & Y pixel

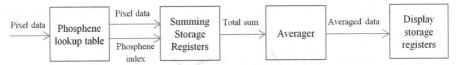

Fig. 5. Averaging sampler implementation

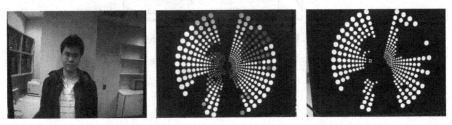

Fig. 6. Thresholding: full resolution image (left), 4-level image (middle), binary image (right)

count values are compared against the mapping lookup table. This lookup table stores the corresponding phosphene index number for each pixel within the central 480 × 480 window of the full camera view. Pixels not belonging to a phosphene are assigned number zero. Once the phosphene index number is determined, the pixel is sampled by adding to a storage register that corresponds to that particular phosphene index number. This process repeats until all pixels have been sampled. Finally, an average is performed on all of the storage registers according to the number of pixels that are within each phosphene, and the results are stored in a separate set of storage registers.

3.3 Thresholding

Various studies [4,8,20] have shown that the modulation of phosphene brightness is possible using a number of different techniques. However, there is some ambiguity in the possible number distinguishable brightness levels.

Our system takes an optimistic approach at simulation of this property, having the option to display at 2, 4 or 8 levels of intensity or greyscale. Since our system uses 10-bit storage registers for pixels, the full greyscale intensity range is 0 to 1023. This range is divided evenly in order to create bands of intensity for 2, 4 and 8 level modes. Results of 2 and 4-level thresholding are shown in Figure 6. It is often difficult to perceive the results of the system in a static image form, therefore we encourage you to view the videos we have listed in the Appendix.

To avoid high frequency oscillation between intensity bands, a hysteresis feature was included. Two threshold values are used to define changes between intensity bands, instead of one value. When a phosphene's intensity is between the two thresholds, no change occurs. Figure 7 shows how hysteresis reduces the oscillation problem.

3.4 Phosphene Modelling

Stimulation of each electrode on the implant will produce a phenomenon in the patient's visual field known as a phosphene, whose appearance is somewhat similar to a bright

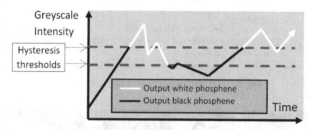

Fig. 7. Binary thresholding with hysteresis

Fig. 8. Phosphene modelling: without Gaussian function (left), with Gaussian function (right)

spot of light [4]. Rather than simply using square pixels that perfectly line up with each other, we attempted to model the output visualisation based on what phosphenes would approximately look like.

In the literature, one common approach is to model the phosphene using a 2D Gaussian mask [7]. The 2D Gaussian function is based on the standard distribution curve, except in two dimensions instead of one. This creates the appearance of a round 'spot' where the centre of the spot has the highest intensity value with the intensity values decreasing radially towards the outside edge of the spot, following the standard distribution curve. A comparison between a phosphene with and without the Gaussian function applied is shown in Figure 8.

3.5   Frame Rate Reduction

The ability of a person to detect motion is very important when it comes to mobility exercises in low resolution vision. A key factor that would limit one's ability to detect motion in the immediate environment is the lack of temporal resolution. It is expected that the temporal resolution of electrode stimulation achievable by the Monash Bionic Eye may be in the range of 5-15 frames per second. In order to simulate this temporal resolution and investigate the possible implications it may have on a patient's ability to move around, we have implemented a frame rate reduction function. The output frame rate of our system can be changed in real-time. Our system has 8 different discrete frame rates available for selection (1, 2, 4, 8, 10, 15, 30 and 60 frames per second). Variable frame rate is achieved by holding the stored frame output data for the specific period of the chosen frame rate.

3.6   Threshold Ramping

It became apparent, through general use of the system, that variations of lighting in the environment resulted in poor scene conversion and representation. This was due to the use of fixed intensity values for greyscale thresholding.

An initial solution to this problem was to allow the user to adjust thresholding levels manually via the remote control. This however, proved rather tedious. Furthermore, it was observed that users seemed to move back and forth between a range of thresholds in order to obtain more structural information about a scene. Based on this observation, an automatic threshold ramping function has been developed.

The function makes use of a histogram in order set upper and lower threshold bounds to move between. This is done in real-time and is dependent on current frame pixel data. After the bounds have been established, the threshold values are then incremented or decremented at set time intervals and following either a saw-tooth or triangular ramping pattern. The preferred pattern can be chosen by the user and the time interval can be adjusted so that a faster or slower ramp can be achieved.

Although effective in certain situations, this function was also found to induce nausea occasionally and so has been excluded from evaluation in the preliminary testing we present.

3.7 Extra Functions

Additional functions have been implemented in our system, such as edge detection, histogram assisted threshold selection and dead electrode simulation. These features have not been evaluated in the preliminary testing we present in this paper; however, would be of importance for future psychophysical research we intend to carry out. Figure 9 demonstrates edge detection and dead electrode simulation.

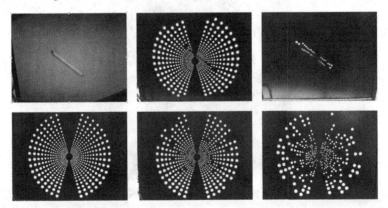

Fig. 9. Edge detection: full resolution (top left), binary thresholding (top middle), edge detection (top right). Dead electrode simulation: 0% (bottom left), 10% (bottom middle), 50% (bottom right).

4 Experimental Setup

After the HatPack was constructed, three different psychophysical preliminary testing experiments were devised by the authors and conducted by a number of Monash Vision Group staff and post-graduate students as volunteers. These experiments were not formalised clinical trials, but rather preliminary trials to test the effectiveness of the system and to examine the effective difference between the modes and parameters set on the system on the end user. The three experiments were a mobility based obstacle avoidance walking maze test, a sit-down contrast discrimination hand-eye co-ordination chess-board placement test, and finally a seated rolling ball interception task that involved lowered frame rates.

4.1   Maze Test

In this test, there were 7 test subjects (6 male, 1 female). The maze test involved subjects walking through a course while avoiding obstacles. The obstacles were large cardboard boxes and office chairs with wheels. The placement of the obstacles was randomised within the maze area and 5 different configurations of obstacle layout were developed, one for each mode tested and kept consistent between subjects. Subjects were not allowed to see the obstacle layout before each test. The starting point was around the corner from the main rectangular maze area, and the end point was at a table at the far end wall of the maze. There is a small black box on the table and the test ends when the subject finds and picks up the box.

Fig. 10. Maze Test obstacle layout

For the test, both time to completion and number of collisions were recorded for all subjects. Subjects were allowed to touch the obstacles in the maze so only unintentional collisions were counted. The 5 modes tested were a control (full resolution, full colour), 4-level thresholding (full frame rate), binary thresholding (full frame rate) and reduced frame rate at 15Hz and 4Hz (both with 4-level thresholding). Subjects were given 2 minutes accommodation time just before the test for each mode where they could adjust to using the system around a cardboard box and two chairs placed away from the actual maze area. Subjects were also given a minimum of 5 minutes break in between each test.

4.2   Chessboard Test

In this test, there were 7 test subjects (6 male, 1 female) and were the same subjects as used in the previous Maze Test. The task required subjects to sit down at a table with a chessboard in front of them and 16 chess pieces (8 black, 8 white) placed in a random pile to the left of the chessboard. The objective was for the subjects to sort and place any black coloured pieces on any white square in the bottom half of the chessboard, and the white pieces on black squares in the top half of the chessboard.

Fig. 11. Chessboard Test finished example

For the test, both time to completion and number of mistakes were recorded for all subjects. For a piece to be considered as correctly placed, at least half of it had to be over the right square. Another aspect to this experiment was to test for learning effects that come from repeated usage of the system. As such, the non-control modes tested were repeated 3 times in this order (all at full frame rate): control (full resolution, full colour), binary thresholding, 4-level thresholding, binary, 4-level, binary, 4-level. Before the testing, subjects were asked to attempt the task without wearing the system in order to familiarise themselves with the task itself. The testing was conducted in a single session, with a minimum 1 minute break in between each test.

4.3   Ball Interception Test

Once again, the same 7 test subjects (6 male, 1 female) were used. The Ball Interception Test required subjects to sit down at a modified table tennis table. The table had a horizontally sliding 'paddle' mounted on the participant's end and at the opposite end 4 fixed half-pipe ramps that balls could be rolled down. The ramps were all the same length, had the same elevation angle, and had a 'fast' and 'slow' roll point for the balls. The ramps were partially covered by a cardboard screen so that the participant could not see which ramp the ball was rolling down until it reached the bottom. The objective was for the subjects to use the sliding paddle to block/intercept/stop the balls (similar to the video game 'Pong' or to 'Air Hockey' in an arcade).

The number of successfully intercepted balls was recorded for all subjects. To be considered as a successful interception, the ball was not allowed to hit the rail that the paddle was sliding along. In cases where there was ambiguity in the outcome (ie. balls getting stuck under paddle or between the paddle and side block), a note was made

Fig. 12. Ball Interception Test Setup

and that particular roll was done again at the end of the trial. For each mode, 60 rolls were done. A predefined randomised roll sequence was used that ensured each of the 4 ramps was used equally and also that there was an equal amount of 'fast' and 'slow' rolls. The aim of this test was to observe the effects of lowered frame rates, and so the modes tested were: a control (full resolution, full colour) and 4-level thresholding at 4Hz, 10Hz, and 60Hz. Subjects were given 5 minutes to familiarise themselves with the task before beginning the trials. The testing was conducted in a single session, with a minimum 1 minute break in between each mode tested.

5   Results and Discussion

5.1   Maze Test

The left graph in Figure 13 details the time to completion (in seconds) for each mode, averaged over the 7 subjects. The order of the modes reflects the order that the subjects were tested in. The error bars show the standard error. 2-way, paired T-Tests were conducted between the control time and each of the non-control modes, as well as between the 4-level thresholding full frame rate mode and the other 3 modes (binary and both reduced frame rates).

The times taken for all the non-control modes were significantly higher ($p < 0.05$ for all) than the time for the control. The binary and reduced frame rate modes were slightly longer than the 4-level thresholding full frame rate mode, but all the non-control modes were within the statistical margin of error ($p > 0.05$ for all).

The right graph in Figure 13 details the number of collisions for each mode, averaged over the 7 subjects. The error bars show the standard error. The average number of collisions was very low, due to a few of the subjects not colliding with anything in any of the modes, but the binary thresholding and reduced frame rate modes had more collisions on average than the 4-level thresholding full frame rate.

5.2   Chessboard Test

The left graph in Figure 14 details time to completion (in seconds) for each mode, averaged over the 7 subjects. The order of the modes reflects the order the subjects were

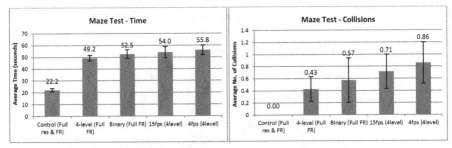

Fig. 13. Maze Test: Left – Mode vs. Average Time (seconds), Right – Mode vs. Average No. of Collisions

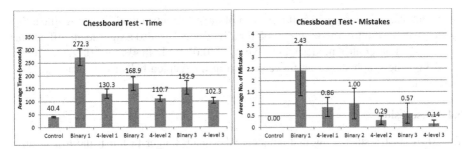

Fig. 14. Maze Test: Left – Mode vs. Average Time (seconds), Right – Mode vs. Average Mistakes

tested in and shows how the same modes were tested repeatedly 3 times to examine learning effects. The error bars show the standard error. 2-way, paired T-Tests were conducted between the control time and each of the non-control modes, as well as between the binary and 4-level thresholding for each pair of repeated tests (eg. 1st binary with 1st 4-level).

The times taken for all non-control modes were significantly longer than control mode ($p < 0.05$ for all). The times taken for the binary modes were significantly longer than the 4-level thresholding for the same repeated number of trial ($p < 0.05$ for the 1st and 2nd pairs of tests, $p = 0.063$ for the 3rd pair). The times for all modes decreases with increasing number of repeated tests.

The right graph in Figure 14 details the number of mistakes for each mode, averaged over the 7 subjects. The error bars show the standard error. The average number of mistakes was quite low due to some subjects not making any mistakes. The trend however clearly looks similar to the Chessboard Time graph with decreasing number of mistakes with repeated trials.

5.3 Ball Interception Test

The graph shown in Figure 15 details the number of successfully intercepted balls for each mode, averaged over the 7 subjects. Once again, the order of the modes reflects the order the subjects were tested in, and the errors bars show standard error. 2-way, paired T-Tests were conducted between the control mode and each of the non-control modes, as well as between the relative non-control modes (4Hz vs 10Hz, 10Hz vs 60Hz).

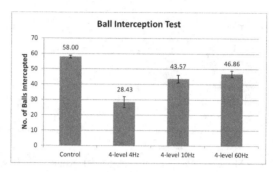

Fig. 15. Maze Test: Mode vs. Average No. of Intercepted Balls

The number of intercepted balls for each of the non-control modes was significantly shorter than the control mode ($p < 0.05$ for all). There was also a significant difference between the non-control modes themselves ($p < 0.05$ for both 4Hz vs 10Hz, and 10Hz vs 60Hz). The graph shows an upward trend in intercepted balls as the frame rate is increased.

5.4 Discussion

The Maze Test results show that subjects take much longer to finish the test in any of the non-control modes compared to the control, and that although the binary and reduced frame rate modes took slightly longer to complete than the 4-level full frame rate mode, the difference was not significant. This trend is also shown in the average number of collisions, but the standard error is very large. From observations made while building and testing the system, reduction in colour depth and frame rate does increase the difficulty of most general tasks including navigational and obstacle avoidance tasks. Possible reasons for this not being made clear in this particular test's results are that the maze area was fairly small and straightforward so the task could be completed in a relatively short amount of time, and the number of test subjects was low, presenting a relatively large error. Also, the obstacles used in this test were large and obvious and so subjects may not have benefited a lot from an increased colour depth and frame rate. Another problem could be the order of the modes in which the subjects were tested was made consistent and that the 'harder' modes were tested later. A learning effect just from repeated testing, even with the changing obstacle placement and accommodation time between tests, could cause a decrease in times for the later tested 'harder' modes and hence reduce differences between them and the earlier test 'easier' modes.

For the Chessboard Test, the results demonstrate that the binary modes were significantly longer than the 4-level thresholding modes for each repeated test. The results also show that there is a clear downwards trend with increasing number of tests for both modes. The average number of mistakes also shows these trends, and that the binary has more mistakes than the 4-level and that both modes decrease over repeated testing, however the standard error is very large. The reason the tests were completed much faster on 4-level compared to binary is likely because this test is based primarily on contrast discrimination and the extra levels of grey available on the 4-level allow the subjects to be able to tell the difference between the dark and light chess pieces and chessboard as well as the grey table more rapidly. This shows that different tasks may benefit differently from various modes. A significant learning effect was evident as times and mistakes would decrease with repeated testing, probably leading to an eventual plateau point where times do not get much faster. It is apparent that as people keep repeating a task they are unfamiliar with, they will improve at it. There should be no reason why it is not the same when using a visual prosthesis simulator, or even a patient with a visual prosthesis implant itself.

The Ball Interception Test results show a clear upward trend in successfully intercepted balls as the frame rate is increased. This was exactly as expected, as a lower frame rate would give a user less time to react and would not allow for smooth tracking of the ball's movement. The differences between the control mode and non-control modes, as well as between the consecutive control modes themselves all showed significant differences in results. It was noticed however, that a considerable number of 'misses' that occurred in the tests were due to misjudgement of where the user thought they had actually moved the paddle to (the paddle would be offset from the balls final location by only a small margin). This is clearly to do with the ability of the participant to coordinate themselves with the paddle placement without actually looking at it. Although, we did allow them some time to 'calibrate' with a simple hand-eye coordination task before each test, the subject's still seemed to have some trouble.

5.5   Limitations of the System

While our system uses a physiologically based model for mapping of phosphenes, it does not represent the gaze-locked nature of a cortical implant. In the case of a real cortical visual prosthesis, the patient will not be able to focus on different points of the visual field with eye movements. In our system however, the user is able to scan the presented pattern voluntarily. To overcome this limitation, an eye-tracker would be required to allow the system to move the pattern along with the movement of the user's eyes, therefore 'locking' the gaze at a specific point (usually at the center) in the presented pattern.

6   Conclusions and Future Work

This paper has presented a simulator for a cortical visual prosthesis. By addressing fundamental limitations in current simulator systems through its portability, and physiologically based phosphene mapping, the system has met expectations and makes a good platform for investigation, improvement and tuning of algorithms for use with a visual prosthesis. The completion of preliminary psychophysical testing has shown that the number of greyscale intensities has a significant effect on results for certain tasks. It was also shown that reducing the frame rate can have a significant effect on the ability of the user to observe and interact with moving objects/things in the environment. Finally, a learning effect was found to be present with repeated trials and will need to be addressed in future work with broader and more rigorous sets of psychophysical testing. It is hoped that, through the use of the HatPack simulator and through further psychophysical testing, valuable insight can be gained and used to improve the implementation of future visual prosthesis devices.

Acknowledgements. Monash Vision Group is funded through the Australian Research Council Research in Bionic Vision Science and Technology Initiative (SR1000006). The authors would like to thank the members of Monash Vision Group that participated in the trials and all those that shared their valuable opinions and advice. Thanks to Dr. Nicholas Price, for help with designing the Ball Interception Test. Finally, thanks to Grey Innovation for help with the physical layout of the integrated simulator system.

References

1. Bak, M., Girvin, J.P., Hambrecht, F.T., Kufts, C.V., Loeb, G.E., Schmidt, E.M.: Visual Sensations Produced by Intracortical Microstimulation of the Human Occipital Cortex. Medical & Biological Engineering & Computing 28, 257–259 (1990)
2. Balasubramanian, M., Polimeni, J.R., Schwartz, E.L.: The V1–V2–V3 complex: quasi-conformal dipole maps in primate striate and extra-striate cortex. Neural Networks 15, 1157–1163 (2002)
3. Bear, M.F., Connors, B.W., Paradiso, M.A.: Neuroscience: Exploring the Brain. Lippincott Williams & Wilkins, Baltimore (2007)
4. Brindley, G.S., Lewin, W.S.: The Visual Sensations Produced by Electrical Stimulation of the Visual Cortex. J. Physiology 196, 479–493 (1968)
5. Canny, J.: A Computational Approach to Edge Detection. IEEE Transactions on Pattern Analysis and Machine Intelligence 8, 679–698 (1986)
6. Chen, S.C., Hallum, L.E., Lovell, N.H., Suaning, G.J.: Visual Acuity Measurement of Prosthetic Vision: A Virtual-Reality Stimulation Study. J. Neural Engineering 2, S135–S145 (2005)
7. Chen, S.C., Suaning, G.J., Morley, J.W., Lovell, N.H.: Simulating Prosthetic Vision: I. Visual Models of Phosphenes. Vision Research 49, 1493–1506 (2009)
8. Dobelle, W.H., Mladejovsky, M.G.: Phosphenes produced by Electrical Stimulation of Human Occipital Cortex, and their Application to the Development of a Prosthesis for the Blind. J. Physiology 243, 553–576 (1974)
9. Dobelle, W.H., Mladejovsky, M.G., Evans, J.R., Roberts, T.S., Girvin, J.P.: 'Braille' Reading by a Blind Volunteer by Visual Cortex Stimulation. Nature 259, 111–112 (1976)
10. Dowling, J.A., Maeder, A.J., Boles, W.: Mobility Enhancement and Assessment for a Visual Prosthesis. In: Proceedings of SPIE Medical Imaging 2004: Physiology, Function and Structure from Medical Images, vol. 5369, pp. 780–791 (2004)
11. Duncan, R.O., Boynton, G.M.: Cortical Magnification within Human Primary Visual Cortex Correlates with Acuity Thresholds. Neuron. 38, 659–671 (2003)
12. Fehervari, T., Matsuoka, M., Okuno, H., Yagi, T.: Real-Time Simulation of Phosphene Images Evoked by Electrical Stimulation of the Visual Cortex. In: Wong, K.W., Mendis, B.S.U., Bouzerdoum, A. (eds.) ICONIP 2010, Part I. LNCS, vol. 6443, pp. 171–178. Springer, Heidelberg (2010)
13. Horton, J.C., Hoyt, W.F.: The Representation of the Visual Field in Human Striate Cortex: A Revision of the Classic Holmes Map. Archives of Opthalmology 109, 816–824 (1991)
14. Humayun, M.S., De Juan, E., Dagnelie, G., Greenberg, R.J., Propst, R.H., Phillips, D.H.: Visual Perception Elicited by Electrical Stimulation of the Retina in Blind Humans. Archives of Opthalmology 114, 40–46 (1996)
15. Lee, J.S.J., Haralick, R.M., Shapiro, L.G.: Morphologic Edge Detection. J. Robotics and Automation 3, 142–156 (1987)
16. Monash Vision Group, http://www.monash.edu.au/bioniceye

17. Polimeni, J.R., Balasubramanian, M., Schwartz, E.L.: Multi-Area Visuotopic Map Complexes in Macaque Striate and Extra-Striate Cortex. Vision Research 46, 3336–3359 (2006)
18. Schira, M.M., Wade, A.R., Tyler, C.W.: Two-Dimensional Mapping of the Central and Parafoveal Visual Field to Human Visual Cortex. J. Neurophysiology 97, 4284–4295 (2007)
19. Schira, M.M., Tyler, C.W., Spehar, B., Breakspear, M.: Modeling Magnification and Anisotropy in the Primate Foveal Confluence. PLoS Computational Biology 6, 1–10 (2010)
20. Schmidt, E.M., Bak, M.J., Hambrecht, F.T., Kufta, C.V., O'Rourke, D.K., Vallabhanath, P.: Feasibility of a Visual Prosthesis for the Blind Based on Intracortical Microstimulation of the Visual Cortex. Brain 119, 507–522 (1996)
21. Schwartz, E.L.: Spatial Mapping in the Primate Sensory Projection: Analytic Structure and Relevance to Perception. Biological Cybernetics 25, 181–194 (1977)
22. Srivastava, N.R., Troyk, P.R., Dagnelie, G.: Detection, Eye-Hand Coordination and Virtual Mobility Performance in Simulated Vision for a Cortical Visual Prosthesis Device. J. Neural Engineering 6, 1–14 (2009)
23. Van Rheede, J.J., Kennard, C., Hicks, S.L.: Simulating Prothetic Vision: Optimizing the Information Content of a Limited Visual Display. J. Vision 10, 1–15 (2010)
24. Veraart, C., Raftopoulos, C., Mortimer, J.T., Delbeke, J., Pins, D., Michaux, G., Vanlierde, A., Parrini, S., Wanet-Defalque, M.: Visual Sensations Produced by Optic Nerve Stimulation Using an Implanted Self-Sizing Spiral Cuff Electrode. Brain Research 813, 181–186 (1998)
25. Wandell, B.A., Dumoulin, S.O., Brewer, A.A.: Visual Field Maps in Human Cortex: Review. Neuron. 56, 366–383 (2007)
26. Zhao, Y., Lu, Y., Tian, Y., Li, L., Ren, Q., Chai, X.: Image Processing Based Recognition of Images with a Limited Number of Pixels Using Simulated Prosthetic Vision. Information Sciences 180, 2915–2924 (2010)

Appendix

Video 1: http://www.youtube.com/watch?v=oAxaNloHVHg
Video 2: http://www.youtube.com/watch?v=2byh1qQfWGQ
Video 3: http://www.youtube.com/watch?v=gIVrnsk04LA

Pathogen Detection Using Magnetoelastic Biosentinels

Howard Clyde Wikle, III , Suiqiong Li, Aleksandr Simonian, and Bryan A. Chin

Materials Research & Education Center, Auburn University, Auburn, AL, U.S.A.
{hcw0002,lisuiqi,simonal,chinbry}@auburn.edu

Abstract. Biosentinels, used to detect, signal, and capture pathogenic bacteria, are discussed. The biosentinel is based on magnetically soft magnetoelastic resonators coated with a selective and specific biorecognition layer. The biosentinels are actuated, monitored, and controlled wirelessly by external magnetic fields. The biosentinels mimic the function of naturally occurring biological defensive systems, such as white blood cells, seeking out and capturing pathogenic bacteria. After binding with the target pathogen, the mass of the biosentinel increases causing the resonant frequency to decrease, providing instantaneous detection of the pathogen. The biosentinels require no on-board power, harvesting electromagnetic energy from the surroundings for propulsion, navigation, and signaling the detection of target pathogens.

Keywords: Pathogen detection, Biosensor, Bio-inspired, Phage, Magnetoelastic, Wireless sensor.

1 Introduction

For centuries, humankind has attempted to mimic the designs of Nature to develop new engineering materials and systems. The human blood system is an excellent example of one of Nature's amazing creations that inspires us in this work. The human blood contains many components that work synergistically to keep us healthy. As part of the immune system, white blood cells are the main defensive mechanism against pathogenic invaders. There are a variety of white blood cell types (neutrophil, eosinophil, lymphocytes, etc.) that target different pathogens. This capability serves as the model for a bio-inspired system of autonomous sentinels for the capture and detection of invasive pathogens. To provide proof-in-principal, research results for bacterial detection using magnetoelastic biosentinels in liquid analytes are presented. Potential short term applications include the capture and detection of bacteria in urine and li-quid food products such as water, juices and milk. A variety of sentinels, similar to different types of white blood cells, may be constructed to target different bacterial pathogens (Figure 1). The envisioned sentinels will autonomously move through a liquid, seeking out and capturing specific invading bacterial pathogens. A sentinel is constructed of a freestanding magnetoelastic (ME) resonator (transducer platform) that is coated with a biorecognition layer (bacteriophage) that specifically captures or binds a single type of pathogen.

J. Gabriel et al. (Eds.): BIOSTEC 2012, CCIS 357, pp. 68–79, 2013.

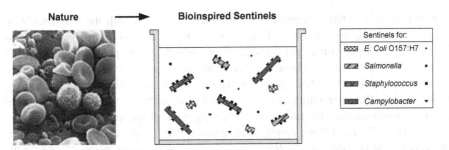

Fig. 1. Bio-inspired sentinels will target different types of bacteria (*E. coli, Salmonella* Typhimurium, etc.) mimicking white blood cells that target different invasive pathogens [1]

2   Theory of the Biosentinel

A biosentinel is constructed of a freestanding magnetoelastic (ME) resonator (transducer platform) that is coated with a biorecognition layer (e.g. antibodies or bacteriophage) that specifically captures or binds a single type of pathogen. The resonator is constructed from an iron-based, amorphous alloy with magnetostrictive properties. Magnetostrictive materials undergo a change in shape when subjected to an applied magnetic field. If the magnetic field is aligned along the length direction of the resonator and varied at the proper frequency, the structure can achieve resonance. The detection principle of the ME sentinels is shown in Figure 2. The freestanding ME resonator serves as the transduction platform, actuated into resonance by the application of an alternating magnetic field. Upon contact with the specific target bacteria, the biorecognition element on the biosentinel's surface captures the target bacterial cells, causing the overall sensor mass to increase which results in a decrease in the resonant frequency. The resonant frequency is remotely and wirelessly measured using a pick-up coil. No onboard power is required by a biosentinel; instead, electromagnetic energy is harvested from the surroundings.

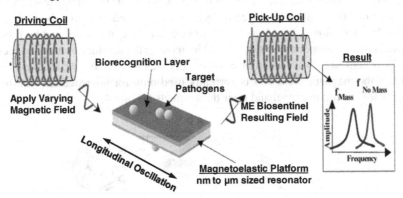

Fig. 2. Detection principle of ME biosentinel. A modulated magnetic field causes the sentinel to resonate. Binding of target bacteria to the sentinel causes the resonant frequency to decrease.

ME biosentinels have unique advantages that stem from both the magnetoelastic resonator platform and the phage biorecognition layer. The biosentinels are wireless devices, enabling in-situ remote detection of multiple target pathogens (Figure 3). Due to its wireless nature, a large number of sentinels can be deployed simultaneously, which significantly enhances the probability of binding with a target pathogen. More importantly, the binding of target pathogens on only one out of many sentinels can be easily detected. By taking advantage of these properties and capabilities of phage-coated ME resonators, a system of sentinels that mimics the functions of white blood cells can be built and deployed for enhanced medical diagnostics, food safety, or water quality applications.

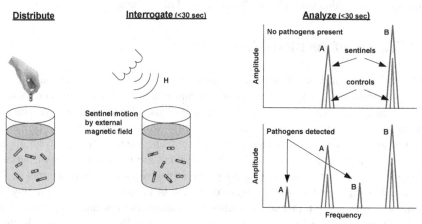

Fig. 3. A large number of sentinels targeting different pathogens may be mixed together and interrogated simultaneously for pathogen detection. Different pathogens may be detected simultaneously since the sentinels are designed to operate in different frequency ranges.

One of the key parameters of these biosentinels is the minimum detection limit. At low bacterial concentrations, the odds of detection are improved either by increasing the number of ME sentinels deployed or by exposing the sentinels to a dynamic environment. For detection in liquid media, dynamic exposure can be achieved by flowing the media past the immobilized sentinels or by moving the sentinels within the media. While flow cells are a viable option, another approach to achieve greater exposure is to harness the magnetic field that is currently used only for interrogation to provide the forces for motion to propel and steer the sentinels (Figure 4).

Fig. 4. The magnetic field generated by the detection system can induce sentinel movement

2.1   The Resonator Platform

Acoustic wave (AW) resonators as biosensing platforms have been widely investigated [2, 3] with quartz crystal microbalances [4–6], microcantilevers [7, 8], and magnetoelastic resonators [9–12] among the platforms. Acoustic resonators are mass sensitive devices where a change in the mass load on the sensor surface causes a change in the sensor's resonant frequency. Acoustic resonators are characterized by two important parameters: 1) the sensitivity (S_m) which represents the resonant frequency shift per unit mass load; and 2) the resonance performance (Q factor), which is defined as the ratio of the energy stored in the resonant structure to the total energy losses per oscillation cycle. A high Q factor means lower attenuation and a sharper resonant peak and thus better resolution in determining the resonant frequency. In a viscous environment, all AW resonators suffer from the damping effects of the media on the vibratory response of the platform. Viscous damping causes a decrease in f, S_m and Q [13]. This is why some AW resonators work well in vacuum or air but not in liquids. The Q value of ME resonators was found to exhibit a much higher Q value (>1000 in air & ~100 in water) than microcantilevers of the same length [14]. This advantage originates from the longitudinal mode of vibration for the freestanding ME resonator's design compared to the transverse vibration mode of cantilevers [15]. The minimum detectable mass for an acoustic sensor platform depends on the ability to resolve resonant frequency shifts as a result of the mass loading.

Magnetoelasticity is the bidirectional coupling of magnetic and elastic fields within a material [16, 17]. With the application of an external magnetic field, elastic distortions arise from the rearrangement of magnetic domains (magnetostrictive behavior) causing the material to change shape. This deformation induces stresses which, conversely, cause the magnetization of the material to change (Villari effect). In the presence of an oscillating external magnetic field (H), the ME resonator will align with the field along its easy axis and will resonate by elastically deforming in response inducing an opposing magnetic field that may be measured with an induction pickup coil. When the frequency of H matches the natural resonant frequency of the platform, the platform will resonate with further enhanced motion. In air, the fundamental resonant frequency f_0 of the longitudinal vibratory motion of a freestanding thin ribbon of magnetoelastic material (with length L) is [18]:

$$f_0 = \frac{1}{2L}\sqrt{\frac{E}{\rho(1-\sigma)}} \qquad (1)$$

where E is Young's modulus of elasticity, ρ is the density, and σ is Poisson's ratio.

For biological detection, the surface of the ME resonator is coated with a biorecognition element such as an antibody or phage to form a biosentinel. This biorecognition element is designed to specifically bind the target of interest. When the ME biosentinel comes into contact with the target pathogens, the biorecognition element will capture/bind the pathogens, creating an additional mass load on the biosentinel resulting in a decrease in the resonant frequency. Therefore, the presence and concentration of any target pathogens can be identified by monitoring the resonant frequency shifts of the biosentinel. For the uniform addition of a small mass Δm to the original mass m_0 (such that $\Delta m \ll m_0$), the resulting mass loaded resonant frequency is [13]:

$$f = \sqrt{\frac{1}{(1 + \Delta m / m_0)}} f_0 \qquad (2)$$

The addition of a small mass results in a shift in resonant frequency Δf such that $\Delta f = f - f_0 < 0$. The mass load on the ME resonator can easily be obtained by simply measuring the resonant frequency shift.

The mass sensitivity (S_m) can be obtained from a truncated Maclauren series expansion of (2), giving:

$$S_m = \frac{df}{dm} \approx -\frac{1}{2}\frac{f_0}{m_0} \qquad (3)$$

The sensitivity of the ME biosentinel is compared with microcantilevers in Figure 5. For ME biosentinels and cantilevers fabricated from the same material and of the same size, the ME sensor exhibits an S_m about 100 times better than the cantilever. Advanced microfabrication processes will enable the optimization of the resonance performance of the ME sentinels which will lead to improved pathogen detection capabilities.

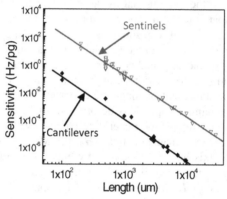

Fig. 5. Sensitivity vs. length for cantilever and biosentinels. Biosentinels are 100 times more sensitive.

Many ME biosentinels may be simultaneously deployed but remain individually distinguishable [19, 20]. Different types of ME biosentinels can be made by changing dimensions, microstructures, heat treatments and/or compositions of the resonators and the different biorecognition elements may be employed to detect different pathogens. Each type of resonator will have its own characteristic resonant frequency. Deploying a large number of ME sensors greatly increases the probability that the sensors will react with the targeted pathogen, thereby reducing detection time. When multiple sensors are employed, the apparent sensitivity can be significantly higher than that of a single sensor.

2.2 Biosentinel Motion

An external magnetic field may also be used to impart motion to the biosentinels, propelling them through a media to gain greater exposure to potential biological targets

[21]. Motion of the biosentinels is the result of forces (F) and torques (T) that are experienced by the biosentinel that arise in the presence of an external magnetic field (H). The forces and torques are given in terms of vector quantities by [22]:

$$F = \mu_0 \int_V (M \cdot \nabla)H dv \qquad (4)$$

$$T = \mu_0 \int_V (M \times H) \, dv \qquad (5)$$

where M is the magnetization of the biosentinel, μ_0 is the permeability of free space, and v is the material volume. In general, M and H vary over the body of the biosentinel. At the micrometer dimensional scales of the biosentinels, the external magnetic field can be assumed to be constant with little error. However, for magnetically soft materials, such as the amorphous magnetoelastic materials used for the biosentinels, the magnetization is a nonlinear function of the external magnetic field and is dependent on the body shape. Only ellipsoids may possess a uniform magnetization throughout the body. For other shapes, geometric effects give rise to demagnetizing fields along different directions (shape anisotropy), causing some directions to be more easily magnetized than others. The modeling of the magnetic fields has typically been achieved with finite element analysis which is unsuitable for real-time control. Recently, analytical models employing equivalent ellipsoids to determine the body magnetization [23] have been published to aid with the magnetic propulsion and navigation of microrobots [24, 25]. Models such as these enable the real-time actuation and control of magnetic microsensors and robots for targeted medical therapeutic and diagnostic applications.

2.3   Fabrication of ME Resonators

Magnetoelastic resonators may be formed by bulk cutting from commercially available ribbons [26–28], electrochemical deposition [29–31], electroless plating [32], and physical vapor deposition [33, 34]. The ME resonators described here are fabricated using standard microelectronic fabrication techniques of photolithography and physical vapor deposition (sputtering). A resonator fabricated by this technique is shown in Figure 6 and the fabrication process is shown schematically in Figure 7. A full description of the process is given in [33]. Magnetoelastic resonators are fabricated on a patterned wafer by co-depositing iron and boron at controlled rates under vacuum. The resonators are coated with gold to protect against oxidation and provide a biocompatible surface to immobilize the biorecognition layer. Fabrication of the sensor platform begins by coating a plain silicon test wafer with a layer of chromium followed by gold, each to a thickness of 30–40 nm. Next, a spin-coated layer of photoresist is deposited on the gold surface, at least twice as thick as the desired ME resonator thickness. The photoresist is then UV exposed using a positive mask that patterns the shape and dimensions of the ME resonators. The wafer is then developed, rinsed, dried, and inspected for pattern integrity and thickness.

To begin forming the resonators, a gold layer is first deposited onto the patterned wafer to a thickness of 30–40 nm. Iron (dc) and boron (rf) are then co-sputtered to form the magnetostrictive alloy. The thickness of the alloy film depends on process conditions, and is generally limited by the thickness of the photoresist layer. Highly magnetostrictive films up to 7 μm thick have been obtained using this dual-cathode method. Finally, another gold layer is deposited on top of the iron-boron film so that the magnetostrictive particles are completely encased in gold. The resonators are freed from the wafer by lift-off using an acetone rinse and collected using a magnet. Approximately 40,000 sentinels can be fabricated on an 8 inch wafer at a cost of approximately $28 USD. Hence the cost of a single ME sensor is less than 1/1000 of a cent.

Fig. 6. SEM micrograph comparing the size of a ME resonator with the Y in "LIBERTY" on a penny. The resonators are microelectronically fabricated and are smaller than a particle of dust. The resonators require no on-board power and their cost is less than 1/1000 of a cent each when fabricated in large numbers.

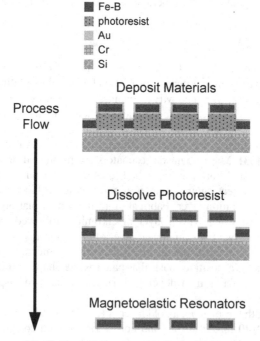

Fig. 7. The ME resonator fabrication process

2.4   Immobilization of the Biorecognition Layer

To form functional biosentinels, a biorecognition layer must be immobilized onto the resonator platform to bind the specific target species. Antibodies [35–38] and bacteriophages [39–45] have been used as the biorecognition layer to detect *Salmonella* Typhimurium cells and *Bacillus anthracis* spores. Filamentous phage (Figure 8) has been genetically engineered to serve as the biorecognition layer [46–48]. In one application, filamentous E2 phage for binding to *S.* Typhimurium was affinity selected from a landscape f8/8 phage library and provided by the Department of Biological Sciences at Auburn University. The clone E2 phage was verified to be highly specific and selective towards *S.* Typhimurium [47]. The phage was immobilized on the ME sensor surface using physical adsorption. Each ME sensor platform was placed in a vial containing 300 µL of E2 phage suspension (5×10^{11} vir/mL in 1× Tris-Buffered Saline (TBS)). These vials were then rotated and incubated on a rotor (running at 8 rpm) for 1 hr. After the immobilization process, the sensors were washed three times with 1× TBS solution and two times with sterile distilled water in order to remove salt and any unbound or loosely bound phage.

In order to reduce nonspecific binding, bovine serum albumin (BSA) solution was then immobilized on the sensor surfaces to serve as a blocking agent. The ME biosentinels were immersed into 1 mg/mL BSA solution for at least 1 hr, followed by a distilled water rinse. In this study, control sensors were fabricated and used to calibrate the effects of environmental changes, such as temperature and non-specific binding. The control sensor is identical to the measurement biosensor except it lacks the E2 phage coating. The control sensors were also treated with BSA to block non-specific binding.

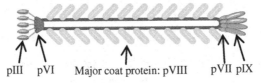

pIII   pVI   Major coat protein: pVIII   pVII pIX

Fig. 8. Filamentous phage

3   Characterization of ME Biosentinel Performance

3.1   Detection of Pathogens by the ME Sentinels

S. Typhimurium culture (ATCC 13311) was provided by the Department of Biological Sciences at Auburn University, Auburn, AL in the form of a suspension at a concentration of 5×10^8 CFU/mL. The suspension was serially diluted in water to obtain test solutions with a range of concentrations from 5×10^1 to 5×10^7 CFU/mL. All test solutions were prepared on the same day as the biosentinel testing. The test solutions were stored at 4 °C and equilibrated to room temperature in a water bath prior to lusing in experiments.

The resonant frequency of the biosentinels was measured using an HP 8751A network analyzer with S-parameter test set (Agilent Technologies, Santa Clara, CA, USA). The ME sentinels (control and measurement) were placed in a tube containing pure water and the resonant frequency of the biosentinel measured. The network analyzer scanned, measured and recorded the resonant frequency spectrum of the ME biosentinel as a function of time. After each 30 min exposure the analyte was changed to the next highest dilution. Figure 9 shows the frequency shift measurements for ME biosentinels 500 × 100 × 4 μm in size. Note that the control sentinel shows a nearly constant frequency (no frequency shift), while the measurement sentinel undergoes a frequency shift of nearly 120 kHz. The ME biosentinel exhibited a sensitivity of nearly -24 kHz/decade and a detection limit less than 50 CFU/mL of *S*. Typhimurium in water.

A scanning electron microscope (SEM) was used to confirm and compare the binding of *S*. Typhimurium on the phage-coated measurement and control sentinels. After the bacterial detection, the ME biosentinels were exposed to osmium tetroxide (OsO$_4$) vapour for 45 min. The sensors were then mounted onto aluminum stubs and examined using the SEM. Figure 10 shows the SEM micrographs for the measurement and control sentinels. The control biosentinel shows only a few cells are bound to the surface while the measurement biosentinel is nearly completely covered with bound *S*. Typhimurium bacteria.

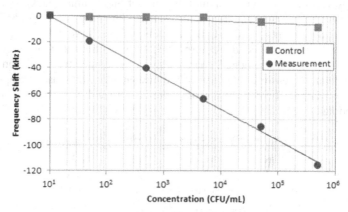

Fig. 9. Response of 500 μm long measurement and control biosentinels exposed to increasingly higher concentrations of *S*. Typhimurium. The detection limit is less than 50 CFU/mL.

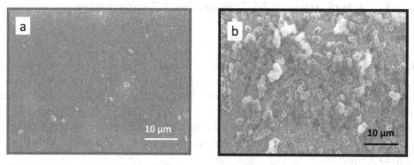

Fig. 10. The SEM images show near zero binding of *Salmonella* cells to the control biosentinel (a) and a large number of bound *Salmonella* cells to the measurement biosentinel (b).

4   Conclusions

New magnetoelastic sentinel technologies under development will lead to a plethora of devices for food safety, biosecurity, environmental monitoring, and medical care. Biosentinels composed of magnetically soft resonator platforms coated with a biorecognition layer such as phage or antibodies can wirelessly and remotely detect the binding and capture of specific pathogenic bacteria. Magnetic propulsion and navigation of magnetoelastic biosentinels offers the potential for targeted medical therapeutics and diagnostics that can open new avenues of treatment, increase the efficacy of existing medical procedures, and improve the overall quality of health care.

References

1. Wetzel, B., Schaefer, H.: National Cancer Institute (1982)
2. Ballantine Jr., D.S., White, R.M., Martin, S.J., Ricco, A.J., Zellers, E.T., Frye, G.C., Wohltjen, H.: Acoustic Wave Sensors: Theory, Design, & Physico-Chemical Applications. Academic Press, San Diego (1996)
3. Vellekoop, M.J.: Acoustic Wave Sensors and their Technology. Ultrasonics 36, 7–14 (1998)
4. Cooper, M.A., Singleton, V.T.: A Survey of the 2001 to 2005 Quartz Crystal Microbalance Biosensor Literature: Applications of Acoustic Physics to the Analysis of Biomolecular Interactions. J. Mol. Recognit. 20, 154–184
5. Becker, B., Cooper, M.A.: A Survey of the 2006-2009 Quartz Crystal Microbalance Biosensor Literature. J. Mol. Recognit. 24, 754–787 (2011)
6. Dixon, M.C.: Quartz Crystal Microbalance with Dissipation Monitoring: Enabling Real-Time Characterization of Biological Materials and their Interactions. J. Biomol. Tech. 19, 151–158 (2008)
7. Hansen, K.M., Thundat, T.: Microcantilever Biosensors. Methods 37, 57–64 (2005)
8. Buchapudi, K.R., Huang, X., Yang, X., Ji, H.-F., Thundat, T.: Microcantilever Biosensors for Chemicals and Bioorganisms. Analyst. 136, 1539–1556 (2011)
9. Li, S.Q., Orona, L., Li, Z., Cheng, Z.-Y.: Biosensor Based on Magnetostrictive Microcantilever. Appl. Phys. Lett. 88, 73507 (2006)
10. Xie, F., Yang, H., Li, S.Q., Shen, W., Wan, J., Johnson, M.L., Wikle, H.C., Kim, D.-J., Chin, B.A.: Amorphous Magnetoelastic Sensors for the Detection of Biological Agents. Intermetallics 17, 270–273 (2009)
11. Huang, S., Hu, J., Wan, J., Johnson, M.L., Shu, H., Chin, B.A.: The Effect of Annealing and Gold Deposition on the Performance of Magnetoelastic Biosensors. Mater. Sci. Eng., C. 28, 380–386 (2008)
12. Shen, W., Mathison, L.C., Petrenko, V.A., Chin, B.A.: Design and Characterization of a Magnetoelastic Sensor for the Detection of Biological Agents. J. Phys. D: Appl. Phys. 43, 15004 (2010)
13. Stoyanov, P.G., Grimes, C.A.: A Remote Query Magnetostrictive Viscosity Sensor. Sens. Actuators, A. 80, 8–14 (2000)
14. Cheng, Z.-Y., Li, S.Q., Zhang, K.W., Fu, L.L., Chin, B.A.: Novel Magnetostrictive Microcantilever and Magnetostrictive Nanobars for High Performance Biological Detection. Adv. Sci. Technol. 54, 19–28 (2008)

15. Castille, C., Dufour, I., Lucat, C.: Longitudinal Vibration Mode of Piezoelectric Thick-Film Cantilever-Based Sensors in Liquid Media. Appl. Phys. Lett. 96, 154102 (2010)

16. Dapino, M.J., Smith, R.C., Faidley, L.E., Flatau, A.B.: A Coupled Structural-Magnetic Strain and Stress Model for Magnetostrictive Transducers. J. Intell. Mater. Syst. Struct. 11, 135–152 (2000)

17. Barandiarán, J.M., Gutiérrez, J., García-Arribas, A.: Magneto-elasticity in Amorphous Ferromagnets: Basic Principles and Applications. Phys. Status Solidi A. 208, 2258–2264 (2011)

18. Liang, C., Morshed, S., Prorok, B.C.: Correction for Longitudinal Mode Vibration in Thin Slender Beams. Appl. Phys. Lett. 90, 221912 (2007)

19. Cheng, Z.-Y.: Applications of Smart Materials in the Development of High Performance Biosensors. MRS Proceedings, 888, 0888-V10-06 (2005)

20. Huang, S., Yang, H., Lakshmanan, R.S., Johnson, M.L., Wan, J., Chen, I.-H., Wikle, H.C., Petrenko, V.A., Barbaree, J.M., Chin, B.A.: Sequential Detection of Salmonella Typhimurium and Bacillus anthracis Spores Using Magnetoelastic Biosensors. Biosens. Bioelectron. 24, 1730–1736 (2009)

21. Nelson, B.J., Kaliakatsos, I.K., Abbott, J.J.: Microrobots for Minimally Invasive Medicine. Annu. Rev. Biomed. Eng. 12, 55–85 (2010)

22. Furlani, E.P.: Permanent Magnet and Electromechanical Devices: Materials, Analysis, and Applications. Academic Press, San Diego (2001)

23. Beleggia, M., Graef, M.D., Millev, Y.T.: The Equivalent Ellipsoid of a Magnetized Body. J. Phys. D: Appl. Phys. 39, 891–899 (2006)

24. Abbott, J.J., Ergeneman, O., Kummer, M.P., Hirt, A.M., Nelson, B.J.: Modeling Magnetic Torque and Force for Controlled Manipulation of Soft-Magnetic Bodies. IEEE Trans. Rob. 23, 1247–1252 (2007)

25. Nagy, Z., Ergeneman, O., Abbott, J.J., Hutter, M., Hirt, A.M., Nelson, B.J.: Modeling Assembled-MEMS Microrobots for Wireless Magnetic Control. In: 2008 IEEE International Conference on Robotics and Automation, pp. 874–879. IEEE (2008)

26. Horikawa, S., Bedi, D., Li, S.Q., Shen, W., Huang, S., Chen, I.-H., Chai, Y., Auad, M.L., Bozack, M.J., Barbaree, J.M., Petrenko, V.A., Chin, B.A.: Effects of Surface Functionalization on the Surface Phage Coverage and the Subsequent Performance of Phage-Immobilized Magnetoelastic Biosensors. Biosens. Bioelectron. 26, 2361–2367 (2011)

27. Shen, W., Lakshmanan, R.S., Mathison, L.C., Petrenko, V.A., Chin, B.A.: Phage Coated Magnetoelastic Micro-Biosensors for Real-Time Detection of Bacillus anthracis Spores. Sens. Actuators, B. 137, 501–506 (2009)

28. Wan, J., Johnson, M.L., Guntupalli, R., Petrenko, V.A., Chin, B.A.: Detection of Bacillus anthracis Spores in Liquid Using Phage-Based Magnetoelastic Micro-Resonators. Sens. Actuators, B. 127, 559–566 (2007)

29. Li, S., Fu, L., Wang, C., Lea, S., Arey, B., Engelhard, M., Cheng, Z.-Y.: Characterization of Microstructure and Composition of Fe-B Nanobars as Biosensor Platform. MRS Proceedings, 962, 0962-P09-14 (2006)

30. Fujita, N., Inoue, M., Arai, K., Izaki, M., Fujii, T.: Uniaxial Magnetic Anisotropy of Amorphous Fe–B Films Deposited Electrochemically in a Magnetic Field. J. Appl. Phys. 85, 4503 (1999)

31. Fujita, N., Inoue, M., Arai, K., Lim, P.B., Fujii, T.: Electrochemical Deposition of Amorphous Films with Soft Magnetic Properties. J. Appl. Phys. 83, 7294 (1998)

32. Fujita, N., Tanaka, A., Makino, E., Squire, P.T., Lim, P.B., Inoue, M., Fujii, T.: Fabrication of Amorphous Iron-Boron Films by Electroless Plating. Appl. Surf. Sci. 113-114, 61–65 (1997)

33. Johnson, M.L., LeVar, O., Yoon, S.H., Park, J.-H., Huang, S., Kim, D.-J., Cheng, Z.-Y., Chin, B.A., Odum, L.: Dual-Cathode Method for Sputtering Magnetoelastic Iron-Boron Films. Vacuum 83, 958–964 (2009)

34. Johnson, M.L., Wan, J., Huang, S., Cheng, Z.-Y., Petrenko, V.A., Kim, D.-J., Chen, I.-H., Barbaree, J.M., Hong, J.W., Chin, B.A.: A Wireless Biosensor Using Microfabricated Phage-Interfaced Magnetoelastic Particles. Sens. Actuators, A. 144, 38–47 (2008)

35. Guntupalli, R., Hu, J., Lakshmanan, R.S., Huang, T.-S., Barbaree, J.M., Chin, B.A.: A Magnetoelastic Resonance Biosensor Immobilized with Polyclonal Antibody for the Detection of Salmonella Typhimurium. Biosens. Bioelectron. 22, 1474–1479 (2007)

36. Guntupalli, R., Lakshmanan, R.S., Hu, J., Huang, T.-S., Barbaree, J.M., Vodyanoy, V.J., Chin, B.A.: Rapid and Sensitive Magnetoelastic Biosensors for the Detection of Salmonella Typhimurium in a Mixed Microbial Population. J. Microbiol. Methods 70, 112–118 (2007)

37. Guntupalli, R., Lakshmanan, R.S., Johnson, M.L., Hu, J., Huang, T.-S., Barbaree, J.M., Vodyanoy, V.J., Chin, B.A.: Magnetoelastic Biosensor for the Detection of Salmonella Typhimurium in Food Products. Sens. Instrum. Food Qual. Saf. 1, 3–10 (2007)

38. Guntupalli, R., Lakshmanan, R.S., Wan, J., Kim, D.-J., Huang, T., Vodyanoy, V.J., Chin, B.A.: Analytical Performance and Characterization of Antibody Immobilized Magnetoelastic Biosensors. Sens. Instrum. Food Qual. Saf. 2, 27–33 (2008)

39. Li, S.Q., Johnson, M.L., Banerjee, I., Chen, I.-H., Barbaree, J.M., Cheng, Z.-Y., Chin, B.A., Li, Y., Chen, H.: Micro-fabricated Wireless Biosensors for the Detection of S. Typhimurium in Liquids. In: Proc. SPIE, vol. 7676 (2010)

40. Li, S.Q., Li, Y., Chen, H., Horikawa, S., Shen, W., Simonian, A., Chin, B.A.: Direct Detection of Salmonella Typhimurium on Fresh Produce Using Phage-Based Magnetoelastic Biosensors. Biosens. Bioelectron. 26, 1313–1319 (2010)

41. Lakshmanan, R.S., Guntupalli, R., Hong, J.W., Kim, D.-J., Cheng, Z.-Y., Petrenko, V.A., Barbaree, J.M., Chin, B.A.: Selective Detection of Salmonella Typhimurium in the Presence of High Concentrations of Masking Bacteria. Sens. Instrum. Food Qual. Saf. 2, 234–239 (2008)

42. Lakshmanan, R.S., Guntupalli, R., Hu, J., Kim, D.-J., Petrenko, V.A., Barbaree, J.M., Chin, B.A.: Phage Immobilized Magnetoelastic Sensor for the Detection of Salmonella Typhimurium. J. Microbiol. Methods 71, 55–60 (2007)

43. Lakshmanan, R.S., Guntupalli, R., Hu, J., Petrenko, V.A., Barbaree, J.M., Chin, B.A.: Detection of Salmonella Typhimurium in Fat Free Milk Using a Phage Immobilized Magnetoelastic Sensor. Sens. Actuators, B. 126, 544–550 (2007)

44. Huang, S., Li, S.Q., Yang, H., Johnson, M.L., Wan, J., Chen, I.-H., Petrenko, V.A., Barbaree, J.M., Chin, B.A.: Optimization of phage-based magnetoelastic biosensor performance. Sens. Transducers J. 3, 87–96 (2008)

45. Huang, S., Yang, H., Lakshmanan, R.S., Johnson, M.L., Chen, I.-H., Wan, J., Wikle, H.C., Petrenko, V.A., Barbaree, J.M., Cheng, Z.-Y., Chin, B.A.: The Effect of Salt and Phage Concentrations on the Binding Sensitivity of Magnetoelastic Biosensors for Bacillus anthracis Detection. Biotechnol. Bioeng. 101, 1014–1021 (2008)

46. Petrenko, V.A., Smith, G.P.: Phages from Landscape Libraries as Substitute Antibodies. Protein Eng., Des. Sel. 13, 589–592 (2000)

47. Sorokulova, I.B., Olsen, E.V., Chen, I.-H., Fiebor, B., Barbaree, J.M., Vodyanoy, V.J., Chin, B.A., Petrenko, V.A.: Landscape Phage Probes for Salmonella Typhimurium. J. Microbiol. Methods 63, 55–72 (2005)

48. Petrenko, V.A.: Landscape Phage as a Molecular Recognition Interface for Detection Devices. Microelectron. J. 39, 202–207 (2008)

Multi-source Harvesting Systems for Electric Energy Generation on Smart Hip Prostheses*

Marco P. Soares dos Santos[1], Jorge A.F. Ferreira[1], A. Ramos[1], Ricardo Pascoal[1],
Raul Morais dos Santos[2,3], Nuno M. Silva[2], José A.O. Simões[1], M.J.C.S. Reis[4],
António Festas[1], and Paulo M. Santos[2]

[1] Department of Mechanical Engineering, University of Aveiro, Aveiro, Portugal
[2] UTAD - University of Trás-os-Montes e Alto Douro, Vila Real, Portugal
[3] Center for the Research and Technology of Agro-Environmental and Biological Sciences, Vila Real, Portugal
[4] Institute of Electronics and Telematics Engineering of Aveiro/UTAD, Vila Real, Portugal
marco.santos@ua.pt

Abstract. The development of smart orthopaedic implants is being considered as an effective solution to ensure their everlasting life span. The availability of electric power to supply active mechanisms of smart prostheses has remained a critical problem. This paper reports the first implementation of a new concept of energy harvesting systems applied to hip prostheses: the multi-source generation of electric energy. The reliability of the power supply mechanisms is strongly increased with the application of this new concept. Three vibration-based harvesters, operating in true parallel to harvest energy during human gait, were implemented on a $Metabloc^{TM}$ hip prosthesis to validate the concept. They were designed to use the angular movements on the flexion-extension, abduction-adduction and inward-outward rotation axes, over the femoral component, to generate electric power. The performance of each generator was tested for different amplitudes and frequencies of operation. Electric power up to 55 μJ/s was harvested. The overall function of smart hip prostheses can remain performing even if two of the generators get damaged. Furthermore, they are safe and autonomous throughout the life span of the implant.

1 Introduction

1.1 Scope of the Problem and Background

Currently, there is no cure for most causes of failure of total joint replacement, except surgical revision [1]. Although drug administration, such as through antimicrobial therapy, suppressive antibiotic therapy, outpatient parenteral antimicrobial therapy and antibiotic prophylaxis, are being used to hinder progressive failure following joint replacement [2], surgical revisions have been the only "medical prescription" for most causes of failure [1]. However, these surgical procedures are not therapeutic methods which are performed to cure or to prevent early failures, but only to relieve pain and to

---

* The authors would like to thank the Portuguese Foundation for Science and Technology (FCT) for their financial support under the Grant PTDC/EME-PME/ 105465/2008.

J. Gabriel et al. (Eds.): BIOSTEC 2012, CCIS 357, pp. 80–96, 2013.

improve joint function [3]. The ordinary methodology to improve prostheses' function has been based on the research of new designs, new materials, new fixation techniques and new surgical techniques [4–6]. Although the everlasting life span is an essential requirement for the next healthcare bio-systems generations, the 20-year revision rate of current orthopaedic prostheses is still higher than 20%. Demographic changes and scientific breakthroughs are the main reasons ascribed to the increase in the number of primary and revision joint replacements [3], as well as the strong demand for joint replacements and revisions predicted for the coming years [7]. After the first revision procedure, the risk of failure increases even more [8]. Furthermore, the increase in the number of inpatients less than 65 years due to joint disorders [9] is also being considered an important reason that supports the hypothesis of developing a new methodology to design prostheses with the ability to control their own life span. Current hip prostheses are passive implants because they are not smart enough to promote maximal bone-implant interaction. They match their design methodology with a design "not to know" and "not to act against".

Instrumented prostheses have been developed since the 60's of the 20th century [10]. Their methodological basis is to perform *in-vivo* measurement and data storage functions to optimize passive implants, surgical procedures, preclinical testing and physiotherapy programs [11, 12]. They have been used to validate models of the physiological environment and customize physiotherapy programs [13, 14]. Contact forces and moments in the joint, temperature distribution along the implant, articular motions, misalignments and detection of hip loosening [15–22] are the main quantities which have been collected by instrumented implants. Telemetric platforms for orthopaedic implants are being optimized to minimize electric energy consumption [22, 23]. Also, activation circuits to wake up deep sleep electronics have already been developed to instrument hip prostheses [24].

Several causes of implant failures were already identified [25]. Loosening, infection, instability, heterotopic ossification or fractures not only can conduct to pain and inability to walk, to self-care and to perform activities of daily living, but also can cause cardiovascular, pulmonary, renal, arterial, nerve or infectious complications, or even malignancy. More than 80% of the non-success surgical procedures are due to loosening of the prosthetic stem and cup [26]. Methods for hip loosening detection in hip implants, as well as to identify the regions impaired by this progressive failure throughout the implant's life span, are currently being proposed [19, 26]. Efficient power management circuits were designed to energize telemetric system of smart hip implants [27]. Even though the number of methods and configurations to transduce energy from the surrounding environment into electric energy is increasing [28–32], few research efforts have been conducted to provide electric power supply for instrumented hip prosthesis. Vibration-based energy harvesting is being considered the most appropriate method to generate electric energy to supply the active elements of instrumented prostheses [27]. In order to enable loosening detection, an electromagnetic power transducer was recently proposed by Morais et al. [27] to harvest electrical energy from the human gait to supply smart hip prostheses. However, it was designed only with a single generator, which decreases the reliability of the electric power generation because it is not a redundant structure for power supplying. No studies have been reported about methods to ensure high reliability of the electric energy generation on instrumented prostheses.

1.2   Method

Three vibration-based energy harvesting systems were designed to implement a reliable electric power supply of a smart hip prosthesis. Linear models were developed in order to analyse their accuracy predicting the energy generation. Each generator was independently tested. Voltage generation was acquired from different rotational and translational movements.

1.3   Paper Contribution

This paper's main contribution is to validate the multi-source generation concept applied to smart hip prostheses. The main goal in the implementation of these harvesting systems is to enable a multifunctional ability of the hip prosthesis, namely to monitor and report failures, and carry out mechanical-based therapeutic prescriptions.

1.4   Main Conclusions

This study shows that it is possible to implement high reliability electric power supplies for active hip prostheses. It was also concluded that linear models of the generators are very inaccurate in this particular application. Experimental results show that they must be optimized in order to maximize their performance during typical walking speeds and to reduce their volume, which demands for accurate non-linear models to predict the energy generation for multi-displacements of the hip prostheses.

1.5   Outline

The new concept of smart hip prosthesis is introduced in section 2. The design of the three power generator prototypes is detailed in section 3. Experimental and simulation results are presented in section 4. Discussion and conclusions are stated respectively in sections 5 and 6.

2   The New Concept of Smart Hip Prosthesis

Passive prostheses are orthopaedic implants without active components implemented to overcome failures that may occur over time. They are designed: (1) without information about themselves, about the physiological environment that surrounds them and about how to fix their own problems; (2) without resources to eliminate causes of failures; (3) without a "true" connection with medical specialists. The ineffectiveness of this method to overcome complications after primary joint replacement is caused by this passivity, because a maximal interaction with the surrounding physiological environment is not taken into account. The concept of smart orthopaedic implant was proposed to be based on a *smartness-to-measure* methodology [33], but is becoming obsolete as the concept of individualized therapy evolves for accommodating patient physiologic idiosyncrasies [34]. Several studies have contributed to identify the function of strain,

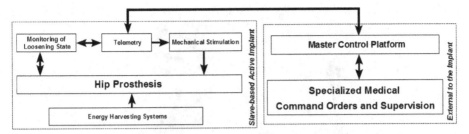

Fig. 1. Block diagram of the new concept of smart orthopaedic implant

load and frequency of mechanical stimulus on the osteogenic responses [35–37], which contributed to hypothesize that the implementation of mechanisms to carry out artificial stimulation programs on bone cells can be an effective and efficient solution to overcome hip loosening. The new concept is based on the *smartness-to-act* methodology in order to prevent and cure failures following primary arthroplasty, avoiding the need of revision procedures. Smart orthopaedic devices must ensure personalized therapy, through: (1) remote communication with external systems outside the human body; (2) monitoring of physical and biological states which can be used to detect early causes of failure; (3) decision-making ability when causes of failure are detected; (4) mechanical actuation-based therapy in the physical and biological states which have a decisive influence on the lifetime of the implant, depending on medical supervision. This long-term survivability biosystem is based on the *tip-of-the-spear* methodology [34], envisioning the application of Paul Ehrlich's *magic bullet* concept [38] to orthopaedic devices: implants must have enough "knowledge" and tools to administer the therapy in the most suitable place and at the most suitable time. To validate the concept, a prototype is being developed, according to the Fig. 1, in order to enable the early detection and cure of aseptic loosening throughout the life span of the hip prosthesis. The methodology is based on the implementation of mechanical micro-stimulation to remodel the bone surrounding the implant in the regions where loosening is detected. Several subsystems were already implemented by this research team toward the validation of a new concept of smart hip prostheses, namely: (1) a generator of electric energy based on double permanent magnet vibration, as well as power management schemes [27]; (2) a telemetric architecture [24, 27, 39]; (3) a piezoresistive-based mechanism to detect prosthetic loosening [26]; (4) a piezoelectric-based stimulation mechanism [35, 40, 41].

3   Material and Methods

3.1   Hip Prosthesis Prototype

A *Metabloc*^{TM} straight stem system (*Zimmer* Corporate, Warsaw, Indiana, EUA), size 10, was hollowed to design a hip prosthesis prototype comprising three vibration-based electric power generators (Fig. 2). Fatigue tests on hollow bone implants were already conducted [16], which have proved the safety of the approach.

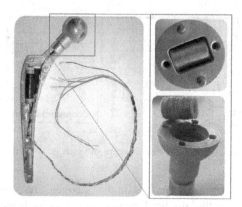

Fig. 2. Multiple Energy harvesting systems for a hip prosthesis prototype

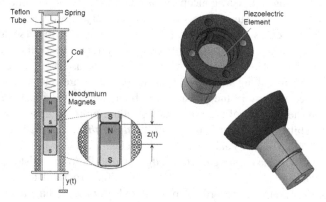

Fig. 3. (Left) Scheme of the TEEH generator; (Right) CAD of the PEH generator

3.2   Translation-Based Electromagnetic Power Generator (TEEH)

Power Generator Prototype Design. An electromagnetic power transducer was de-signed in the body of the hip prosthesis, as shown in Fig. 3 (left). It transduces mechan-ical movements, from the abduction-adduction and flexion-extension axes, into electric energy. The generator prototype comprises an extension coil spring ($K = 2.45$ N/m, 5 mm of diameter and 0.2 mm^2 of wire section) and 2 neodymium disc magnets N35 (6 mm of diameter, 6 mm of height and 1.22 T of magnetic field). These magnets are suspended inside a Teflon tube ($c_m = 0.04$) where enamelled copper wire (0.1 mm of diameter, 27 mm of length and 1.72×10^{-8} Ωm of electrical resistivity) was wound ($N = 2000$ turns, 124.4 Ω of total wire resistance), which in turn was attached to the hip prosthesis fixture. The coil and the prosthesis make up the body frame. A relative displacement $z(t)$ between the magnets and the frame comes up due to the hip displace-ments $y(t)$, which are transmitted by the body frame.

Linear Model of the TEEH System. Linearly, vibration-based generators can be modelled as second-order mass-spring-damper systems [29, 42, 43], as described by equation (1). The mechanical structure can be modelled as an inertial frame where a suspended mass is coupled to a spring, which in turn is coupled to a damping element. The mass represents a set of magnets and the damper represents the sum of the comprising parasitic losses and the electrical energy extracted by the transducer [42].

$$m\ddot{z}(t) + c\dot{z}(t) + kz(t) = -m\ddot{y}(t) \tag{1}$$

m is the mass of the magnets, k the stiffness of the spring and c is the damping coefficient. When this damping system is excited by an external sinusoidal vibration $y(t) = Y\sin(\omega t)$, the solution of (1) is given by (2). $\omega_n = (k/m)^{1/2} = 4.98$ Hz is the natural frequency and $\zeta = c/2m\omega_n = 0.256$ its total damping ratio.

$$z(t) = \frac{Y\omega^2}{\left(\left(\omega_n^2 - \omega^2\right)^2 + \left(2\zeta\omega_n\omega\right)^2 \right)^{1/2}} \sin\left(\omega t - arctg\left(\frac{2\zeta\omega_n\omega}{\omega_n^2 - \omega^2} \right) \right) \tag{2}$$

3.3   Rotation-Based Electromagnetic Power Generator (REEH)

Power Generator Prototype Design. An electromagnetic transducer was designed using the modular ball head of the hip prosthesis and the acetabular component. Energy is harvested from the rotation around the flexion-extension axis and around the inward-outward. The ball head was hollowed to comprise a circular winding of enamelled copper wire (AWG 42, 0.063 mm of diameter), which was coiled ($N = 4710$ turns, 682 Ω of total wire resistance, 117.1 m of total length of the coil, 7.92 mm of average diameter) around a Teflon tube (5.8 mm of diameter, 12 mm of length), whose core was designed to be a steel cylinder (4 mm of diameter, 14 mm of length, 100 of relative permeability). 24 neodymium disc magnets N52 (6 mm of diameter, 2 mm of height and 1.48 T of magnetic field) were placed on the structure of an acetabular component, in order to set the magnetic field lines over the volume of the upper half of the ball head, as presented by Fig. 4 (right). 6 groups of 2 magnets, positioned equidistantly, were settled symmetrically in the acetabulum with 6 other groups of 2 magnets, which were also equidistantly positioned. Figure 4 (left) illustrate a scheme of the REEH generator.

Linear Model of the REEH System. The total harvested energy is the total sum of the energy that can be harvested from the rotation around the flexion-extension axis and the energy acquired from the rotation around the inward-outward axis, according to equation (3).

$$V_{emf}^{\hat{x}\hat{z}} = -\pi R^2 N B \frac{d\alpha_{\hat{z}}}{dt} sin(\alpha_{\hat{z}}) - \pi R^2 N B \frac{d\alpha_{\hat{x}}}{dt} sin(\alpha_{\hat{x}}) \tag{3}$$

3.4   Piezoelectric Power Generator (PEH)

Power Generator Prototype Design. A piezoelectric power generator was designed to harvest energy from the axial load over the hip joint. Figure 3 (right) provides a CAD

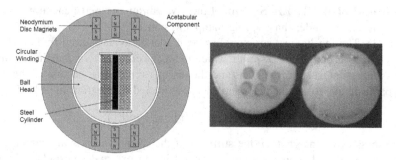

Fig. 4. (left) Scheme of the REEH transducer; (right) Acetabulum of high density polyethylene

modulation of the PEH generator. A piezoelectric ceramic diaphragm (ref. 7BB-12-9, *muRata* Corporate, Kyoto, Japan) with 9 mm of diameter and 0.22 mm of thickness (12 mm of plate size, 0.1 mm of plate thickness and 9.0 ± 1.0 kHz of resonant frequency) was placed on the lower half of ball head of the hip prosthesis.

Neural Network Model of the PEH System. Piezoelectric energy harvesters can be modelled as mechanical damping systems [29, 43]. Because the piezoelectric element is attached to the hip prosthesis structure, the mechanical damping ratio and proof mass are very difficult to find due to the geometry of the prosthesis. An artificial neural network (ANN) model was developed to predict the power and energy generation [44]. A multilayer "feed-forward" ANN was trained to perform the matching between a series of pairs of the frequency and amplitude of sinusoidal axial forces, and the average power and peak-to-peak voltage generation. The ANN consists of one input layer, with two neurons, two hidden layers, with seven neurons each, and one output layer, with two neurons, as shown in figure 5 and predicted by equation (4). The Levenberg-Marquardt's algorithm was used as the training algorithm and the mean square error of 1.0×10^{-20} as the convergence criteria for the network training. Sigmoid functions (Tansig) for the hidden layers and linear function (Purelin) for the output layer were used as the transfer functions.

$$\mathbf{y}_N = f_L\left(\mathbf{LW}_2 f_S\left(\mathbf{LW}_1 f_S\left(\mathbf{IW}_1 \mathbf{i}_N + \mathbf{b}_1\right) + \mathbf{b}_2\right) + \mathbf{b}_3\right) \tag{4}$$

Here, $\mathbf{y}_N$ is the output 2×1 matrix, $\mathbf{i}_N$ is the 2×1 input matrix, $\mathbf{IW}_1$ is a input weight 7×2 matrix, $\mathbf{LW}_1$ and $\mathbf{LW}_2$ are respectively layer weight 7×7 and 2×7 matrices, and $\mathbf{b}_1$, $\mathbf{b}_2$ and $\mathbf{b}_3$ are respectively bias 7×1, 7×1 and 2×1 matrices. f_L and f_S are linear and sigmoid functions, respectively.

4   Experimental and Simulation Results

The experimental average and peak power, energy and peak-to-peak voltage generation were analysed. These experimental results were compared with the linear models reported in sections 3.2, 3.3 and 3.4.

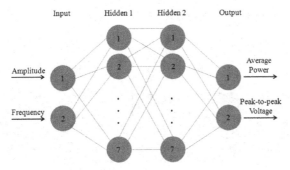

Fig. 5. ANN used to model the average power and peak-to-peak voltage of the PEH generator

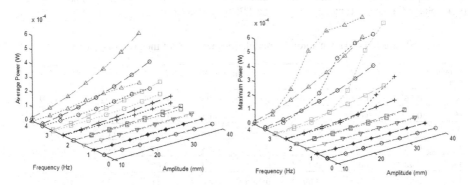

Fig. 6. Experimental (dotted line) and simulated (dashed line) average (in the left) and maximum (in the right) power harvested from the TEEH transducer

4.1 TEEH Results

A load resistance of 979 Ω was used to enable the energy transfer when sinusoidal input vibrations, with amplitudes in the range 10 mm to 40 mm and frequencies in the range 0.5 Hz to 4 Hz, were applied to this generator. Figure 6 shows the results of the experimental and simulated average and maximum power, whereas Fig. 7 highlights the results of the experimental and simulated energy and peak-to-peak voltage. Tables 1 to 4 report the modulation errors. The maximum energy harvested was 53.7 μJ/s when the sinusoidal function has an amplitude of 40 mm and a frequency of 4 Hz. This harvester is able to provide 567.4 μW of instantaneous peak power when the input is excited with an amplitude of 40 mm and a frequency of 3 Hz.

4.2 REEH Results

A load resistance of 8.98 kΩ was used to enable energy transfer of this generator when sinusoidal rotations in the flexion-extension axis, with amplitudes in the range 50° to 70° and frequencies in the range 0.5 Hz to 2.5 Hz, were applied to the generator. Using the magnetic field in the winding measured at the ends of the winding (80 mT), the experimental and simulated results are reported in Fig. 8 and Fig. 9. Tables 5 to 8 report

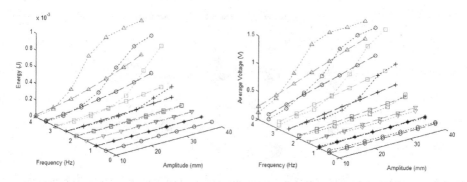

Fig. 7. Experimental (dotted line) and simulated (dashed line) energy (in the left) and peak-to-peak voltage (in the right) harvested from the TEEH transducer

Table 1. TEEH generator — modulation error (W) of the average power

Amplitude (mm)	Frequency (Hz)							
	0.5	1	1.5	2	2.5	3	3.5	4
10	$0.06^{(1)}$	$0.009^{(2)}$	$0.05^{(2)}$	$0.02^{(3)}$	$0.04^{(3)}$	$0.08^{(3)}$	$0.02^{(4)}$	$0.03^{(4)}$
15	$0.02^{(1)}$	$0.02^{(2)}$	$0.11^{(2)}$	$0.04^{(3)}$	$0.08^{(3)}$	$0.18^{(3)}$	$0.03^{(4)}$	$0.05^{(4)}$
20	$0.13^{(1)}$	$0.04^{(2)}$	$0.20^{(2)}$	$0.06^{(3)}$	$0.15^{(3)}$	$0.31^{(3)}$	$0.05^{(4)}$	$0.09^{(4)}$
25	$0.26^{(1)}$	$0.06^{(2)}$	$0.31^{(2)}$	$0.10^{(3)}$	$0.23^{(3)}$	$0.47^{(3)}$	$0.08^{(4)}$	$0.12^{(4)}$
30	$0.43^{(1)}$	$0.09^{(2)}$	$0.44^{(2)}$	$0.14^{(3)}$	$0.32^{(3)}$	$0.62^{(3)}$	$0.10^{(4)}$	$0.18^{(4)}$
35	$0.62^{(1)}$	$0.12^{(2)}$	$0.60^{(2)}$	$0.19^{(3)}$	$0.39^{(3)}$	$0.71^{(3)}$	$0.14^{(4)}$	$0.26^{(4)}$
40	$0.85^{(1)}$	$0.15^{(2)}$	$0.78^{(2)}$	$0.24^{(3)}$	$0.48^{(3)}$	$0.85^{(3)}$	$0.19^{(4)}$	$0.35^{(4)}$

$^{(1)} \times 10^{-7}$; $^{(2)} \times 10^{-5}$; $^{(3)} \times 10^{-4}$; $^{(4)} \times 10^{-3}$.

Table 2. TEEH generator — modulation error (W) of the maximum power

Amplitude (mm)	Frequency (Hz)							
	0.5	1	1.5	2	2.5	3	3.5	4
10	$0.70^{(1)}$	$0.03^{(2)}$	$0.01^{(2)}$	$0.01^{(3)}$	$0.001^{(4)}$	$0.01^{(4)}$	$0.01^{(4)}$	$0.02^{(4)}$
15	$0.30^{(1)}$	$0.02^{(2)}$	$0.06^{(2)}$	$0.03^{(3)}$	$0.003^{(4)}$	$0.01^{(4)}$	$0.02^{(4)}$	$0.02^{(4)}$
20	$0.54^{(1)}$	$0.01^{(2)}$	$0.15^{(2)}$	$0.05^{(3)}$	$0.01^{(4)}$	$0.02^{(4)}$	$0.0002^{(4)}$	$0.08^{(4)}$
25	$0.53^{(1)}$	$0.0002^{(2)}$	$0.27^{(2)}$	$0.07^{(3)}$	$0.02^{(4)}$	$0.003^{(4)}$	$0.06^{(4)}$	$0.26^{(4)}$
30	$0.48^{(1)}$	$0.05^{(2)}$	$0.35^{(2)}$	$0.11^{(3)}$	$0.004^{(4)}$	$0.09^{(4)}$	$0.23^{(4)}$	$0.29^{(4)}$
35	$0.37^{(1)}$	$0.04^{(2)}$	$0.39^{(2)}$	$0.13^{(3)}$	$0.09^{(4)}$	$0.32^{(4)}$	$0.25^{(4)}$	$0.22^{(4)}$
40	$0.69^{(1)}$	$0.12^{(2)}$	$0.64^{(2)}$	$0.03^{(3)}$	$0.17^{(4)}$	$0.44^{(4)}$	$0.22^{(4)}$	$0.14^{(4)}$

$^{(1)} \times 10^{-6}$; $^{(2)} \times 10^{-5}$; $^{(3)} \times 10^{-4}$; $^{(4)} \times 10^{-3}$.

the modulation errors. The maximum energy harvested was 0.77 μJ/s. The "plus" sign refers to peak-to-peak amplitudes in the range $-10°$ to $60°$, $-10°$ to $50°$ and $-10°$ to $40°$; the "square" sign refers to peak-to-peak amplitudes in the range $-20°$ to $50°$, $-20°$ to $40°$ and $-20°$ to $30°$.

Table 3. TEEH generator — modulation error (J) of the energy

Amplitude (mm)	Frequency (Hz)							
	0.5	1	1.5	2	2.5	3	3.5	4
10	0.17[1]	0.01[2]	0.03[2]	0.01[3]	0.003[4]	0.01[4]	0.01[4]	0.02[4]
15	0.16[1]	0.01[2]	0.10[2]	0.03[3]	0.01[4]	0.01[4]	0.02[4]	0.02[4]
20	0.14[1]	0.02[2]	0.18[2]	0.06[3]	0.01[4]	0.02[4]	0.02[4]	0.11[4]
25	0.13[1]	0.04[2]	0.29[2]	0.08[3]	0.02[4]	0.01[4]	0.08[4]	0.40[4]
30	0.11[1]	0.07[2]	0.42[2]	0.12[3]	0.01[4]	0.07[4]	0.29[4]	0.50[4]
35	0.09[1]	0.10[2]	0.53[2]	0.14[3]	0.06[4]	0.29[4]	0.44[4]	0.47[4]
40	0.06[1]	0.14[2]	0.69[2]	0.07[3]	0.13[4]	0.49[4]	0.45[4]	0.41[4]

[1] $\times 10^{-6}$; [2] $\times 10^{-5}$; [3] $\times 10^{-4}$; [4] $\times 10^{-3}$.

Table 4. TEEH generator — modulation error (V) of the peak-to-peak voltage

Amplitude (mm)	Frequency (Hz)							
	0.5	1	1.5	2	2.5	3	3.5	4
10	0.05	0.02	0.003	0.01	0.003	0.06	0.08	0.11
15	0.03	0.02	0.01	0.04	0.01	0.05	0.08	0.06
20	0.04	0.01	0.03	0.06	0.07	0.08	0.04	0.19
25	0.03	0.01	0.04	0.05	0.06	0.02	0.16	0.51
30	0.03	0.01	0.05	0.08	0.05	0.25	0.47	0.52
35	0.03	0.01	0.04	0.06	0.30	0.60	0.55	0.41
40	0.03	0.02	0.06	0.04	0.38	0.74	0.50	0.31

Table 5. REEH generator — modulation error (W) of the average power [error ("plus" test) | error ("square" test)]

Amplitude (°)	Frequency (Hz)				
	0.5	1	1.5	2	2.5
50	0.11[1] | 0.06[1]	0.04[2] | 0.25[1]	0.09[2] | 0.06[2]	0.18[2] | 0.10[2]	0.33[2] | 0.22[2] |
60	0.16[1] | 0.11[1]	0.06[2] | 0.49[1]	0.22[2] | 0.10[2]	0.40[2] | 0.19[2]	0.54[2] | 0.29[2] |
70	0.30[1] | 0.17[1]	0.11[2] | 0.78[1]	0.24[2] | 0.14[2]	0.44[2] | 0.41[2]	0.32[2] | 0.22[2] |

[1] $\times 10^{-7}$; [2] $\times 10^{-6}$.

Table 6. REEH generator — modulation error (W) of the maximum power [error ("plus" test) | error ("square" test)].

Amplitude (°)	Frequency (Hz)				
	0.5	1	1.5	2	2.5
50	0.07[2] | 0.36[1]	0.24[2] | 0.09[2]	0.05[3] | 0.18[2]	0.08[3] | 0.03[3]	0.17[3] | 0.09[3] |
60	0.13[2] | 0.45[1]	0.31[2] | 0.20[2]	0.11[3] | 0.32[2]	0.23[3] | 0.08[3]	0.24[3] | 0.13[3] |
70	0.22[2] | 0.75[1]	0.58[2] | 0.36[2]	0.11[3] | 0.59[2]	0.22[3] | 0.18[3]	0.18[3] | 0.12[3] |

[1] $\times 10^{-7}$; [2] $\times 10^{-6}$; [3] $\times 10^{-5}$.

Table 7. REEH generator — modulation error (J) of the energy [error ("plus" test) | error ("square" test)]

Amplitude (°)	Frequency (Hz)				
	0.5	1	1.5	2	2.5
50	$0.16^{(1)}$ | $0.09^{(1)}$	$0.06^{(2)}$ | $0.04^{(2)}$	$0.14^{(2)}$ | $0.08^{(2)}$	$0.28^{(2)}$ | $0.15^{(2)}$	$0.49^{(2)}$ | $0.33^{(2)}$
60	$0.24^{(1)}$ | $0.16^{(1)}$	$0.09^{(2)}$ | $0.07^{(2)}$	$0.34^{(2)}$ | $0.15^{(2)}$	$0.59^{(2)}$ | $0.29^{(2)}$	$0.81^{(2)}$ | $0.43^{(2)}$
70	$0.44^{(1)}$ | $0.26^{(1)}$	$0.16^{(2)}$ | $0.12^{(2)}$	$0.35^{(2)}$ | $0.21^{(2)}$	$0.65^{(2)}$ | $0.61^{(2)}$	$0.48^{(2)}$ | $0.33^{(2)}$

$^{(1)} \times 10^{-6}$; $^{(2)} \times 10^{-5}$.

Table 8. REEH generator — modulation error (V) of the peak-to-peak voltage [error ("plus" test) | error ("square" test)]

Amplitude (°)	Frequency (Hz)				
	0.5	1	1.5	2	2.5
50	0.03 | 0.02	0.05 | 0.03	0.08 | 0.04	0.11 | 0.05	0.14 | 0.09
60	0.04 | 0.02	0.06 | 0.04	0.12 | 0.05	0.16 | 0.07	0.15 | 0.09
70	0.05 | 0.02	0.08 | 0.06	0.11 | 0.06	0.13 | 0.11	0.09 | 0.06

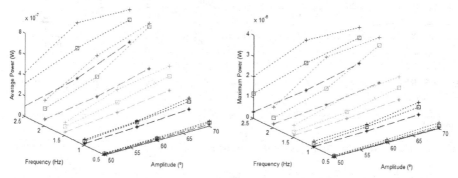

Fig. 8. Experimental (dotted line) and simulated (dashed line) average (in the left) and maximum (in the right) power harvested from the REEH generator

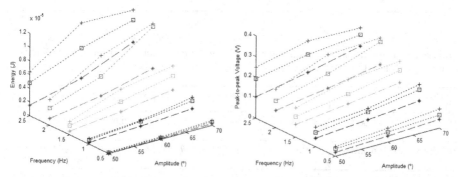

Fig. 9. Experimental (dotted line) and simulated (dashed line) energy (in the left) and peak-to-peak voltage (in the right) harvested from the REEH generator

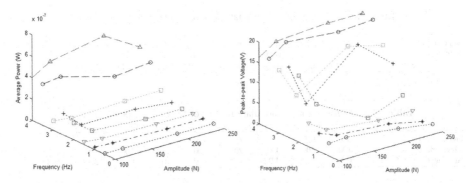

Fig. 10. Experimental average power and peak-to-peak voltage results harvested from the PEH

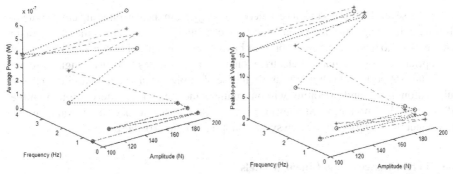

Fig. 11. Validation of the average power harvested from the PEH generators (dash-dot line refers to the network output)

4.3 PEH Results

External sinusoidal forces, with amplitudes in the range 100 N to 250 N and frequencies in the range 0.5 Hz to 4 Hz, were applied to the generator. A load of 1 MΩ was used to enable the energy transfer. Figure 10 reports the experimental average power and peak-to-peak voltage results, whereas Fig. 11 introduces the validation results of the "feed-forward" ANN, using only data not used in the training process. Tables 9 to 10 report the modulation errors. The maximum energy harvested was 0.6 μ J/s.

5 Discussion

5.1 Need for Multi-source Harvester Systems

The high reliability of the electric energy generation system is a necessary condition to ensure high reliability of the therapy based on mechanical stimulation. It is technologically possible to implement active implants with the ability to monitor failures, to communicate the states of its surrounding physiological environment to the medical specialist and carry out mechanical-based therapeutic prescriptions ordered by the specialist. These operations demand a full availability of electric energy.

Table 9. PEH generator — modulation error of the average power

Frequency (Hz)	0.5	0.5	1	1	1.5	3	3.5	4	4
Amplitude (N)	100	200	125	200	200	125	200	100	200
Absolute error (W)[1]	0.0004	0.003	0.0003	0.003	0.002	0.23	0.11	0.03	0.13

[1] $\times 10^{-6}$.

Table 10. PEH generator — modulation error of the peak-to-peak voltage

Frequency (Hz)	0.5	0.5	1	1	1.5	3	3.5	4	4
Amplitude (N)	100	200	125	200	200	125	200	100	200
Absolute error (W)	0.32	1.18	1.22	1.18	0.80	10.19	0.98	3.49	0.88

5.2 Choice of the Energy Harvesting Method

There are several methods to harvest electric energy from the surrounding environment. Biofuel cells, magnetic induction, thermoelectric and vibration are some of the main sources used to harvest energy. Vibration-based generation is being considered the most appropriate solution to harvest electrical energy on prostheses [45]. Although there are no studies supporting this hypothesis, their ease of implementation, their ability of being fully autonomous and operating without maintenance, ensuring safety throughout the life span of the implant, were relevant features taken into consideration to validate this new concept of energy harvesting systems applied to hip prostheses.

5.3 Performance of the Linear Models

Linear models can only ensure accuracy within a narrow window of the systems' operating range. The modulation errors, presented in sections 4.1, 4.2 and 4.3, confirm the inaccuracy of the linear models of the generators, especially those used to model the behaviour of the TEEH generator. The real translational and rotational hip displacements, the friction and gravity forces acting on these systems, the non-linearity behavior of inductors and transduction damping coefficients were jeopardized from linear models. Because experimental results show a higher performance of the TEEH generator in this particular application, it is mandatory the development of non-linear models for high accuracy prediction of electric energy generation considering a broad specter of the generators' operation.

5.4 Optimization of the Multi-source Harvester System

Each transducer must be optimized in order to maximize electric generation during typical walking speeds, namely in the range between 0.5 Hz and 2 Hz. The optimization of the TEEH and PEH generators demands for a "perfect" match between the frequency of the hip kinematics and their resonant frequencies. The implementation of a continuous-matching system is very complex [46] because the duration and frequency of every-day human activities are unpredictable [47]. New methods must be developed to ensure high performance tracking of the hip kinematics' frequency. Each generator must be optimized to supply as much electric energy as demanded by the active mechanics of smart prostheses, even though they may require different periods to generate the same amount of energy.

6 Conclusions

A therapeutic methodology to cure failures of hip prostheses following primary artroplasty, based on a personalized approach, would be of great importance for the quality of life of many patients. Multi-source electric energy generation systems for orthopaedic implants contribute forward the implementation of individualized medicine approaches in the scope of embedded smart bone implants. This new concept was validated with the use of three vibrational energy harvesters: two electromagnetic generators and a piezoelectric. Experimental results confirm the inaccuracy of the generators' linear models operating on hip prostheses fixtures, which invalidates their use in optimization routines.

There is ongoing work to:

1. Develop non-linear models of vibrational energy harvesters. The main goal is to reduce the volume of each generator and maximize their performance;
2. Identify the most appropriate method to generate electrical energy on hip implants;
3. Design a new method to ensure an effective tracking of the frequency of the hip kinematics. The set of generators must be synchronized with the hip dynamics in order to ensure all energy requirements demanded by active elements of the smart prostheses;
4. Design an energy management system for a multitude of power sources;
5. Design of energy harvesting systems which are independent of failures due to contact stresses (for instance, magnetically levitated generators);
6. Design of a redundant multi-source harvester structure. Such a redundant ability, along with the requirements introduced in the previous number and in section 5.4, are sufficient to ensure reliability of the electric energy generation system.

References

1. Ren, W., Blasier, R., Peng, X., Shi, T., Wooley, P.H., Markel, D.: Effect of oral erythromycin therapy in patients with aseptic loosening of joint prostheses. Bone 44(4), 671–677 (2009)
2. Esposito, S., Leone, S.: Prosthetic joint infections: microbiology, diagnosis, management and prevention. International Journal of Antimicrobial Agents 32(4), 287–293 (2008)
3. Dreinhöfer, K.E., Dieppe, P., Günther, K.P., Puhl, W.: EUROHIP - Health Technology Assessment of Hip Arthroplasty in Europe. Springer, New York (2009)
4. Ramos, A., Completo, A., Relvas, C., Simões, J.: Design process of a novel cemented hip femoral stem concept. Materials and Design 33, 313–321 (2012)
5. Jun, Y., Choi, K.: Design of patient-specific hip implants based on the 3D geometry of the human femur. Advances in Engineering Software 41(4), 537–547 (2010)
6. Simões, J.A., Marques, A.T.: Design of a composite hip femoral prosthesis. Materials and Design 26(5), 391–401 (2005)

7. Fehring, T.K., Odum, S.M., Troyer, J.L., Iorio, R., Kurtz, S.M., Lau, E.C.: Joint replacement access in 2016: A supply side crisis. The Journal of Arthroplasty 25(8), 1175–1181 (2010)

8. Ong, K.L., Lau, E., Suggs, J., Kurtz, S.M., Manley, M.T.: Risk of subsequent revision after primary and revision total joint arthroplasty. Clinical Orthopaedics and Related Research 468(11), 3070–3076 (2010)

9. Kurtz, S.M., Lau, E., Ong, K., Zhao, K., Kelly, M., Bozic, K.J.: Future young patient demand for primary and revision joint replacement. Clinical Orthopaedics and Related Research 467(10), 2606–2612 (2009)

10. Rydell, N.W.: Forces acting on the femoral head-prosthesis. A study on strain gauge supplied prostheses in living persons. Acta Orthopaedica Scandinavica 37(88), 1–132 (1966)

11. Damm, P., Graichen, F., Rohlmann, A., Bender, A., Bergmann, G.: Total hip joint prosthesis for in vivo measurement of forces and moments. Medical Engineering & Physics 32(1), 95–100 (2010)

12. Boyle, C., Kim, I.Y.: Comparison of different hip prosthesis shapes considering micro-level bone remodeling and stress-shielding criteria using three-dimensional design space topology optimization. Journal of Biomechanics 44(9), 1722–1728 (2011)

13. Nikooyan, A.A., Veeger, H.E., Westerhoff, P., Graichen, F., Bergmann, G., van der Helm, F.C.: Validation of the delft shoulder and elbow model using in-vivo glenohumeral joint contact forces. Journal of Biomechanics 43(15), 3007–3014 (2010)

14. Rohlmann, A., Gabel, U., Graichen, F., Bender, A., Bergmann, G.: An instrumented implant for vertebral body replacement that measures loads in the anterior spinal column. Medical Engineering & Physics 29(5), 580–585 (2007)

15. Puers, R., Catrysse, M., Vandevoorde, G., Collier, R., Louridas, E., Burny, F., Donkerwolcke, M., Moulart, F.: A telemetry system for the detection of hip prosthesis loosening by vibration analysis. Sensors and Actuators A: Physical 85(1-3), 42–47 (2000)

16. Heinlein, B., Graichen, F., Bender, A., Rohlmann, A., Bergmann, G.: Design, calibration and pre-clinical testing of an instrumented tibial tray. Journal of Biomechanics 40, S4–S10 (2007)

17. Almouahed, S., Gouriou, M., Hamitouche, C., Stindel, E., Roux, C.: Design and evaluation of instrumented smart knee implant. IEEE Transactions on Biomedical Engineering 58(4), 971–982 (2011)

18. Graichen, F., Bergmann, G., Rohlmann, A.: Hip endoprosthesis for in vivo measurement of joint force and temperature. Journal of Biomechanics 32(10), 1113–1117 (1999)

19. Marschner, U., Grätz, H., Jettkant, B., Ruwisch, D., Woldt, G., Fischer, W.J., Clasbrummel, B.: Integration of a wireless lock-in measurement of hip prosthesis vibrations for loosening detection. Sensors and Actuators A: Physical 156(1), 145–154 (2009)

20. Rowlands, A., Duck, F.A., Cunningham, J.L.: Bone vibration measurement using ultrasound: Application to detection of hip prosthesis loosening. Medical Engineering & Physics 30(3), 278–284 (2008)

21. Bergmann, G., Graichen, F., Rohlmann, A., Westerhoff, P., Heinlein, B., Bender, A., Ehrig, R.: Design and calibration of load sensing orthopaedic implants. Journal of Biomechanical Engineering 130(2), 021009 (2008)

22. Graichen, F., Arnold, R., Rohlmann, A., Bergmann, G.: Implantable 9-channel telemetry system for in vivo load measurements with orthopedic implants. IEEE Transactions on Biomedical Engineering 54(2), 253–261 (2007)

23. Valdastri, P., Rossi, S., Menciassi, A., Lionetti, V., Bernini, F., Recchia, F.A., Dario, P.: An implantable ZigBee ready telemetric platform for *in vivo* monitoring of physiological parameters. Sensors and Actuators A: Physical 142(1), 369–378 (2008)

24. Morais, R., Frias, C.M., Silva, N.M., Azevedo, J.L.F., Serôdio, C.A., Silva, P.M., Ferreira, J.A.F., Simões, J.A.O., Reis, M.C.: An activation circuit for battery-powered biomedical implantable systems. Sensors and Actuators A: Physical 156(1), 229–236 (2009)

25. Kim, P.R., Beaulé, P.E., Laflamme, G.Y., Dunbar, M.: Causes of early failure in a multicenter clinical trial of hip resurfacing. The Journal of Arthroplasty 23(6), 44–49 (2008)

26. Alpuim, P., Filonovich, S.A., Costa, C.M., Rocha, P.F., Vasilevskiy, M.I., Lanceros-Mendez, S., Frias, C., Marques, A.T., Soares, R., Costa, C.: Fabrication of a strain sensor for bone implant failure detection based on piezoresistive doped nanocrystalline silicon. Journal of Non-Crystalline Solids 354(19-25), 2585–2589 (2008)

27. Morais, R., Silva, N.M., Santos, P.M., Frias, C.M., Ferreira, J.A.F., Ramos, A.M., Simões, J.A.O., Baptista, J.M.R., Reis, M.C.: Double permanent magnet vibration power generator for smart hip prosthesis. Sensors and Actuators A: Physical 172(1), 259–268 (2011)

28. Wei, X., Liu, J.: Power sources and electrical recharging strategies for implantable medical devices. Frontiers of Energy and Power Engineering in China 2(1), 1–13 (2008)

29. Kaźmierski, T.J., Beeby, S.: Energy Harvesting Systems - Principles, Modeling and Applications. Springer, New York (2011)

30. Kerzenmacher, S., Ducrée, J., Zengerle, R., von Stetten, F.: Energy harvesting by implantable abiotically catalyzed glucose fuel cells. Journal of Power Sources 182(1), 1–17 (2008)

31. Carmo, J.P., Ribeiro, J.F., Silva, M.F., Goncalves, L.M., Correia, J.H.: Thermoelectric generator and solid-state battery for stand-alone microsystems. Journal of Micromechanics and Microengineering 20(8), 1–8 (2010)

32. Lu, M., Guang Zhang, G., Fu, K., Hao Yu, G., Su, D., Feng Hu, J.: Gallium nitride schottky betavoltaic nuclear batteries. Energy Conversion and Management 52(4), 1955–1958 (2011)

33. Burny, F., Donkerwolcke, M., Moulart, F., Bourgois, R., Puers, R., Schuylenbergh, K.V., Barbosa, M., Paiva, O., Rodes, F., Bégueret, J.B., Lawes, P.: Concept, design and fabrication of smart orthopedic implants. Medical Engineering & Physics 22(7), 469–479 (2000)

34. Sakamoto, J.H., van de Ven, A.L., Godin, B., Blanco, E., Serda, R.E., Grattoni, A., Ziemys, A., Bouamrani, A., Hu, T., Ranganathan, S.I., Rosa, E.D., Martinez, J.O., Smid, C.A., Buchanan, R.M., Lee, S.Y., Srinivasan, S., Landry, M., Meyn, A., Tasciotti, E., Liu, X., Decuzzi, P., Ferrari, M.: Enabling individualized therapy through nanotechnology. Pharmacological Research 62(2), 57–89 (2010)

35. Frias, C., Reis, J., e Silva, F.C., Potes, J., Simões, J., Marques, A.T.: Polymeric piezoelectric actuator substrate for osteoblast mechanical stimulation. Journal of Biomechanics 43(6), 1061–1066 (2010)

36. Tanaka, S.M., Li, J., Duncan, R.L., Yokota, H., Burr, D.B., Turner, C.H.: Effects of broad frequency vibration on cultured osteoblasts. Journal of Biomechanics 36(1), 73–80 (2003)

37. Bacabac, R.G., Smit, T.H., Loon, J.J.W.A.V., Doulabi, B.Z., Helder, M., Klein-Nulend, J.: Bone cell responses to high-frequency vibration stress: does the nucleus oscillate within the cytoplasm? The FASEB Journal 20(7), 858–864 (2006)

38. Winau, F., Westphal, O., Winau, R.: Paul Ehrlich - in search of the Magic Bullet. Microbes and Infection 6, 786–789 (2004)

39. Morais, R., Silva, N., Santos, P., Frias, C., Ferreira, J., Ramos, A., Simões, J., Baptista, J., Reis, M.: Permanent magnet vibration power generator as an embedded mechanism for smart hip prosthesis. Procedia Engineering 5, 766–769 (2010)

40. Frias, C., Reis, J., e Silva, F.C., Potes, J., Simões, J., Marques, A.T.: Piezoelectric actuator: Searching inspiration in nature for osteoblast stimulation. Composites Science and Technology 70(13), 1920–1925 (2010)

41. Reis, J.C.: The Bone/Implant Interface: New Approaches to Old Problems. PhD thesis, University of Évora (2010)

42. Beeby, S.P., Tudor, M.J., White, N.M.: Energy harvesting vibration sources for microsystems applications. Measurement Science and Technology 17, R175–R195 (2006)

43. Priya, S., Inman, D.J.: Energy Harvesting Technologies. Springer, New York (2009)
44. Kalogirou, S.: Applications of artificial neural-networks for energy systems. Applied Energy 67, 17–35 (2000)
45. Yuen, S.C., Lee, J.M., Li, W.J., Leong, P.H.: An AA-sized vibration-based microgenerator for wireless sensors. IEEE Pervasive Computing 6(1), 64–72 (2007)
46. Zhu, D., Tudor, M.J., Beeby, S.P.: Strategies for increasing the operating frequency range of vibration energy harvesters: a review. Measurement Science and Technology 21(2), 1–29 (2010)
47. Morlock, M., Schneider, E., Bluhm, A., Vollmer, M., Bergmann, G., Müller, V., Honl, M.: Duration and frequency of every day activities in total hip patients. Journal of Biomechanics 34(7), 873–881 (2001)

An Integrated Portable Device for the Hand Functional Assessment in the Clinical Practice

Danilo Pani[1], Gianluca Barabino[1], Alessia Dessì[1], Matteo Piga[2], Iosto Tradori[2], Alessandro Mathieu[2], and Luigi Raffo[1]

[1] DIEE - Dept. of Electrical and Electronic Engineering,
University of Cagliari, 09123 Cagliari, Italy
[2] Chair of Rheumatology and Rheumatology Unit,
University and AOU of Cagliari, Italy
{danilo.pani,alessia.dessi,
gianluca.barabino,luigi}@diee.unica.it,
matteopiga@alice.it, iosto.tradori@tiscali.it,
mathieu@medicina.unica.it

Abstract. The functionality of the human hand is of paramount importance for the daily life activity of a subject. Several chronic diseases can have localized lesions on the hands, causing disability, as for the Systemic Sclerosis and Rheumatoid Arthritis. In these cases the evaluation of the hand functionality is a necessary step for setting up the therapeutic and rehabilitation program. This research presents a novel device tackling this problem, allowing the evaluation of hand dexterity and strength on 4 simple rehabilitation exercises. Real-time controlled by a wirelessly connected PC where a C++ physician graphical interface enables a user-friendly management of the assessment, the device provides hitherto unavailable measurements. A first evaluation of the device in a real outpatient rheumatology clinic has been performed and the preliminary results reveal the potentialities of the approach.

Keywords: Functional Hand Assessment, Biomedical Embedded System, Rheumatic Diseases, Rehabilitation.

1 Introduction

Human hand is one the principal instruments enabling the interaction of a subject with the surrounding environment. For such a reason, hand disability is perceived as more invalidating from a functional perspective compared to other disabilities involving different districts. As a matter of fact, hand disability hampers the execution of normal daily life activities such as hair brushing, dressing or cooking. Both hand strength and fine movements are often compromised. From a medical viewpoint, hand functionality assessment represents a necessary evaluation to be performed both at the first examination and during the follow-up of the patient in order to quantify the disability level, to properly set up the pharmacological therapy and to define a personalized rehabilitation program.

J. Gabriel et al. (Eds.): BIOSTEC 2012, CCIS 357, pp. 97–110, 2013.

In particular, rheumatic diseases, such as Rheumatoid Arthritis (RA) and Systemic Sclerosis (SSc) could led to lesions localized on the hands, thus producing the afore-mentioned invalidating effects, requiring an integrated therapy including kinesitherapy and a pharmacological protocol. Differently from RA, where movement limitation is prevalently caused by arthritis, patients with SSc suffer a skin thickening whose conse-quence is a limited mobility which in turn exacerbates the problem. For these patients, the functional deficit is evaluated in a clinical setting using physician assessed biome-chanical measurements such as the Range Of Movement (ROM), the hand extension and strength. Furthermore, physician-administered questionnaires are used to obtain the subject's perception of its disability, i.e. the Dreiser test [4] and the Health Assessment Questionnaire (HAQ) [11]. Only for some evaluations (e.g. grip and pinch strength or ROM) some digital devices are able to provide one-shot measurements but there is a lack of commercial functional evaluation tools able to measure the interesting param-eters associated to the execution of typical rehabilitation exercises prescribed to these patients. They would probably help in the assessment procedures, especially if all the devices are integrated on the same hardware platform.

This chapter presents a system for the real-time assessment of the hand functional-ity on rehabilitation exercises usually exploited with patients affected by RA and SSc. Compared to the typical procedures at the state of the art for such evaluation, exploit-ing either subjective scores or one-shot measurements, the proposed system extracts the relevant information from the real-time monitoring of exercise repetitions with hitherto unavailable precision, reducing the background noise due to fatigue and distraction. The microcontroller-based device exploits 4 sensorized custom-made aids for the evaluation of the hand agility (finger tapping and dynamic rotation) and strength/mobility (isomet-ric rotation and hand extension with counter resistance) and can be easily controlled by a PC connected via a Bluetooth link. A stand-alone C++ graphical user interface (GUI) has been implemented exploiting the Qt 4.7 framework to provide a license-free physician support for the real-time control of the device. The device has been evaluated in a real outpatient rheumatology clinic on 6 voluntary subjects affected either by SSc or RA and the preliminary results reveal the potentialities of the proposed approach.

The remainder of this paper is organized as follows. In Sect. 2 a brief analysis of the related works is provided. Sect. 3 provides an overview of the proposed device and the relative hardware implementation details, whereas the physician GUI is presented in Sect. 4. Sect. 5 presents the results of the device application in an outpatient clinic. Conclusions are presented in Sect. 6.

2 Related Works

For functional assessment of the human hand, the most common evaluation techniques involve pinch and grip exercises. Both the Jamar dynamometer (isometric) and the Vig-orimeter (dynamic) represent well established instruments for the clinical evaluation of the grip strength [10]. Commercial devices such as Pablo by Tyromotion GmbH or the H500 Hand Kit by Biometrics Ltd. allow monitoring also the single finger pinch force. In principle, isometric wrist dynamometer can be also used to estimate the torque applied with the finger when the wrist is in a fixed position, in order to evaluate the hand performance with respect to this task. Usually the digital versions of these devices are able to provide maximum, average and standard deviation of the force, but without any temporal analysis within a series except for the systems exploiting additional electromyographic signals [13]. In [5], a grip measurement device is presented, able to perform also some time measurements but only on a single 4.4s grip exercise for the performance assessment in rheumatic patients. A similar work has been presented in [1] for the parkinsonian patients. In both cases the aim is a one-shot functional assessment rather than the evaluation of a series of exercises. An interesting device for rehabilitation mixing torque and grip force has been presented in [7], but is not intended for performance assessment.

The hand agility (severely affected by rheumatoid arthritis and scleroderma) can be in principle evaluated by means of finger tapping tests, originally conceived to assess both motor speed and control in neuropsychology. From the first mechanical devices, other approaches for the monitoring of this kind of exercise have arisen. Approaches including a passive marker-based motion analyzer [6] present a very complex setup not suited for a fast evaluation. Other approaches, based on sensorized gloves [3], are uncomfortable for patients with hand deformities caused by arthritis. In [9], a touch system based on a 4-finger active sensor (injecting on the hand a small sinusoidal current at 1.5 kHz) has been presented along with its support software. An App (Digital Finger Tapping Test 1.0) with limited functionalities is also available for IPhone users. An approach based on the detection of the exerted force in the tapping activity is presented in [8].

To the best of our knowledge, the realization of a low cost device for the quantitative monitoring of both agility and strength exercises for hand functional evaluation on rheumatic patients during real rehabilitation exercises has not been presented in literature until now.

3   The Hand Functional Assessment Device

The proposed device is conveniently packaged in a lightweight metal briefcase, as shown in Fig. 1. The patient can perform 4 exercises with a single hand at a time, with as many sensorized devices. By using a GUI installed on his PC, the physician can choose which exercise to execute, evaluating in real-time how the patient executes it not only in terms of correct position but also looking at barely perceptible execution parameters that the digital device is able to reveal. For instance, a real-time updated plot discloses sensors wave shape while numerical data such as peak and running-average values are displayed on the GUI, allowing a finer monitoring compared to a traditional visual inspection. From this point of view the device can be conveniently used as a kinesitherapic monitoring system. The physician can stop the execution at any time but an upper bound is imposed by the predefined number of repetitions of the same exercise hard coded in the device firmware. During the exercise execution, the device automatically extracts the relevant measures from the signals and updates their statistics (min, max, avg, std, etc.) in order to provide at every time the parameters needed for the quantitative hand functionality assessment.

From the patient viewpoint, the interaction with the device is very simple. The hand to use is indicated by a led and the movements to be performed exploiting the sensorized embedded aids are well defined by the rehabilitation protocol, as follows.

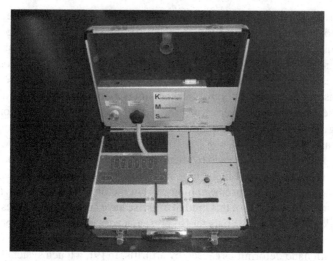

Fig. 1. The prototypical device for the real-time hand functionality assessment

On the vertical panel there are two knobs. The outer one allows the evaluation of the patient manipulation dexterity (exercise of *dynamic rotation*). The patient must rotate as fast as possible the knob using his fingers, shaped in a pinch grasp, without any wrist rotation and maintaining the forearm on the horizontal plane. The adjacent knob allows to evaluate the clockwise and anticlockwise rotation torque (*isometric rotation* exercise) with the same grasp type and restrictions of the previous exercise.

On the horizontal panel of the device it is possible to perform the other two exercises. One is a revised version of the *finger tapping* exercise, which must be performed on the exposed printed circuit board (PCB). The patient must touch key-shaped pads on the PCB following a specific sequence (little finger, ring finger, middle finger, first finger and thumb) as playing the piano. It is allowed to have multiple fingers on the keys provided that the sequence is correctly performed and closed with a thumb tapping. The last exercise allows evaluating the *hand extension* ability. The patient must rest the hand between the two L-shaped aluminium profiles, touching them with the thumb and the little finger. Then he must open and close the hand (always on the horizontal plane) in rhythm, allowing the aid to appreciate opening and closing agility. A constant counter-resistance is applied.

The correct position of the patient's hand for the 4 exercises is depicted in Fig. 2.

3.1   Hardware Architecture

From an hardware perspective, the device leverages a mother board hosting the main MSP430FG4618 microcontroller unit (MCU) @1MHz, the analog front-end for the sensorized aids, all the power supply circuitry and the visible/audible feedback devices. The analog and digital sensorized aids are tightly connected to the device whereas the Bluetooth module for the PC connection is detachable being connected to the device by a 25-pole female D connector, guaranteeing also access to the JTAG ports to program the 2 MCUs embedded in the device. A single power supply at 3.3V is available on board, obtained from a single-cell rechargeable Li-ion battery.

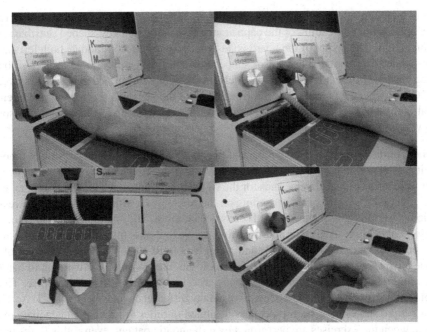

Fig. 2. From top left, clockwise: dynamic rotation, isometric rotation, finger tapping and hand extension exercises on the proposed hand functional evaluation device

The Sensorized Aids. A precision multi-turn potentiometer, equipped with a 30mm aluminium knob, has been used as dynamic rotation aid. The potentiometer (Vishay 534, 20kΩ, 2W) is able to perform 10 turns opposing a torque of 0.006 Nm. The resistance varies linearly with the rotation of the knob so that it suffices to measure the voltage on the wiper to detect the angular position at any instant. Due to the low opposing torque, the exercise can be considered without any load.

On the contrary, the isometric rotation aid is composed of a 5-lobe 50mm plastic knob able to slightly turn on its own axis pulling along with it a T bar nut able to press one of two thin-film force sensors (the low-cost Tekscan FlexiForce A201, max 110N), for clockwise and anticlockwise rotations. These sensors linearly vary their conductance in response to the applied force. Being an isometric exercise, thanks to the aforementioned design, the knob cannot spin.

The hand extension exercise is dynamic but it introduces a counter-resistance. It is evaluated by means of an analogue draw wire position sensor (LX-PA-15 by TME) mounted on a roller (CES30-88-ZZ by Rollon) free to move on a 40cm linear zinc plated guide (TES30-1040 by Rollon): the wire coming out from the sensor is attached to a second roller mounted on the same guide. The two rollers are attached to as many L-shaped aluminium profiles actuated by the patient opening and closing his hand. The sensor is characterized by a nominal wire rope tension of 3.9N, which must be overcome by the patient in order to extend his hand.

Lastly, the finger tapping aid exploits a capacitive touch board. Compared to the one presented in [9], the capacitive approach is still able to provide a detection of the touch without any counter-resistance from the measuring device but also avoids any direct current injection in the patient's hand. The touch board is based on the MSP430F2013 MCU, managing the reading of the capacitance associated to 8 key-shaped sensible areas on a PCB. The layout of the board is designed to easily accommodate both left and right-handed exercises and the sensor shape allows to find a comfortable hand position taking care of different hand sizes and deformities. This sensorized aid provides over an I2C bus, whenever required, the current status of the keys in a single byte: the interpretation of the data in the light of the exercise to execute is up to the main processor firmware.

Signal Conditioning. Given the nature of the involved signals, which are slowly time-varying, it is possible to operate at rather low sampling frequencies, with consequent benefits in terms of real-time bounds for the signal processing algorithms. In order to let the signal processing algorithms running on the MCU work with an adequate time resolution, a sampling frequency of 150Hz has been chosen. The analogue interface block is essentially composed of four non-inverting, active low-pass filters, implemented with an operational amplifier (TLV2375) and a single pole RC net. The value of its cut-off frequency has been set to about 48Hz to exploit the filter as anti-alias with guard band of about 25Hz under the Nyquist frequency, also limiting the 50Hz mains noise. The outputs of the four filters are connected to as many different channels of the MCU ADC. All the input stages have been especially designed in order to provide an adequate response when the exercises are performed by a rheumatic patient, even if this limits the operating range of the sensor.

Two different configurations have been employed. The first configuration is used for the Tekscan FlexiForce sensor, which has been connected between ground and the operational amplifier inverting input, making the stage a variable gain amplifier. Using a fixed input, provided by a voltage reference at 0.5V, the output varies linearly with the force applied to the sensor (between 0.5 and 3.3V). The second configuration, used for the potentiometric sensors, has a fixed gain and a variable input voltage. The potentiometric sensor is inserted in a voltage divider, with the wiper connected to the stage input, so that the output value is proportional to the voltage present at the wiper.

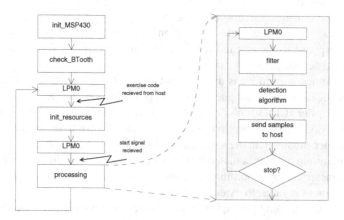

Fig. 3. Embedded control software flow diagram

Patient Interface. The device includes low-level user interface elements and some patient feedbacks, motivating him and aiding a correct execution of the exercises. Two leds indicate which hand must be used to execute the exercise and another led gives a time reference blinking at 1Hz, which is useful for sustained position tests. They are placed on the front panel for improved visibility. Moreover, a buzzer chimes whenever the device detects a successful event, letting the user know that the device has effectively captured his action. The device has been also provided with a double digit 7 segments display, which has different functions depending on the exercise, providing:

- the percentage of the effort with respect to the maximum bound (extension and torque),
- the number of correct sequences performed (finger tapping),
- the percentage of rotation over 10 turns (dynamic rotation).

Two buttons, placed on the horizontal plane and connected to two different external interrupt pins of the MCU, provide a way for the patient to interact with the device. The first one starts the exercise when the patient is ready, allowing to correctly position the hand, whereas the second one can be used to skip a single repetition of an exercise.

3.2 Embedded Control Software

The operation of the MCU is controlled by the firmware loaded onto its flash memory, written in C and developed under the CCS v4.0 IDE by Texas Instruments. The firmware flow is depicted in Fig. 3: as soon as all the initializations have been carried out, the MCU enters the low power mode (LPM), where both CPU and MCLK are disabled. The rest of the processing is then managed asynchronously by interrupt service routines (ISR). All the resources present on the board, as operational amplifiers, finger tapping MCU and Bluetooth module, are initially held in reset. Then the Bluetooth module is set up and configured by setting the operating mode, device name and password. The firmware enters an endless loop, where each iteration corresponds to the execution of an entire exercise. Inside the loop the MCU goes immediately in LPM, waiting for the execution code of the exercise to launch coming from the PC. Receiving the exercise code triggers the USCI A port ISR, which wakes up the MCU. Depending on the selected exercise, some initializations are carried out, the ADC input channel is set to the corresponding input pin (except for the finger tapping exercise) and the device patient interface is set accordingly. After that the MCU goes in LPM again, waiting the start signal (by pushing the white button), which unlocks the execution.

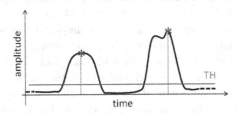

Fig. 4. Peak detection for the extension and isometric rotation exercises

From now on the processing is timed by a timer. In the corresponding ISR, either the value at the ADC input is sampled and stored or the new digital word from the tapping board is read. A global counter is incremented, to keep track of the number of samples gathered and hence to extract time measurements from it. Then the actual signal processing takes place on a sample-by-sample basis, in different ways depending on the specific exercise, as explained in the following. The current sample is sent to the host machine through the Bluetooth link, and only every second a vector containing statistics which characterize the execution is sent too. If the stop condition which identifies the end of an exercise is not met, the MCU enters the LPM again from which it will be released by the acquisition of a new sample, otherwise the processing steps back to the main loop, entering in LPM until the device gets triggered again from the GUI.

For all the exercises but the tapping one, the samples are first low-pass filtered by an 8-tap moving average filter in order to smooth the signal. For the extension and isometric rotation evaluations the algorithm simply detects the signal peaks corresponding respectively to a hand extension or a torque application, as showed in Fig. 4. This is done by comparing each sample with a threshold, which is set to a specific value by inverting each sensor calibration curve. In particular the minimum acceptable values are 1 cm for the extension exercise and 0.8 Kg for the isometric rotation one. The mean value of the first acquired samples are used as the zero value of the subsequent measures. The peak event is validated only if at least 75 consecutive samples are above the threshold and only as soon as the samples go under the threshold again. The peak maximum value, its duration and position are determined and used to compute their incremental mean values as:

$$\bar{m}_N = \frac{(\bar{m}_{N-1}(N-1) + s)}{N} \tag{1}$$

where $\bar{m}_i$ is the mean value computed over i samples, and s is the value of the new sample. The device also stores the absolute maximum and minimum values for the peak amplitude within the sequence. It is worth to underline that the variables which hold the average values are float numbers, though the MCU is a 16 bit platform and floating point is not supported in hardware. Nevertheless all these operations are translated by the compiler in the proper microcode without additional coding effort.

The algorithm is different in the case of the dynamic rotation, since different signal features must be detected. The typical signal has a terraced wave shape as showed in Fig. 5, where the edges correspond to the spinning of the potentiometer whereas the plateaus indicate that the transducer is still. The duration of both edges and plateaus, and the amplitude of each edge, are computed. To detect both onset and end of an edge, a simple detection mechanism based on thresholds has been designed, exploiting the smoothness of the filtered signal. A FIFO buffer of 14 samples is linearly updated at every new sample. The mean value of the oldest 4 samples is computed and compared with the most recent sample. If the difference is greater than an empirically determined threshold, the algorithm detects an edge and marks the onset n samples before the most recent one. When the difference falls back under the threshold, the edge end is marked and the processing is repeated, until the potentiometer reaches the limit. By using absolute values, the processing is the same for both clockwise and counter-clockwise exercises.

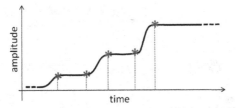

Fig. 5. Detection of the edges onset and end of the typical dynamic rotation signal for segmentation

The finger tapping exercise differs from the others because there are no analogue signals involved. The MCU on the main board acts as the master of the I2C channel, requesting the 8-bit word (one bit for each key in the touchpad) provided by the sensorized aid whenever the sampling timer expires. For this exercise, the timer has been set differently for a sampling frequency of 50Hz, which is in line with the state of the art [6] and allows the complete scanning of the 8 keys in a sampling period, in the worst case (when all the keys are touched). As a new word is received, it is mirrored, if necessary, in order to have the least significant bit always referred to the thumb key. When the first not null data is received, the algorithm detects the less significant bit set to 1 and creates a mask used, at the next touch, to check if the next key tapped corresponds to a less significant bit or not. If this is true, the mask is updated and the processing goes on, otherwise an error flag is set. The sequence terminates when the thumb touch is detected (lsb = 1). If the number of touches is equal to five the valid sequence counter is incremented or, if either the error flag is set or the sequence length differs from five, the bad sequence counter is. This processing is performed in real-time and when the exercise is complete, an additional routine computes the relevant statistics, including average touch duration for each finger, average distance between them, total consecutive touches and total duration of the exercise.

4   The Advanced Stand Alone Physician Interface

By means of a standalone GUI based on the Qt 4.7 framework (a C++ graphic framework), the physician can monitor in real-time on a host PC the execution quality of the exercises, also extracting the information needed for the hand functionality assessment. Since the Bluetooth device driver exports a serial interface towards the user applications, it can be managed using the QtExtSerialPort class which is not included in the framework by default but can be integrated with minor effort. Once the link has been established the device sends its calibration values to the host, which will be used to perform the scaling of the received data on the PC, in order to lighten the processing on the device microcontroller. The communication is handled by means of a simple protocol made of 8 bit wide control codes. In the main window it is possible to choose the exercise and the hand to use whereas the exercise progress can be analyzed in a different window, specific for the selected exercise, which pops up as soon as the exercise is started. Both windows are depicted in Fig. 6. In every exercise-specific window there is the possibility both to stop the execution and to go back to the main window, where the physician can select a new exercise.

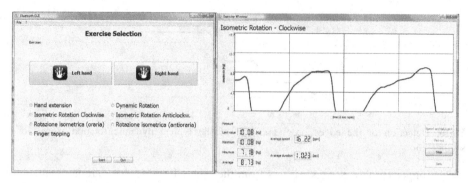

Fig. 6. Physician interface main window (left) and exercise monitoring (right)

Even though the signal is sent to the PC on a sample-by-sample basis, its plot is refreshed only every 75 input samples, shifting towards left the previous blocks in the plot linear buffer, for the sake of efficiency. All the received samples are logged thus, at the end of the execution, the user can visualize a static plot of the whole signal including the relevant delineation markers extracted in real-time by the device (Fig. 7). The GUI receives the functional assessment relevant parameters (e.g. speed of execution, position and the amplitude of the last peak, the maximum, the minimum and the mean value of the executions), to be presented on the GUI, every 150 samples of the signal.

Beyond the whole plot of the acquired signal, the interface also enables the visualization of the "Speed and Value plot" (Fig. 7), which overprints to a bar graph showing the peak values, a line graph representing the frequency of the repetitions. This information can be useful to evaluate how much the performance is dependent by the execution speed, being important to know if smaller values achieved by the patient are caused by a higher execution speed or by fatigue. It should be noted that traditional assessment techniques do not consider time as discriminative factor, thus reducing the informative content of the measurements.

5   Device Application in an Outpatient Clinic

In a rheumatologic clinical setting, 6 volunteers were enrolled with the aim to test the portable prototypical system presented above. All the patients were enrolled from the outpatient clinic of the Chair of Rheumatology, Department of Medical Sciences, University of Cagliari, Italy. They were evaluated in order to participate to the clinical test if they fulfilled the following inclusion criteria: age 18 − 75 years, ability to give informed consent, clinical remission of the inflammatory disease phase, no change in antirheumatic treatment in the three previous months, need to perform a rehabilitation program due to limitation in ability to perform usual self-care, vocational, and avocational activities because of an inactivity periods that preceded the clinical remission of inflammatory phase.

All patients are female, underwent a clinical examination and were assessed according to international guidelines. Three of them present SSc and suffered from flexion contractures, caused by retraction of skin, subcutaneous tissues and tendon sheats.

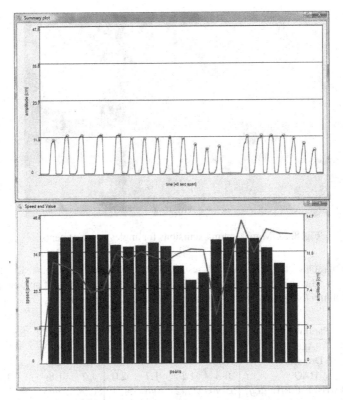

Fig. 7. Marked whole signal plot (top) and "speed and value" plot (bottom) for the extension test

Three of them are affected by RA and suffered from muscular hypotrophy, capsular and tendon sheats fibrosis of the hands and wrists without deformities. Before starting the rehabilitation phase, they underwent a functional assessment through the traditional tools and the portable prototypical device to test its ergonomics and functionality.

Traditional assessments of hand function were performed using the Dreiser test, the HAQ, the ROM. For the latter, the movements leading to the hand positions presented in Fig. 8 have been considered, namely wrist flex-extension, wrist lateral-lateral and finger lateral-lateral. Hand extension ability was evaluated through the experimental device and by traditional tools. The patient dexterity (exercise of dynamic rotation and finger tapping exercise) and the rotation torque (isometric rotation exercise) were assessed only by the experimental device since no instruments are currently available for such evaluations. Demographic characteristics and results are shown in Tab. 1 and Tab. 2.

Although it is not possible to compare the results obtained with the traditional tools against those recorded using the experimental device, because of the low number of subjects, it is worth mentioning that the latter seems to fit with the former. As an example, the SSc2 patient, who showed the highest Dreiser's and HAQ scores and the poorest ROM and traditionally evaluated extension performances, due to high disability levels, had the poorest performances at the finger tapping, dynamic rotation and extension exercises evaluated through the experimental device. Since previous studies

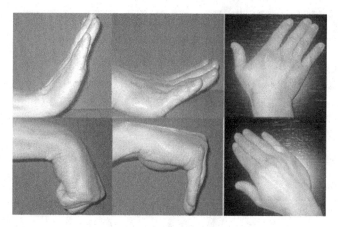

Fig. 8. ROM maximum excursions for angles measurements

Table 1. Demographic characteristics and results of the traditional assessment. RA 1 to 3 are patients with RA whereas SSc 1 to 3 are patients with SSc. Normal values, if any, are presented in the *range* column.(R:right, L:left, F-E: flex-extension, L-L: lateral-lateral)

Parameter	Range	RA 1	RA 2	RA 3	SSc 1	SSc 2	SSc3
Age		47	53	58	45	43	47
Dreiser	0-30	15	16	20	18	21	21
HAQ	0-3	1.2	1.7	2.0	1.2	1.8	1.3
ROM wrist F-E R/L [deg]	65-90	90/42	70/60	50/40	80/90	75/65	90/75
ROM wrist L-L R/L [deg]	90	35/25	25/15	25/30	70/60	45/60	55/70
ROM fingers F-E R/L [deg]	90	65/85	60/60	70/65	88/88	65/80	90/95
Extension R/L [cm]		6.5/4.5	4.8/5.5	8.5/8	8/8.2	5.5/5.3	10.5/10.5

have reported that ROM was related to some kinds of hand function in patients with SSc [12,2] and ROM seems related to the performances recorded by the experimental device, it is conceivable that in the future the latter might represent an instrument to quantify the hand function, or disability, in SSc patients. As another example, patient RA3, who showed the poorest wrists but the best fingers ROM performances, because of a prevalent anatomical damage at wrists level, had a good performances at the dynamic rotation exercise which do not involve wrist movement. Moreover, in the same patient, the low values recorded at the extension exercises evaluated using the experimental device as compared to those traditionally recorded might be ascribed to the counter resistance or to a difficulty in the execution of the exercise, as demonstrated by the low extension speed recorded. Therefore, the experimental device appeared able to differentiate the anatomic level of disability as well as ROM but differently from Dreiser or HAQ which are general indicators. Moreover, allowing the registration of speed parameters, the experimental device might estimate the quality of the exercise or the difficulty in performing it.

Table 2. Results of the assessment through the experimental device. All the parameters are given for Right/Left hand. RA 1 to 3 are patients with RA whereas SSc 1 to 3 are patients with SSc. Data expressed in sec represent the mean time interval between consecutive repetitions. (FT: Finger Tapping, ICR: Isometric Clockwise Rotation, ICcR: Isometric Counterclockwise Rotation, DR: Dynamic Rotation)

Exercise	RA 1	RA 2	RA 3	SSc 1	SSc 2	SSc3
Extension $[cm]$	7.1/4.4	4.0/5.0	3.6/3.2	8.2/7.9	5.2/5.3	12.2/11.7
Extension $[s]$	1.8/1.3	2.2/5.0	2.9/2.2	1.5/1.4	2.9/2.3	1/0.8
Extension speed $[ppm]^\dagger$	22.1/24.2	40/40	8.7/10.9	28.7/27.6	10.9/16	38.4/50.2
FT correct $[\sharp]$	20/20	11/18	20/13	20/20	3/9	2/20
FT wrong $[\sharp]$	0/4	19/12	2/7	10/10	27/21	28/9
FT speed $[tps]^{\dagger\dagger}$	2.4/2.8	1.4/1.7	1.8/1.6	2.6/3.1	0.8/1.4	0.4/2.5
ICR $[Kg]$	2.0/2.5	3.8/3.3	2.4/3.3	2.6/5.0	3.9/3.6	1.2/3
ICR $[s]$	1.1/1.3	1.1/1.1	1.5/0.8	0.9/1.2	2.0/2.3	1.1/0.8
ICR speed $[ppm]^\dagger$	28.3/23.0	26/19.0	21.3/35.0	34.0/21.9	12.1/21.0	26.3/32.0
ICcR $[Kg]$	3.6/2.3	1.2/3.3	2.5/2.7	6.3/4.1	5.2/4.7	2.9/2.9
ICcR $[s]$	1.3/1.3	1.4/1.1	1.1/0.8	1.1/1.2	1.5/1.1	0.8/0.8
ICcR speed $[ppm]^\dagger$	27.8/24.8	25.0/26.0	24.6/28.2	30.0/31.3	22.0/30.0	39.0/41.0
DR $[deg]$	195/81	195/113	183/272	224.5/229	119/139	163/110
DR speed $[deg/s]$	340/472	257/189	381/388	334/387	544/869	646/297

† ppm = peaks per minute; †† tps = touches per second

6 Conclusions

The device developed in this research is a valuable aid for the assessment of the hand functionality in chronically ill subjects requiring a quantitative evaluation for the proper set up of the personalized pharmacological and rehabilitation programs. Compared to other devices at the state of the art, it embeds the sensorized aids necessary to execute different kinds of exercises in a single low-cost framework, allowing the extraction of the relevant parameters from the analysis of several repetitions of the same movement performed in real rehabilitation exercises typically prescribed to RA and SSs patients. In this way the noise given by fatigue and distraction can be better identified compared to one-shot measurements, and also the temporal features of the acquired signals can be accurately recorded in order to enable a more complete analysis. The device, easily controllable exploiting a stand-alone software interface on the physician PC, has been preliminary evaluated in the outpatient clinic of the Chair of Rheumatology of the University of Cagliari, Italy. All patients completed the assessment through the experimental device without complaining of pain or discomfort. Furthermore, at this stage of development, the prototypical system is easy to use and well perceived by both patients and physicians representing a promising new tool in a clinical field that apparently lacks of devices allowing the real-time quantitative assessment of the hand functionality. Further studies in larger population are needed to evaluate if such device might be considered a reliable instrument to quantify the hand function in patients affected by rheumatic diseases.

Acknowledgements. The research leading to these results has received funding from the Region of Sardinia, Fundamental Research Programme, L.R. 7/2007 "Promotion of the scientific research and technological innovation in Sardinia" under grant agreement CRP2_584 Re.Mo.To. Project. The authors wish to thank V. Lussu, L. Piras, I. Secci, N. Zaccheddu, F. Boi and M. Crabolu. Alessia Dess gratefully acknowledges Sardinia Regional Government for the financial support of her PhD scholarship (P.O.R. Sardegna F.S.E. Operational Programme of the Autonomous Region of Sardinia, European Social Fund 2007-2013 - Axis IV Human Resources, Objective 1.3, Line of Activity 1.3.1.)

References

1. Andria, G., Attivissimo, F., Giaquinto, N., Lanzolla, A., Quagliarella, L., Sasanelli, N.: Functional evalu ion of handgrip signals for parkinsonian patients. IEEE Transactions on Instrumentation nd Measurement 55(5), 1467–1473 (2006)
2. Badley, I , Wagstaff, S., Wood, P.: Measures of functional ability (disability) in arthritis in relation o impairment of range of joint movement. Ann. Rheum. Dis. (43), 563–569 (1984)
3. Busta' ante, P., Grandez, K., Solas, G., Arrizabalaga, S.: A low-cost platform for testing activ es in parkinson and ALS patients. In: 12th IEEE International Conference on e-Health Net orking Applications and Services (Healthcom), pp. 302–307 (2010)
4. D' ser, R., Maheu, E., Guillou, G., Caspard, H., Grouin, J.: Validation of an algofunctional j lex for osteoarthritis of the hand. Rev. Rheum. Engl. 6(suppl. 1), 43S–53S (1995)
5. lelliwell, P., Howe, A., Wright, V.: Functional assessment of the hand: reproducibility, acceptability, and utility of a new system for measuring strength. Ann. Rheum. Dis. 46, 203–208 (1987)
6. Jobbágy, A., Harcos, P., Karoly, R., Fazekas, G.: Analysis of finger-tapping movement. Journal of Neuroscience Methods 141, 29–39 (2005)
7. Lambercy, O., Dovat, L., Gassert, R., Burdet, E., Teo, C.L., Milner, T.: A haptic knob for rehabilitation of hand function. IEEE Transactions on Neural Systems and Rehabilitation Engineering 15(3), 356–366 (2007)
8. Macellari, V., Morelli, S., Giacomozzi, C., Angelis, G.D., Maccioni, G., Paolizzi, M., Giansanti, D.: Instrumental kit for a comprehensive assessment of functional recovery (2006)
9. Muir, S.R., Jones, R.D., Andreae, J.H., Donaldson, I.M.: Measurement and analysis of single and multiple finger tapping in normal and parkinsonian subjects. Parkinsonism Related Disorders 1(2), 89–96 (1995)
10. Peters, M.J.H., van Nes, S.I., Vanhoutte, E.K., Bakkers, M., van Doorn, P.A., Merkies, I.S.J., Faber, C.G.: Revised normative values for grip strength with the jamar dynamometer. Journal of the Peripheral Nervous System 16, 47–50 (2011)
11. Poole, J., Steen, V.: The use of the health assessment questionnaire to determine physical disability in systemic sclerosis. Arthritis Care. Res. 4, 27–31 (1991)
12. Poole, J., Steen, V.: Grasp pattern variations seen in the scleroderma hand. Am. J. Occup. Ther. (48), 46–54 (1994)
13. Seo, N.J., Rymer, W.Z., Kamper, D.G.: Delays in grip initiation and termination in persons with stroke: Effects of arm support and active muscle stretch exercise. J. Neurophysiol. 101(6), 3108–3115 (2009)

Part II

Bioinformatics Models, Methods and Algorithms

Forests of Latent Tree Models to Decipher Genotype-Phenotype Associations

Christine Sinoquet[1,*], Raphaël Mourad[2], and Philippe Leray[3]

[1] LINA, UMR CNRS 6241, Université de Nantes, 2 rue de la Houssinière, BP 92208, 44322 Nantes Cedex, France
[2] Center for Computational Biology and Bioinformatics, Department of Molecular and Medical Genetics, Indiana University, Indianapolis, IN, 46002, U.S.A.
[3] LINA, UMR CNRS 6241, Ecole Polytechnique de l'Université de Nantes, rue Christian Pauc, BP 50609, 44306 Nantes Cedex 3, France
{christine.sinoquet,philippe.leray}@univ-nantes.fr

Abstract. Genome-wide association studies have revolutionized the search for genetic influences on common genetic diseases such as diabetes, obesity, asthma, cardio-vascular diseases and some cancers. In particular, together with the population aging concern, increasing health care costs require that further investigations are pursued to design scalable and efficient tools. The high dimensionality and complexity of genetic data hinder the detection of genetic associations. To decrease the risks of missing the causal factor and discovering spurious associations, machine learning offers an attractive framework alternative to classical statistical approaches. A novel class of probabilistic graphical models (PGMs) has recently been proposed - the forest of latent tree models (FLTMs) - , to reach a trade-off between faithful modeling of data dependences and tractability. In this chapter, we assess the great potentiality of this model to detect genotype-phenotype associations. The FLTM-based contribution is first put into the perspective of PGM-based works meant to model the dependences in genetic data; then the contribution is considered from the technical viewpoint of LTM learning, with the vital objective of scalability in mind. We then present the systematic and comprehensive evaluation conducted to assess the ability of the FLTM model to detect genetic associations through latent variables. Realistic simulations were performed under various controlled conditions. In this context, we present a procedure tailored to correct for multiple testing. We also show and discuss results obtained on real data. Beside guaranteeing data dimension reduction through latent variables, the FLTM model is empirically proven able to capture indirect genetic associations with the disease: strong associations are evidenced between the disease and the ancestor nodes of the causal genetic marker node, in the forest; in contrast, very weak associations are obtained for other latent variables. Finally, we discuss the prospects of the model for association detection at genome scale.

Keywords: Probabilistic Graphical Model, Bayesian Network, Latent Tree Model, Detection of Genetic Association, Latent Variable, Data Dimension Reduction, Scalability.

---

* Christine Sinoquet and Raphaël Mourad are joint first authors.

J. Gabriel et al. (Eds.): BIOSTEC 2012, CCIS 357, pp. 113–134, 2013.

1   Introduction

Thanks to their ability to capture (conditional) independences and dependences between variables, probabilistic graphical models (PGMs) offer an adapted framework for a fine modeling of relationships between variables in an uncertain data framework. A PGM is a probabilistic model relying on a graph encoding conditional dependences within a set of random variables. A PGM provides a compact and natural representation of the joint distribution of the set of variables. Bayesian networks (BNs) are a commonly used branch of PGMs.

Despite the fact that the observed variables are often sufficient to describe their joint distribution, sometimes, additional unobserved variables, also named latent variables, have a role to play. In this context, hierarchical Bayesian networks such as latent tree models (LTMs), formerly named hierarchical latent class models, were proposed. LTMs are tree-shaped BNs where leaf nodes are observed while internal nodes are not. LTMs generalize latent class models (LCMs), defined as containing a unique latent variable and edges only connecting the latent variable to all the observed variables. In LTMs, multiple latent variables organized in a hierarchical structure allow to depict a large variety of relations encompassing local to higher-order dependences (see Figure 1). LCMs enforce observed variables to be independent, conditionally on the latent variable. In contrast, LTMs relax this local independence assumption which is often violated for observed data.

Few algorithms have been developed to learn such models and still fewer for applications in association genetics [1]. Forests of LTMs have been recently proposed as *potentially* useful for association studies [2,3]. In the biomedical research domain, association studies rely on the description of DNA variants at characterized genome loci - or genetic markers - for all subjects in case and control cohorts. Such studies attempt to identify any putative dependence - or association - between one or possibly some genetic markers and the affected/unaffected status. In the case of a single causal locus, a putative association is revealed if the distribution of variants between cases and controls shows an accumulation of the former with respect to some variant(s). From now on, we will refer to the most popular genetic markers, that is, Single Nucleotide Polymorphisms (SNPs).

One of the first motivations to propose this novel model - the forest of LTMs (FLTM) - is to take account of linkage disequilibrium (LD) in the most possible faithful way. Linkage disequilibrium occurs because DNA variants close on the chromosome are scarcely separated by the shuffling of chromosomes (recombination) that takes place during sex cell formation. Such variants are therefore transmitted together (as an haplotype) from parent to child. Such patterns are at the basis of the so-called haplotype block structure [4]: "blocks" where statistical dependences between loci are high alternate with shorter regions characterized by low statistical dependences, the recombination hotspots. LD is crucial for association studies since a causal locus not sharply coinciding with a SNP is nevertheless expected to be flanked by SNPs highly likely to be shown (indirectly) associated with the phenotype. Besides, benefitting from high correlations is appealing to implement data dimension reduction.

Data dimension reduction exploiting LD is not new to genetics. However, tackling this issue through adapted Bayesian networks has but recently been proposed [3].

Besides, FLTM models seem appealing to enhance association studies: due to their hierarchical structure, FLTM models would help pointing out a region containing a genetic factor associated with a studied disease. However, bottom-up information fading is likely to be observed in the hierarchical structure. The impact on downstream analyses such as association studies remains questionable. The very point is to check whether latent variables covering a causal region are found associated with the disease. This chapter conducts a systematic and comprehensive evaluation of the ability of the FLTM model to help evidence genetic associations through latent variables. In this framework, we address the case of the single causal genetic factor.

The organization of the chapter is as follows. After the Section "Definitions", Section 3 provides the motivation for the FLTM model proposal, together with the context of this proposal: the FLTM-based contribution is first put into the perspective of PGM-based works addressing LD modeling; then the contribution is considered from the viewpoint on LTM learning. The next Section describes the specific FLTM learning algorithm developed. Section 5 briefly outlines the advantages of FLTM, confirmed by evaluation. In particular, this Section highlights three advantages of the FLTM model, which are crucial to detect genetic associations: scalability, faithfulness in LD modeling, high data dimension reduction. In Section 6, the notion of "indirect association" is defined; then is detailed the protocol implemented for the methodical evaluation of FLTM latent variables' ability to capture indirect associations. Section 6 discusses the results of intensive tests run on realistic simulated data; finally, tests applied on real genotypic data are also shown.

2 Definitions

In the context of this chapter, we restrain our concern to discrete and finite variables (either observed or latent).

Definition 1 (Conditional Independence). *Given a subset of variables $S \subseteq X \backslash \{X_i, X_j\}$, conditional independence between X_i and X_j given S ($X_i \perp\!\!\!\perp X_j \mid S$) is defined as: $\mathbb{P}(X_i, X_j \mid S) = \mathbb{P}(X_i \mid S)\, \mathbb{P}(X_j \mid S)$. The non-equality entails that both variables are conditionally dependent given S.*

Definition 2 (Bayesian Network). *BNs are defined by a directed acyclic graph $G(X, E)$ and a set of parameters θ. The set of nodes $X = \{X_1, ..., X_p\}$ represents p random variables and the set of edges E captures the conditional dependences between these variables (i.e. the structure). The set of parameters θ describes conditional probability distributions $\theta_i = [\mathbb{P}(X_i / Pa_{X_i})]$ where Pa_{X_i} denotes node i's parents. If a node has no parent, then it is described by an a priori probability distribution. The variables are described for n observations. X is a BN with respect to G if it satisfies the local Markov property stating that each variable is conditionally independent of its non-descendants given its parent variables: $X_i \perp\!\!\!\perp X \backslash \operatorname{desc}(X_i) \mid Pa_{X_i}$ for all $i \in \{1, ..., p\}$ where $\operatorname{desc}(X_i)$ is the set of descendants of X_i.*

Due to the local Markov property, the joint probability distribution writes as a product of individual distributions, conditional on the parent variables: $\mathbb{P}(X) = \prod_{i \in \{1, ..., p\}} \theta_i$.

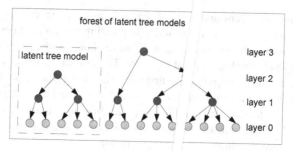

Fig. 1. Latent tree model and forest of latent tree models. The light shade indicates the observed variables whereas the dark shade points out the latent variables.

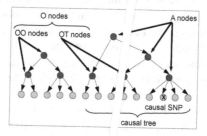

Fig. 2. Illustration of key terms specific to our approach. A nodes: ancestor nodes of the causal SNP; O nodes: other latent nodes categorized in OT nodes (in causal tree) and OO nodes (outside the causal tree). See Figure 1 for node nomenclature.

For further understanding, we now briefly recall some definitions, including that of another branch of PGMs, the Markov random fields, which will be mentioned in Section 3.

Definition 3 (Markov Random Field). *Given an undirected graph $G(X, E)$, the set of random variables X form a Markov random field (MRF) with respect to G if it satisfies the local Markov property, stating that a variable is conditionally independent of all other variables given its neighbours.*

In this case, the joint distribution can be factorized over the cliques of the graph: $\mathbb{P}(X) = \prod_{C \in cl(G)} \phi_C(X_C)$, where $cl(G)$ is the set of cliques of G and the functions ϕ_C are the so-called potentials.

A commonly used class of MRFs, the decomposable MRFs, represents those MRFs whose graph is triangulated (i.e. no cycle of length strictly greater than 3 is allowed).

Definition 4 (Entropy, Mutual Information). *The entropy of variable X writes as:*

$$\mathcal{H}(X) = -\sum_{i=1}^{n} \mathbb{P}(x_i) \log \mathbb{P}(x_i)$$

where $\mathbb{P}(x_i)$ is the probability mass function of outcome x_i.

Given two variables X_i and X_j, the mutual information measures the dependence of the two variables, expressing the difference of entropies between the independent model

$\mathbb{P}(X_i) \; \mathbb{P}(X_j)$ *and the dependent model* $\mathbb{P}(X_i|X_j) \; \mathbb{P}(X_j)$: $\mathcal{I}(X_i, X_j) = \Big(\mathcal{H}(X_i) +$ $\mathcal{H}(X_j) \Big) - \Big(\mathcal{H}(X_i|X_j) + \mathcal{H}(X_j) \Big) = \mathcal{H}(X_i) - \mathcal{H}(X_i|X_j)$. *The larger the difference between entropies, the higher is the dependence.*

Definition 5 (Allele, SNPs, Haplotype). *Due to the presence of pairs of chromosomes in the human genome, the DNA at a given chromosome locus (SNP) may either be described through a pair of variants (alleles or phased data) at the finer description level or through a unique variant (unphased data). As SNPs are biallelic, only two alleles are encountered at the corresponding loci (instead of the 4 possible nucleotides A,T,C,G). Thus, SNPs are discrete variables whose three possible values may be coded as, say, 0, 1 and 2, to respectively account for aa, {Aa, aA} (usually not distinguishable) and AA, where A and a are the two alleles. An haplotype is defined as a sequence of alleles.*

3 Motivation and Related Work

3.1 Motivation

To tackle the difficult problem of disease association detection, several algorithms coming from the machine learning domain have been proposed. Some of them use PGMs [5,6]. Recently, forests of latent tree models have been investigated for LD modeling purpose [3]. A forest of latent tree models (FLTM) is a forest whose trees are LTMs (see Figure 1). FLTMs generalize LTMs, since the variables are not constrained to be dependent upon one another, either directly or indirectly. Thus, FLTMs can describe a larger set of configurations than LTMs.

When modeling such highly correlated variables as those in genotypic data, the challenge is all the more crucial for downstream analyses such as study and visualization of linkage disequilibrium, mapping of disease susceptibility genetic patterns and study of population structure. Most notably, the benefits of using FLTMs to model LD rely on their ability to account for multiple degrees of SNP dependences and to naturally deal with the fuzzy nature of LD block boundaries. As will further be emphasized, this latter advantage results from the FLTM learning algorithm, which does not impose that the SNPs subsumed by the same latent variable be neighbouring SNPs (along the genome).

3.2 Probabilistic Graphical Models to Model Linkage Disequilibrium

The FLTM-based model is meant as an improving alternative over other PGM-based works addressing LD modeling. Besides learning of parameters (θ), that is *a priori* and conditional probabilities for Bayesian networks, and probability distributions for cliques and separators for Markov random fields, the most challenging task in PGM learning is structure inference. Thomas and Camp pioneered the use of PGMs to model LD [7]. To reach this aim, their approach relies on the general class of *decomposable Markov Random Fields* (DMRF). Decomposable graphs allow the efficient computation of the likelihood of the structure, given the data. Thus, structure learning is easily performed navigating the structure space while optimizing a log-likelihood-based score. To explore the DMRF space, operations based on connection or disconnection of randomly-selected nodes were designed to build the neighbourhood of an incumbent graph G. Whatever the method used for model estimation (downhill search or simulated annealing), a severe issue alleviates the efficiency of the model learning algorithm: the operational characterization of general decomposable graphs is not possible. Therefore, starting from the incumbent graph G, one has to check *a posteriori* whether a proposal G' in the neighbourhood of G is decomposable. In the BN-based approach of Lee and Shatkay, hill climbing with random restarts is constrained by the greedy sparse candidate procedure, a standard used to accelerate structure learning [8].

Models with latent variables have also been investigated to model and exploit LD. Their common characteristic is the use of the SEM (structural expectation maximization) procedure. SEM successively optimizes parameters θ, conditional to the structure S, then optimizes S conditional to θ. In this line, Greenspan and Geiger efficiently infer ancestral haplotypes using BNs with latent variables and taking account of recombination hotspots, bottlenecks, genetic drift and mutation [9]. Their model is defined by a partition of the region under analysis into blocks, with one or more ancestor haplotypes defined for each block (latent variables). A Markov chain expresses the dependences between the block genealogies, reflecting the fact that LD exists between blocks as well as within them. First, an ensemble of local models which are locally optimal are inferred. For this purpose, the search space of models is explored through Gibbs-style iterations. All model parameters are optimized at each stage of this process, using the local search and an adapted EM algorithm. The Latent Class Model (LCM) (see Introduction) is the basic component of Kimmel and Shamir's model [10]. The latter is a collection of LCMs. The principle of the modeling lies in assigning each locus to a cluster, *i.e.* an LCM. In this category, Zhang and Ji's approach improves over the previous block-based approach: a cluster-based approach allows non contiguous SNPs in the same cluster [1].

None of the previous approaches is scalable when more than one thousand SNPs and one thousand individuals are considered. To cope with genome-wide data, various leads have been explored. Hidden Markov models (HMMs) are a particular class of PGMs, in which the latent variables are ruled by a Markov chain. Daly *et al.* developed an HMM model for LD mapping [4]. In this line, Scheet and Stephens proposed an HMM-based model where latent states correpond to ancestral haplotypes [11]. To capture the features inherent to LD, *i.e.* that cluster-like patterns tend to be local in nature, an HMM is used: it models the fact that alleles at nearby markers are likely to arise from the same cluster. Besides, the model can deal with gradual decline of LD with distance. The corresponding well-known software program, fastPHASE, is accurate and enables to handle very large datasets (one thousand individuals and hundreds of thousands of SNPs). A variable-length markov chain (VLMC) was proposed by Browning and Browning [12]. In VLMCs, the memory length can vary along the chain, depending on the context. Thus, the memory length will be larger for regions of high LD, whereas it will be small for hotspots, where recombination entails low LD. In contrast to HMMs, no prespecification of the structure is required to learn a VLMC, such as the number of ancestral haplotypes, and model learning is performed through a fast heuristic. The widely used Beagle software is expected to scale to one thousand individuals and one hundred thousand SNPs.

Additional investigations on DMRFs led to two scalable proposals. To escape the computationally demanding checking of the decomposability property, appropriate moves were tailored by Verzilli *et al.* to preserve this property while browsing the search space of DMRFs [5]. To move from a triangulated graph to another one, these authors designed moves that involve changes in the sets of cliques and separators. In the merge move, a randomly selected clique is merged with another one. The split move splits up a non-singleton clique selected at random. More recently, Abel and Thomas used rules coming from graph theory, to keep the graph decomposability property when sampling in the neighbourhood of a DMRF [13]. The authors are able to enumerate all the DMRF neighbours of an incombent DMRF. Central to this enumeration is the concept of junction tree (JT): sampling in the space of JTs is substituted to sampling in the space of DMRFs [14].

To reach scalability in LD modeling, the recent BN-based proposal of Mourad *et al.* constrains the model structure to be a forest of latent tree models. Thus, a specific algorithm may be tailored to efficiently learn the structure. We recapitulate the approaches abovementioned in Table 1.

Table 1. Probabilistic graphical models dedicated to the modeling of linkage disequilibrium

Scalability	Model	Main aim	Model learning	Details	Reference	Software
No	Markov random field	LD modeling	Hill climbing		Thomas and Camp (2004)	HapGraph
	Bayesian network	Selection of tagging SNPs	Hill climbing	Greedy sparse candidate procedure	Lee and Shatkay (2006)	–
	Bayesian network with latent variables	High density LD mapping	SEM	Block partitioning, local Gibbs sampling, Markov chain	Greenspan and Geiger (2004)	Haploclock
		LD modeling and phase inference		Block-based collection of latent class models	Kimmel and Shamir (2005)	–
		LD modeling		Cluster-based collection of latent class models	Zhang and Ji (2009)	–
Yes	Hidden Markov Model	LD modeling	EM		Daly et al. (2001)	–
		Phase inference, missing data imputation			Scheet and Stephens (2006)	fastPHASE
	Variable Length Markov Chain	Genome-wide association study	Specific heuristic	Probabilistic automaton	Browning and Browning (2007)	Beagle
	Markov random field		Monte Carlo Markov chain		Verzilli et al. (2006)	
		Genome-scale LD modeling	Hill climbing	Sampling in the space of junction trees	Abel and Thomas (2011)	
	Forest of latent tree models	LD modeling, dimension reduction	Specific procedure	Ascending clustering of variables	Mourad et al. (2010)	CFHLC+

3.3   Learning Latent Tree Models

As for general BNs, besides learning of parameters (θ), *i.e. a priori* and conditional probabilities, one of the tasks in LTM learning is structure inference. This task generally remains the most challenging due to the complexity of the search space. Regarding LTM learning, the proposals published in the literature fall into two categories. The first category relies on standard Bayesian network learning techniques. The second category is based on the clustering of the variables. To learn θ, both categories rely on the expectation maximization (EM) algorithm or an EM variant.

In the first category, the algorithms explore the search space through a local search strategy and optimize a score, such as the BIC score [15]. Zhang proposed a greedy algorithm, to navigate in the structure search space [16]. This algorithm is coupled with a hill climbing procedure, to adjust the cardinality of the latent variables. A more efficient variant of this algorithm has recently been proposed [17]. In this first category, structural expectation maximization (SEM) has also been adapted to the case of Bayesian networks with latent variables. SEM successively optimizes θ, conditionally on the structure S, then optimizes S conditionally on θ. Parameter learning being a time-consuming step, in this framework, Zhang and Kocka have adapted a procedure, called local EM, to optimize the variables whose connection or cardinality has been modified in the transition from former to current model [18]. In one of the two LTM-based approaches dedicated to LD modeling, Zhang and Ji use a set of LCMs and apply a SEM strategy [1]. The number of LCMs has to be specified. To avoid getting trapped in local optima while running the EM algorithm to learn a set of latent models, these authors have adapted a simulated annealing approach.

The above score-based approaches require the computation of the maximum likelihood in presence of latent variables, a prohibitive task regarding computational burden. Thus, various methods based on the clustering of variables have been implemented. They all construct the model following an ascending strategy; they all rely on the mutual information (MI) criterion, to identify clusters of dependent variables. In their turn, these methods may be sub-categorized into binary- and non binary-based approaches.

Hwang and co-workers' learning algorithm is dedicated to binary trees and binary latent variables [19]. It has to be noted that the trees are possibly augmented with connections between siblings, that is nodes sharing the same parent into the immediate upper layer. Also confining themselves to binary trees, Harmeling and Williams have proposed two learning algorithms [20]. One of them approximates MI between a latent variable H and any other variable X, based on a linkage criterion (single, complete or average) applied for X and the variables in the cluster subsumed by H. A variant of this first algorithm locally infers the data corresponding to any latent variable; therefore it is possible to achieve an exact computation of the MI criterion between a latent variable and any other variable.

Two approaches have been proposed to circumvent binary tree-based structures. Wang and co-workers first build a binary tree; then they apply regularization and simplification transformations which may result in subsuming more than two nodes through a latent variable [21]. In their approach devoted to LD modeling, Mourad and collaborators implement the clustering of variables through a partitioning algorithm [3]; the latter yields cliques of pairwise dependent variables. Besides, this method imposes the control of information fading as the level increases in the hierarchy, which generally results in the production of a forest of LTMs (instead of a single LTM).

Two of the above cited methods have been shown to be tractable. For these two non binary-based approaches, some reports are available: Hwang and co-workers' approach was able to handle 6000 variables and around 60 observations. The scalability of the FLTM construction by Mourad *et al.* has been shown for benchmarks describing 10^5 variables and 2000 individuals.

Table 2. Representations based on latent tree models and their corresponding learning algorithms. Scalability has been shown for Hwang *et al.*'s approach, to a certain extent, and for Mourad *et al.*'s model, to genome-scale.

Method	Model	Specificity of Model Learning	Application	Reference
Local search	General LTM	Two-level algorithm (adjustment of optimal cardinalities conditional on structure, structure optimization)	Multidimensional clustering of categorical data	Zhang (2004)
		Grow-restructure-thin strategy, optimized EM (restricted likelihood)		Chen *et al.* 2011
	Cluster-based collection of latent class models	SEM	Inference of latent structure	Zhang and Kocka (2004)
			LD modeling	Zhang and Ji (2009)
Hierarchical ascending clustering of variables	Binary tree-based LTM augmented with possible connections between sibling nodes, binary latent variables		Large-scale data analysis	Hwang *et al.*(2006)
	Binary trees	Variant with local exact computations preceded by imputation of latent variables	Inference of latent structure	Harmeling and Williams (2011)
	General arity of trees and latent variables	Processing of an initial binary tree	Approximation of the underlying Bayesian network through an LTM	Wang *et al.*(2008)
		Partitioning of variables	LD modeling	Mourad *et al.* (2011)

From a methodological viewpoint, the recapitulation provided in table 2 puts the contribution of Mourad and collaborators into the perspective of LTM-based approaches. This table emphasizes the pioneering aspect of an LTM-based method dedicated to genome-scale LD modeling. The next Section thoroughly describes the learning algorithm for Mourad *et al.*'s approach.

4   The Learning Algorithm

Two versions of the FLTM learning algorithm were designed by the authors. This Section depicts the version used to test the ability of the FLTM model to detect genetic associations. Five details of the algorithm are highlighted. The Section ends with a short recapitulation regarding the evaluation of FLTM with respect to the LD modeling aspect.

4.1   Outline of the Algorithm

The learning is performed through an adapted agglomerative hierarchical clustering procedure. At each iteration, a partitioning method is used to assign variables into non-overlapping clusters. The partitioning is based on the identification of cliques of strongly dependent variables in the complete graph of pairwise dependences. Amongst the clusters, each cluster of size at least two is a candidate for subsumption into a latent variable H. To acknowledge or reject the creation of H, a prior task considers the LCM rooted in this latent variable candidate and whose leaves are all the variables of the cluster. Parameter learning using the EM algorithm is performed for this LCM. Then probabilistic inference allows missing data imputation for the latent variable. Once all the data are known for this LCM, a validation step checks whether the latent variable captures enough information from its children. If a latent variable is validated, its child variables are then replaced with the latent variable. In contrast, the nodes in unvalidated clusters are kept isolated for the next iteration. Iterating these steps yields a hierarchical structure. In other words, latent variables capture the information borne by underlying observed variables (*e.g.* genetic markers). In their turn, these latent variables, now playing the role of observed variables, are synthesized through additional latent variables, and so on.

The FLTM learning algorithm is now more formaly depicted (see Algorithm 1). The ascending hierarchical clustering (AHC) process is initiated from the first layer consisting of univariate models. Each such univariate model is built for any observed variable (lines 2 and 3). The termination of the AHC process arises if each cluster identified is reduced to a singleton (line 7) or if no cluster of size at least 2 could be validated (line 18). At each step, an LCM is first learnt for each cluster containing at least two nodes (line 12); the validity of the proposed subsumption is then checked (line 13 to 16). For simplification, the cardinality of the latent variable is estimated as an affine function of the number of variables in the corresponding cluster (line 11). After validation, the LCM is used to enrich the FLTM model (line 14): a node corresponding to the new latent variable L_{i_k} is created and connected to the child nodes; the *prior* distributions of the child nodes are replaced with distributions conditional on the latent variable. L_{i_k} is added to the set of latent variables, whereas its imputed values, $D[L_{i_k}]$, are stored (line 15). All variables in C_{i_k} are then dismissed and replaced with the latent variable (line 15). In contrast, the nodes in unvalidated clusters are kept isolated for the next step.

4.2   Details of the Algorithm

Five points of this algorithm are now detailed. For a start, light is shed on the two points establishing the difference between the initial version in [3] and the novel version, CFHLC+.

Window-Based Data Scan versus Straightforward Data Scan. First, the reader is reminded that the initial observed variables are SNPs, which are located along the genome in a sequence of "neighbouring" (but generally non contiguous) genetic markers. To meet the scalability criterion, a divide-and-conquer procedure was implemented in [3]: the data are scanned through contiguous windows of identical fixed sizes. However, such splitting is questionable. It entails a bias in the processing of the variables located in the neighbourhood of the artificial window frontiers. Managing overlapping windows would not have led to a practicable algorithm. Therefore, a first notable difference with the algorithm in [3] lies in that the novel version does not require data splitting. Instead, a simple principle is implemented: not all pairs of variables are processed by the partitioning algorithm. Beyond a physical distance on the chromosome, δ, specified by the geneticist, variables are not allowed in the same cluster. Setting the δ constraint actually corresponds to implementing a *sliding* window approach.

Partitioning of Variables into Cliques. Standard agglomerative hierarchical clustering considers a similarity matrix. As a latent variable is intended to connect pairwise dependent variables, the standard agglomerative approach was adapted accordingly. Within each window, the previous version runs a clique partitioning algorithm on the *complete* graph of pairwise dependences. In the novel version, no complete matrix is

---

Algorithm 1. FLTM model learning

INPUT: $\mathbf{X_{obs}}$, a set of p observed variables $(X = X_1, ..., X_p)$,

$\quad\quad$ $\mathbf{D[X]}$, the corresponding data for n subjects,

$\quad\quad$ **PartitionProc**, a procedure to partition variables into non-overlapping clusters of variables,

$\quad\quad$ $\tau_{\mathbf{pairwise}}$, a threshold used to guide $PartitionProc$,

$\quad\quad$ **Criter**, a criterion to estimate information fading through the bottom-up model construction,

$\quad\quad$ $\tau_{\mathbf{latent}}$, a threshold used to constrain information fading,

$\quad\quad$ α_1, α_2 and $\mathbf{card_{max}}$, parameters used to estimate the cardinality of latent variables.

OUTPUT: $\mathscr{F}$ and θ, respectively the graphical structure and the parameters of the FLTM model,

$\quad\quad$ **L**, the whole set of latent variables identified through the construction $(L = \{L_1, ..., L_m\})$,

$\quad\quad$ $\mathbf{D[L]}$, the corresponding data imputed for the n subjects.

Convention: $\mathbf{D[V]}$, data relative to the set of variables V, and describing the n subjects.

1: $\mathscr{F} \leftarrow \emptyset$; $\theta \leftarrow \emptyset$; $L \leftarrow \emptyset$; $D[L] \leftarrow \emptyset$

2: $\{\cup_i DAG_{univ_i}, \cup_i \theta_{univ_i}\} \leftarrow learnUnivariateModels(X)$ $\quad$ /* processing of layer 0 */
3: $\mathscr{F} \leftarrow \cup_i DAG_{univ_i}$; $\theta \leftarrow \cup_i \theta_{univ_i}$

4: $step \leftarrow 1$; $X \leftarrow X_{obs}$
5: **while true**
6: $\quad$ $\{C_1, ..., C_c\} \leftarrow partitionIntoClusters(X, D[X], PartitionProc, \tau_{pairwise})$
7: $\quad$ **if** all clusters C_q are singletons **then break end if**

8: $\quad$ $\{C_{i_1}, ..., C_{i_{c_2}}\} \leftarrow selectNonSingletonClusters(C_1, ..., C_c)$
9: $\quad$ $nbValidClusters \leftarrow 0$

10: $\quad$ **for** k = 1 **to** c_2
11: $\quad\quad$ $card_{LV} \leftarrow min(\alpha_1 \times numberOfVariables(C_{i_k}) + \alpha_2, card_{max})$
12: $\quad\quad$ $\{DAG_{i_k}, \theta_{i_k}, L_{i_k}, D[L_{i_k}]\} \leftarrow learnLatentClassModelThroughEM(C_{i_k}, D[C_{i_k}], card_{LV})$

13: $\quad\quad$ **if** $(Criter(DAG_{i_k}, D[C_{i_k}] \cup D[L_{i_k}]) \geq \tau_{latent})$ $\quad$ /* validation of current cluster i_k */
14: $\quad\quad\quad$ $incr(nbValidClusters)$; $\mathscr{F} \leftarrow mergeStructures(\mathscr{F}, DAG_{i_k})$;
$\quad\quad\quad\quad$ $\theta \leftarrow mergeParameters(\theta, \theta_{i_k})$
15: $\quad\quad\quad$ $L \leftarrow L \cup L_{i_k}$; $D[L] \leftarrow D[L] \cup D[L_{i_k}]$; $X \leftarrow (X \setminus C_{i_k}) \cup L_{i_k}$
16: $\quad\quad$ **end if**
17: $\quad$ **end for**

18: $\quad$ **if** $(nbValidClusters = 0)$ **then break end if**
19: $\quad$ $incr(step)$
20: **end while**

---

required anymore. The physical constraint δ leads to consider a sparse matrix of pairwise dependences, where only computed values are stored.

The clique partitioning algorithm CAST devoted to the clustering of variables is employed [22]. The dependence between two variables, evaluated through pairwise mutual information, is used to derive a binary similarity measure (requested by CAST), depending on a threshold $\tau_{pairwise}$. The algorithm CFHLC+ automatically fixes this threshold, based on a given quantile value of the mutual information values in the whole matrix (*e.g.* the median value). The CAST algorithm has been adapted to take account of the physical constraint δ. The physical constraint imposed by the sliding window size δ allows to adjust the variable bandwidth of the sparse dependence matrix.

It has to be noted that, unlike SNPs, latent variables are not characterized by a physical location on the chromosome. In this specific case, the locations of the SNPs subsumed by a given latent variable are averaged to provide the location of this latent variable.

Data Imputation for Latent Variables. Data imputation is processed locally, that is considering the LCM rooted in the latent variable and whose leaves are the variables in the cluster. For simplification, the cardinality of the latent variable is estimated as an affine function of the number of leaves. Parameter learning is first performed in this LCM, through the EM algorithm. This step yields the marginal distribution of the latent variable and the conditional distributions of the child variables. Therefore, (linear) probabilistic inference can be carried on, based on the following principle:

$$\mathbb{P}(H = c | \mathbf{x}^j) = \frac{\Pi_{i=1}^{p} \mathbb{P}(x_i^j | H = c) \, \mathbb{P}(H = c)}{\sum_{c=1}^{k} \Pi_{i=1}^{p} \mathbb{P}(x_i^j | H = c) \, \mathbb{P}(H = c)},$$

with k the cardinality of latent variable H, c a possible value for H, j an observation, *i.e.* an individual, and $\mathbf{x}^j$ the vector of values $\{x_1^j, ..., x_p^j\}$ corresponding to the variables in the cluster $\{X_1, ..., X_p\}$.

Local Parameter Learning. In parallel with structure growing, the parameters of the forest of LTMs are learned locally (see Subsection 4.2). At a given iteration, for any variable identified as a leaf node in an LCM (corresponding to a cluster), the current marginal distribution of this variable is replaced with its conditional distribution learnt in the LCM. Thus, during the bottom-up construction of the FLTM, marginal distributions are successively replaced with conditional distributions.

Validation of Latent Variables. The subsumption of the candidate cluster into the latent variable H is validated through a criterion averaging a normalized dependence measure between H and each of H's child nodes:

$$Criter = \frac{1}{|C_H|} \sum_{i \in C_H} \frac{\mathcal{I}(X_i, H)}{\min\left(\mathcal{H}(X_i), \mathcal{H}(H)\right)} \geq \tau_{latent},$$

with $|C_H|$ the size of cluster C_H.

4.3 Role of Parameters

In the forest of LTMs, the subsumption process is controlled through thresholds $\tau_{pairwise}$ and τ_{latent}, and constraint δ. No latent variable is allowed to subsume variables which are not highly pairwise dependent ($\tau_{pairwise}$) or which are in regions too far from one another (δ); τ_{latent} controls bottom-up information fading through the hierarchy. $\tau_{pairwise}$, τ_{latent} and δ thus monitor the number of connected components (trees) and the number of layers in the forest. These three parameters rule the trade-off between faithfulness to the underlying reality and tractability of the modeling.

5   Evaluation

The application software, CFHLC+, is available at http://sites.google.com/site/ raphaelmourad/Home/programmes. It is developed in C++ and relies on the ProBT library dedicated to Bayesian networks (http://bayesian-programming.org).

The algorithm was tested on datasets describing 10^5 SNPs for 2000 individuals. With the first version, the running time was around 15 hours for an arbitrary window size of 100 SNPs. When setting the sliding window size δ to $0.5 \, Mb$, a reasonable choice to capture LD, the novel algorithm now runs in less than 12 hours. It has to be emphasized that as the algorithm runs EM with 10 restarts, a significant improvement has been brought with respect to the initial version. Finally, the algorithm is shown quasi linear with the number of SNPs and linear with the sliding window size. Such experimentations are reported in [3], together with the examination of the robustness with respect to parameter adjustment.

FLTM was shown to faithfully model linkage disequilibrium. Due to its hierarchical structure, the multiple layers of an FLTM are expected to describe various degrees of LD strength. To check this property, the principle was the following: for some given genomic region, two matrices were compared. The standard triangular matrix M_c of pairwise dependences (r^2 coefficient) between SNPs was first calculated. Then, for each pair of SNPs, the latent variable representing the lowest common ancestor (LCA) was identified. On the other hand, it is easy to compute the mean r^2 over all latent variables located in the same level in the FLTM hierarchy. Thus, each cell of the second matrix, M_d, was assigned the mean r^2 measure associated with the LCA level. For a visual comparison, a color palette where shade darkens whith increasing dependence was assigned to M_c, whereas a discretized palette was affected to M_d. The visual comparison of the two plots brilliantly showed that the FLTM faithfully reflects LD strength variety (see [3]).

In complement, it was also shown that FLTM provides a compact and interpretable view of LD for the geneticist. Low-level latent variables represent short-range LD and are interpreted as haplotype shared ancestry. High-level latent variables correspond to long-range LD, induced by population admixture or natural selection. The flexibility of FLTM was highlighted in [23], where short-, long- and chromosome-wide linkage disequilibrium was modeled and visualized.

Equally important for the genetic association purpose is the dimension reduction aspect, with its consequence, possible bad subsumption. Drastic reductions are observed as a rule (about 85%) (see [3]). However, the quality of the information about the child variables is expected to decrease in a bottom-up fashion, for latent variables. Now the soundness of FLTM for LD modeling is assessed, a demonstration of the ability to capture genetic associations is still requested.

6   Protocol to Assess the Suitability of FLTM to Association Genetics

The objective of the study is to investigate how information about causality fades from bottom to top in the hierarchy and what are the trends regarding the ratios of latent variables erroneously associated with the disease. Therefore, realistic simulated data designed to harbour a causal SNP must be generated. We name *indirect genetic association* any dependence between a causal SNP ancestor node (abbreviated as A) in the FLTM and the disease. This dependence is due to the fact that an A node is likely to capture the information of the causal SNP. If indirect genetic association may be evidenced for A nodes, the identification of A nodes will allow pointing out potentially causal markers since the latter are leaf nodes of the trees rooted in A nodes (see Figure 2 which clarifies the meaning of specific key terms further used). The purpose is to examine the difference between causal SNP ancestors (As) and other latent nodes (abbreviated as Os). The behaviour of Os in causal trees (OTs) and of Os outside causal trees (OOs) will also be examined.

6.1 Simulation of Realistic Genotypic Data

Conducting a systematic analysis under controlled conditions requires that we are able to simulate both realistic SNP data and an association between one of these SNPs and the disease status (affected/unaffected). For this purpose, one of the most widely used software applications was chosen, namely HAPGEN (http://www.stats.ox.ac.uk/~marchini/software/gwas/hapgen.html) [24]. The reader well acquainted with such HAPGEN simulations may skip the two following paragraphs, which describe the simulation in the case of a single causal SNP.

Generating realistic genotypic simulation lies in the ability to mimic linkage disequilibrium. HAPGEN relies on the haplotypes (or sequence of alleles) of a population of reference, to generate new haplotypes as mosaics of the known haplotypes, for a user-specified number of cases and controls. The genotype of any individual is generated based on the two haplotypes simulated for this individual.

HAPGEN selects at random the causal SNP, checking for the minor allele frequency to be within a user-specified range. Assuming causality under a specific disease model and effect sizes, it is straightforward to calculate the genotype frequencies in cases at that locus. On this basis, any case individual is simulated by first simulating the alleles at the causal locus and then working outwards in each direction to construct the two haplotypes. Note that the same mechanism governs the construction of haplotypes, whatever the status of the individual (case or control). The only distinction lies in that the locus from which the extension is started is chosen at random, for controls. For cases, the extension is initiated from the causal locus. The extension processes conditionally on reference haplotypes and is ruled by the fine-scale knowledge of recombination rates and the physical distance between loci, to calculate the probability of breaks in the mosaic pattern as one moves along the region. Moreover, partial copies (of haplotype subregions) are blurred by simulated mutations.

To control the simulation conditions, three ingredients are combined: *minor allele frequency* (MAF) of the causal SNP, severity of the disease expressed as *genotype relative risks* (GRRs) for various *disease models*. The range of the MAF at the causal SNP is specified to be 0.1-0.2, 0.2-0.3 or 0.3-0.4. Various genotype relative risks are considered and the disease model is specified amongst additive, dominant, multiplicative or recessive (*add, dom, mul, rec*). These choices are justified as standards used for simulations in association genetics.

For short, together with GRRs, the disease models allow specifying the probability to be affected, depending on the genotype at the causal locus: $GRR = \frac{\mathbb{P}(affected|Aa)}{\mathbb{P}(affected|aa)}$, where A is the disease allele. The specification of the disease model amongst add, dom, mul and rec allows the adjustment of the probability to be affected when carrying the two disease alleles AA, with respect to the probability to be affected when carrying Aa (or aA). Thus various effect sizes may be simulated (see Table 3).

Table 3. The genotype relative risks for four standard disease models. The value 1 stands for the effect when no disease allele (A) is present at the causal locus (aa). The effect sizes for the carriers of one disease allele (Aa or aA) and two disease alleles (AA) are indicated for all four disease models.

	Genotype Relative Risk		
	Major Homozygous	Heterozygotous	Minor Homozygotous
	aa	aA or (Aa)	AA
additive	1	$1 + \frac{\alpha}{2}$	$1 + \alpha$
dominant	1	$1 + \alpha$	$1 + \alpha$
multiplicative	1	$1 + \alpha$	$1 + \alpha^2$
recessive	1	1	$1 + \alpha$

HAPGEN was run on the widely used reference haplotypes of the HapMap phase II coming from U.S. residents of northern and western European ancestry (CEU) (http://hapmap.ncbi. nlm.nih.gov/). The disease prevalence (percentage of cases observed in a population) specified to HAPGEN was set to 0.01, a standard value used for disease locus simulation. The simulated data were generated for 1000 unaffected subjects and 1000 affected subjects and consist of unphased genotypes relative to a 1.5 Mb region containing around 100 SNPs. Combining all previous conditions leads to testing 36 scenarii (3 × 3 × 4). To derive significant trends, each scenario was replicated 100 times. Together with the objective of a comprisive study, the necessity of replication explains the choice of the number of variables (100 SNPs). Standard quality control for genotypic data was carried out: SNPs with MAF less than 0.05 and SNPs deviant from the so-called Hardy-Weinberg Equilibrium (not detailed) with a p-value below 0.001 were removed.

6.2 Choice of the Association Test

The G^2 standard test of independence was preferred over the well-known Chi^2 test. For relatively small sample sizes (below 300 subjects) as in the real dataset analyzed in SubSection 7.2, G^2 is recommended: $G^2 = 2 \sum_{ij} o_{ij} \cdot \ln(o_{ij}/e_{ij})$, where o_{ij} and e_{ij} are observed and expected frequencies (in absence of genotype-phenotype association) in the cells of table $genotypes \times phenotypes$. Various p-values were obtained through successive tests of the phenotype Y against, respectively, the causal SNP, the causal SNP ancestor nodes (A nodes) and other nodes (abbreviated as Os) in the FLTM's graph. The phenotype Y is the affected/unaffected status.

6.3 Adapted Correction for Multiple Testing

To measure the significance of associations, it is necessary to adapt a permutation procedure dedicated to the computation of the per-test error rate α' (type I error), in order to control the family-wise error rate α (type I error) at, say 5%. Namely, α' defines the significance threshold for each association test. α controls the probability to make one or more false discoveries among all hypotheses when performing multiple association tests.

An advantage of the FLTM strategy relies on the fact that there are less variables in the highest layers than in the lowest ones. Thus, α' is expected to increase with the layer level. That is the reason why a permutation procedure need be adapted to the calculation of layer-specific thresholds α'.

The procedure implemented to obtain the α' threshold specific to each layer is described in Algorithm 2. This procedure performs n_p permutations (line 4). For each permutation, and each layer ℓ of the FLTM, independence tests are run for any variable X_v belonging to this layer and the target variable Y (line 9). Then, for each permutation, the minimum of the p-values over all variables belonging to layer ℓ is added to the distribution of minimum p-values for this layer (line 12). Given a specified family-wise error rate α, this distribution then allows to extract the corresponding α' threshold (line 16). This α' value, specific to each layer, is to be compared with the p-value resulting

---

Algorithm 2. PermutationProcedure $(X, D_X, Y, D_Y, n_p, \alpha)$

INPUT: X, D_X, a set of n_v candidate variables (observed or latent) $X = X_1, ..., X_{n_v}$ and the corresponding data observed or imputed for n individuals,

Y, D_Y, a target variable Y and the corresponding data observed for n individuals,

n_p, the number of permutations,

α, the family-wise error rate.

OUTPUT: $\{\alpha'(1), ..., \alpha'(n_\ell)\}$, the set of per-test error rates respectively computed for layers l to n_ℓ.

1: **for** $\ell = 1$ **to** n_ℓ
2:     $distrib_{minPValues}(\ell) \leftarrow \emptyset$
3: **end for**

4: **for** $p = 1$ **to** n_p
5:     $D_{Y_p} \leftarrow permuteLabels(D_Y)$
6:     **for** $\ell = 1$ **to** n_l
7:         $pValues(p, \ell) \leftarrow \emptyset$
8:         **for each** variable X_v in layer ℓ
9:             $pValue_{p, \ell, v} \leftarrow runAssociationTest(X_v, D_{Y_p})$
10:            $pValues(p, \ell) \leftarrow pValues(p, \ell) \cup pValue_{p, \ell, v}$
11:        **end for**
12:        $distrib_{minPValues}(\ell) \leftarrow distrib_{minPValues}(\ell) \cup min_{X_v}(pValues(p, \ell))$
13:    **end for**
14: **end for**

15: **for** $\ell = 1$ **to** n_l
16:     $\alpha'(\ell) \leftarrow quantile(distrib_{minPValues}(\ell), \alpha)$
17: **end for**

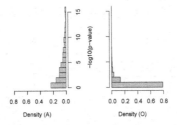

Fig. 3. Histograms of $-log_{10}$(p-value) values resulting from association tests of the phenotype with the causal SNP ancestor nodes (As) and the other latent nodes (Os). General results compiling all simulated scenarii (see 6.1): MAF (0.1-0.2, 0.2-0.3 and 0.3-0.4), heterozygous GRR (1.4, 1.6 and 1.8) and disease model (dominant, recessive, additive and multiplicative)

from the association test between variable X_v (belonging to layer ℓ) and Y. Thus can be assessed association significance, corrected for family-wise type I error, or should one write instead, controlled for *layer-wise* type I error.

7 Detection of Indirect Genetic Associations

7.1 Verification for Simulated Data

Over all 3600 runs (36 scenarii × 100 replicates), up to 7 layers were generated. Results obtained for layers above third layer (comprised) are not reported: such layers do not provide sufficient data to compute representative medians or draw informative boxplots. On average, over all 3600 FLTMs, the percentages of nodes are distributed as follows: 89.1% in layer 0, 9.5% in layer 1, 1.2% in layer 2 and 0.2% in layer 3.

General Trends. Instead of p-values, we will consider the $-log_{10}$(p-value) values. Values near 0 point out independence and the previous indicator increases with the strength of the dependence. Figure 3 compares the histograms of $-log_{10}$(p-value) values resulting from association tests of Y with A nodes and O nodes, respectively. The comparison of these two histograms reveals a large dissimilarity between the two distributions. The majority (70%) of $-log_{10}$(p-value) values relative to A nodes is greater than 1, whereas it is the case for only 19% of O nodes. We observe that large $-log_{10}$(p-value) values (*e.g.*, greater than 5) are common for the former and are very rare for the latter. A non-parametric test, the Wilcoxon rank-sum test, shows a p-value less than 10^{-16}, thus confirming that A and O p-values follow two different distributions.

Figure 4(a) more thoroughly describes the $-log_{10}$(p-value) values observed for the different layers of the FLTM in the cases of tests relative to A and O nodes. The layer 0 refers to the association tests between the phenotype and the causal SNP and serves as the reference value. In this figure, we observe that the association strength for A nodes slowly decreases when the layer number increases, whereas the association for O nodes sticks to $-log_{10}$(p-value) values below 0.4, corresponding to p-values greater than 0.4. Although O nodes reveal false positive associations (less than 10% have a p-value below 0.01), these results clearly highlight a general trend: indirect associations are captured by the A nodes while it is not the case for a large majority of O nodes.

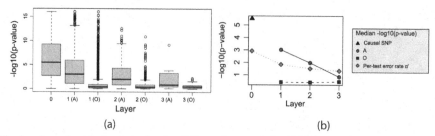

Fig. 4. $-log_{10}$(p-value) values for the different layers of the FLTM, resulting from association tests of the phenotype with the causal SNP ancestor nodes (As) and with the other latent nodes (Os) - simulated data. (a) Boxplots. (b) Median values. Layer 0 shows the results of the association tests between the phenotype and the causal SNP (over all simulated scenarii). See Figure 3 for details about the scenarii, see Subsection 6.3 for the definition of error rate α'

Figure 4(b) emphasizes the general trend of $-log_{10}$(p-value) observed for A and O nodes, and compares the median $-log_{10}$(p-value) value obtained for each layer to the value corresponding to the significance threshold α' specific to this layer (see Subsection 6.3, second paragraph). The remarkable conclusion drawn from this figure is the following: up to the second layer, *significant* associations are identified for A nodes; in contrast, regarding O nodes, for all layers, median $-log_{10}$(p-value) values are smaller than the corresponding $-log_{10}(\alpha')$ values. Focusing on the O distribution, we observe that the percentage of p-values lower than α' (false positives) is 4.7%.

In the following, we refer to the terminology defined in Figure 2. The causal tree is the tree containing the causal SNP. The O nodes divide between nodes located in the causal (OTs) tree and nodes outside the causal tree (OOs) (Os = OTs (21%) + OOs (79%)). The true positives are all A nodes. The distribution for O nodes shows 4.7% of false positives (FPs) on average. But examining separately the OT distribution and the OO distribution reveals respective FP percentages of 16% and 1.6%. Turning it the other way, focusing on FPs shows the following allotment: FPs = FP Os = FP OTs (73%) + FP OOs (27%). Thus 73% of FPs are in the causal tree, representing less than 21% of O nodes. In conclusion, the false positives are mainly confined in the causal tree; in this case, the false discovery is explained by the presence of indirect dependences between the causal SNP and the OTs.

Behaviour under Thirty-Six Genetic Scenarii. The distributions relative to A and O nodes are now compared for each scenario described in Subsection 6.1 (see Figure 5). Globally, similar tendencies are observed over all scenarii: the association strength drops continuously from bottom to top layer; in the case of O nodes, an overwhelming majority of results points out the absence of association, whichever the FLTM's layer concerned.

When considering the easiest case (MAF range = 0.3-0.4, GRR = 1.8 and multiplicative model), over all layers, the A nodes present strong associations ($-log_{10}$(p-value) > 7). Regarding a less ideal but more plausible configuration (MAF range = 0.2-0.3, GRR = 1.6 and additive model), the median $-log_{10}$(p-value) value computed for A nodes decreases from 8.3 at layer 0, to reach 4.6, 3.2 and 2.2 at layers 1, 2 and 3, respectively. On the contrary, when the model is recessive, the association with the causal SNP is low and the A nodes cannot capture anything (similar results are

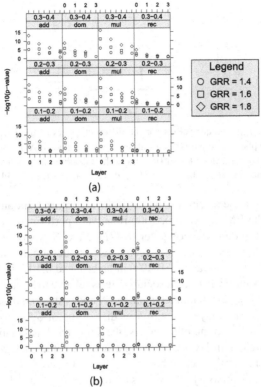

Fig. 5. Median $-log_{10}$(p-value) values for the different layers of the FLTM, resulting from association tests between the phenotype and latent nodes - simulations under thirty-six conditions. (a) Causal SNP ancestor nodes (As). (b) Other latent nodes (Os). The different windows represent possible genetic scenarii. At the top of each window, the range of the simulated causal SNP's minor allele frequency and the disease model assumption (additive, dominant, multiplicative or recessive) are indicated. The three different symbols used refer to the genotype relative risks considered for the simulated causal SNP (see Legend and 6.1). Layer 0 refers to the association tests between the phenotype and the causal SNP (over all 100 replications).

obtained with most of the methods dedicated to association studies). As regards the O nodes, null associations are reported in all configurations.

7.2   Confirmation on Real Data

The ability of FLTM to capture the indirect associations was also evaluated on real data. The dataset used is the 890 *kb* region flanking the *CYP2D6* gene on human chromosome 22q13. This gene has a confirmed role in drug metabolism [25]. The dataset consists of 32 SNP markers genotyped for 268 individuals and was downloaded from the R package graphminer developed by Verzilli and collaborators [5]. The SNP 19 at the position 550 *kb* is the closest marker to *CYP2D6* gene (at 525.3 *kb*). For this reason, SNP 19 is considered as the causal marker.

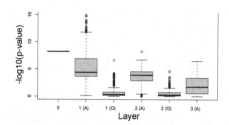

Fig. 6. Boxplot of $-log_{10}$(p-value) values for the different layers of the FLTM, resulting from association tests of the phenotype with the causal SNP ancestor nodes (As) or with the causal SNP non-ancestor nodes (Os) - real data. Layer 0 refers to the association test between the phenotype and the causal SNP (marker 19). In layer 3, no O nodes are observed in the FLTMs.

To take into account the stochastic nature of the algorithm (random initialization of parameters during the EM algorithm), 1000 runs were performed. Each run takes on average $5.4\ s$ on a standard PC computer ($3\ GHz$, $2\ GB$ RAM). On average, over all 1000 FLTMs (1000 replicates), the percentages of nodes are distributed as follows: 82.62% in layer 0, 16.89% in layer 1, 0.39% in layer 2 and 0.10% in layer 3. Figure 6 shows the $-log_{10}$(p-value) values of association tests relative to As and Os. As expected in view of experiments led on simulated data, the A nodes succeed in capturing indirect association, in particular in layer 1, with a median value of 5.5, corresponding to p-values lower than 5.10^{-6}. In the other layers, the strength of associations is lower but remains relatively high as in layer 2 showing a median value of 4, equivalent to a p-value of 10^{-4}. As previously seen, when we focus on O nodes, we observe very few strong associations. The majority of p-values (over 80%) is greater than 0.01.

8   Conclusions

Based on both simulated and real data analyses, this chapter promotes the use of FLTMs as a simple and useful framework for disease association detection in human genetics. Efficient capture of indirect genetic association is achieved through two major reasons: (i) the causal SNP ancestor nodes succeed in capturing indirect associations with the phenotype; (ii) at the opposite, the other latent nodes globally show very weak associations. In other words, this property allows to distinguish between true and false indirect genetic associations.

The numbers of SNPs in the benchmarks used for the simulations were limited. Nonetheless, this limitation is not a bias to the sound characterization of the fading of information in the FLTM hierarchies: bottom-up information decays does concern the forest depth and does not interfere with the forest width. It must be underlined that the tests were not designed to meet the small n, large p condition (many more variables (SNPs) than subjects) as in genome-wide association studies (GWASs). Again, this is not a bias to the study: over thirty-six various scenarii, it was shown that the overwhelming part (about three quarters) of false positives confines in a unique tree, namely the one harbouring the causal SNP (causal tree). In the conditions of a GWAS, the forest width may well be far larger than those observed in our tests, the false positives are expected to remain confined in the causal tree, for the major part.

The scalable FLTM learning algorithm allows to reach orders of magnitude consistent with GWAS demands (10^5 variables, 2000 individuals). In addition to scalability, data dimension reduction advocates the use of FLTM-based modeling in GWASs: the issue of multiple hypothesis testing in GWASs would be resolved by testing a low number of latent variables instead of a large number of observed variables. In the methodical investigation presented in this chapter, a permutation procedure was necessary to correct for multiple testing. In the context of a GWAS, only would be explored the trees rooted in latent variables shown to be significantly associated with the phenotype. Thus the permutation procedure remains necessary to compute the significance threshold specific to each layer. Finally, before envisaging an FLTM-based GWAS, an inescapable prerequisite was testing whether the bottom-up information fading through the forest would nevertheless allow reliable association detection. No less unavoidable was the close examination of ratios of latent variables erroneously associated with the disease. In such an exhaustive analysis of latent variables as above described, the high concentration of (false) associations in a tree pinpointed the causal tree. However, in a GWAS implementation, a mere best-first search in the FLTM would not allow the identification of this high concentration. Therefore, the question remains open to design an optimized procedure where some variant of the best-first FLTM traversal strategy, dimension reduction and conditional dependence testing have a role to play.

References

1. Zhang, Y., Ji, L.: Clustering of SNPs by a Structural EM Algorithm. In: International Joint Conference on Bioinformatics, Systems Biology and Intelligent Computing, pp. 147–150 (2009)
2. Mourad, R., Sinoquet, C., Leray, P.: Learning Hierarchical Bayesian Networks for Genome-Wide Association Studies. In: Lechevallier, Y., Saporta, G. (eds.) 19th International Conference on Computational Statistics (COMPSTAT), pp. 549–556 (2010)
3. Mourad, R., Sinoquet, C., Leray, P.: A Hierarchical Bayesian Network Approach for Linkage Disequilibrium Modeling and Data-Dimensionality Reduction Prior to Genome-wide Association Studies. BMC Bioinformatics 12, 16+ (2011)
4. Daly, M.J., Rioux, J.D., Schaffner, S.F., Hudson, T.J., Lander, E.S.: High-Resolution Haplotype Structure in the Human Genome. Nature Genetics 29(2), 229–232 (2001)
5. Verzilli, C.J., Stallard, N., Whittaker, J.C.: Bayesian Graphical Models for Genome-Wide Association Studies. The American Journal of Human Genetics 79, 100–112 (2006)
6. Han, B., Park, M., Chen, X.-W.: A Markov Blanket-Based Method for Detecting Causal SNPs in GWAS. BMC Bioinformatics 11(suppl. 3), S5+ (2010)
7. Thomas, A., Camp, N.J.: Graphical Modeling of the Joint Distribution of Alleles at Associated Loci. The American Journal of Human Genetics 74, 1088–1101 (2004)
8. Lee, P.H., Shatkay, H.: BNTagger: Improved Tagging SNP Selection Using Bayesian Networks. Bioinformatics 22(14), 211–219 (2006)
9. Greenspan, G., Geiger, D.: High Density Linkage Disequilibrium Mapping Using Models of Haplotype Block Variation. Bioinformatics 20, 137–144 (2004)
10. Kimmel, G., Shamir, R.: GERBIL: Genotype Resolution and Block Identification Using Likelihood. Proceedings of the National Academy of Sciences of The United States of America (PNAS) 102(1), 158–162 (2005)

11. Scheet, P., Stephens, M.: A Fast and Flexible Statistical Model for Large-Scale Population Genotype Data: Applications to Inferring Missing Genotypes and Haplotypic Phase. The American Journal of Human Genetics 78(4), 629–644 (2006)

12. Browning, S.R., Browning, B.L.: Rapid and Accurate Haplotype Phasing and Missing-data Inference for Whole-Genome Association Studies by Use of Localized Haplotype Clustering. The American Journal of Human Genetics 81(5), 1084–1097 (2007)

13. Abel, H.J., Thomas, A.: Accuracy and Computational Efficiency of a Graphical Modeling Approach to Linkage Disequilibrium Estimation. Statistical Applications in Genetics and Molecular Biology 10(1), Article 5 (2011)

14. Thomas, A., Green, P.J.: Enumerating the Junction Trees of a Decomposable Graph. Journal of Computational and Graphical Statistics 18(4), 930–940 (2009)

15. Schwartz, G.: Estimating the Dimension of a Model. The Annals of Statistics 6(2), 461–464 (1978)

16. Zhang, N.L.: Hierarchical Latent Class Models for Cluster Analysis. Journal of Machine Learning Research 5, 697–723 (2004)

17. Chen, T., Zhang, N.L., Liu, T., Poon, K.M., Wang, Y.: Model-Based Multidimensional Clustering of Categorical Data. Artificial Intelligence 176(1), 2246–2269 (2011)

18. Zhang, N.L., Kocka, T.: Efficient Learning of Hierarchical Latent Class Models. In: 16th IEEE International Conference on Tools with Artificial Intelligence (ICTAI), pp. 585–593 (2004)

19. Hwang, K.-B., Kim, B.-H., Zhang, B.-T.: Learning Hierarchical Bayesian Networks for Large-Scale Data Analysis. In: King, I., Wang, J., Chan, L.-W., Wang, D. (eds.) ICONIP 2006, Part I. LNCS, vol. 4232, pp. 670–679. Springer, Heidelberg (2006)

20. Harmeling, S., Williams, C.K.I.: Greedy Learning of Binary Latent Trees. IEEE Transactions on Pattern Analysis and Machine Intelligence 33(6), 1087–1097 (2011)

21. Wang, Y., Zhang, N.L., Chen, T.: Latent Tree Models and Approximate Inference in Bayesian Networks. Machine Learning 32, 879–900 (2008)

22. Ben-Dor, A., Shamir, R., Yakhini, Z.: Clustering Gene Expression Patterns. In: 3rd Annual International Conference on Computational Molecular Biology, pp. 33–42 (1999)

23. Mourad, R., Sinoquet, C., Dina, C., Leray, P.: Visualization of Pairwise and Multilocus Linkage Disequilibrium Structure Using Latent Forests. PLoS ONE 6(12), e27320 (2011)

24. Spencer, C.C., Su, Z., Donnelly, P., Marchini, J.: Designing Genome-Wide Association Studies: Sample Size, Power, Imputation, and the Choice of Genotyping Chip. PLoS Genetics, 5, e1000477+ (2009)

25. Hosking, L.K., Boyd, P.R., Xu, C.F., Nissum, M., Cantone, K., Purvis, I.J., Khakhar, R., Barnes, M.R., Liberwirth, U., Hagen-Mann, K., Ehm, M.G., Riley, J.H.: Linkage Disequilibrium Mapping Identifies a 390 kb Region Associated with CYP2D6 Poor Drug Metabolising Activity. Pharmacogenomics Journal 2(3), 165–175 (2002)

Laser Doppler Flowmeters Prototypes: Monte Carlo Simulations Validation Paired with Measurements

Edite Figueiras[1], Anne Humeau-Heurtier[2], Rita Campos[1], Ricardo Oliveira[1],
Luís F. Requicha Ferreira[1], and Frits de Mul[3]

[1] Instrumentation Center (GEI-CI), Physics Department,
Faculty of Sciences and Technology of Coimbra University,
Rua Larga, 3004-516, Coimbra, Portugal
[2] Laboratoire d'Ingénierie des Systèmes Automatisés (LISA), Université d'Angers,
62 avenue Notre Dame du Lac, 49000 Angers, France
[3] previously at University of Twente, Department of Applied Physics,
Biomedical Optics Group, Enschede, Netherlands
{edite.figueiras,ffmdemul}@gmail.com,
anne.humeau@univ-angers.fr,

Abstract. Two new laser Doppler flowmeter prototypes are herein validated with Monte Carlo simulations paired with measurements. The first prototype is a multi-wavelength laser Doppler flowmeter with different spaced detection fibres that will add depth discrimination capabilities to laser Doppler flowmetry skin monitoring. The other prototype is a self-mixing based laser Doppler flowmeter for brain perfusion estimation. Monte Carlo simulations in a phantom consisting of moving fluid as well as in a skin model are proposed for the first prototype validation. We obtain a good correlation between simulations and measurements. For the second prototype validation, Monte Carlo simulations are carried out on a rat brain model. We show that the mean measurement depth in the rat brain with our probe is 0.15 mm. This positioning is tested *in vivo* where it is shown that the probe monitors the blood flow changes.

Keywords: Laser Doppler Flowmetry, Monte Carlo simulations, Microcirculation.

1 Introduction

Laser Doppler flowmetry (LDF) is a Doppler effect based technique used for microcirculation blood flow monitoring where monochromatic light, guided by optical fibres, is transmitted to the tissues under study. In the tissues the laser light can be reflected, absorbed, transmitted or scattered. The photons scattered by moving particles, like red blood cells (RBCs), are frequency shifted in accordance with the Doppler effect. These photons get red blood cells velocity information. If they are detected, together with photons scattered by static particles, they will produce a modulated photocurrent in the photodetector. This photocurrent is related with the velocity and concentration of the RBCs [1] [2].

J. Gabriel et al. (Eds.): BIOSTEC 2012, CCIS 357, pp. 135–149, 2013.

LDF signals recorded from human skin lack in estimating the sampling depth. These difficulties lead to ambiguities in the discrimination of the fraction of light scattered from superficial and deeper blood microcirculation skin layers [3]. Besides this, commercial available flowmeters use different signal processing algorithms and calibration procedures making impossible the comparison of their results [2]. Concerning LDF invasive measurements, the smallest commercial probes available (with 450 μm diameter) are too large for research studies in small organs of animals as rat brain, causing damage in an extension that may negatively impact local measurements [3].

Monte Carlo methods are statistic methods used in stochastic simulations with applications in several areas as physics, mathematics, and biology. Monte Carlo simulations of light transport are very helpful in photon propagation studies in turbid media, as skin. They have been widely used in LDF area (see for example [4]). In Monte Carlo methods, the light transport in turbid media is based on the simulation of the photon trajectories, where separate photons travel through the tissues. Several phenomena as scattering, absorption and refraction can be simulated based on the scattering functions, Fresnel relations, etc.

We present herein Monte Carlo simulations results for validation of two new laser Doppler flowmeter prototypes. These prototypes have been built in order to eliminate two drawbacks existing in the LDF technique. The first prototype aims at giving depth perfusion measurements information (non invasive prototype) for human skin. The second prototype aims at reducing the size of LDF invasive probes for rat brain measurements (invasive prototype). For the first prototype validation, Monte Carlo simulations in a phantom consisting of moving fluid at six different depths are herein proposed. Simulations in a skin model are also presented. Measurements made in the phantom built and in the human skin are herein presented and compared with the simulation results. For the second prototype validation, Monte Carlo simulations are carried out on a rat brain model paired with *in vivo* measurements. In what follows, we first present the two prototypes, the three simulated models and the measurements protocols. Finally, the Monte Carlo simulations are detailed and the computed signals are presented and compared with the measurements.

2   Materials and Methods

2.1   Prototypes

Skin microcirculation is present in the dermis and is organized into two horizontal plexuses: the most superficial is situated in the papillary dermis at 0.4 - 0.5 mm below the skin surface; the second plexus is located at the dermal subcutaneous interface at 1.9 mm from the skin surface where arteriovenous anastomoses can be found [5]. A new laser Doppler flowmeter prototype with depth discrimination capabilities is being built in order to determine the sampling depth of the backscattered photons used to compute the LDF signal [3]. This prototype is a non-invasive and multi-wavelength prototype device, with 635, 785, and 830 nm laser wavelengths. The probe used is from Perimed AB and has a central emitting fibre and collecting fibres located at 0.14 (F0.14), 0.25 (F0.25) and 1.20 (F1.20) mm from the emitting fibre [3].

A self-mixing based prototype with a miniaturized laser Doppler probe is also being built in order to monitor blood flow changes in rat deep brain structures without causing significant damage to the tissue [3]. In the self-mixing method, the monitor photodiode at the rear face of the laser diode is used for signal detection; a single optical fibre is therefore used for light emission and detection. Pigtailed laser diodes, with 785 and 1308 nm laser wavelengths and with single mode optical fibre are used. Standard single mode optical fibres have 125 and 250 μm of cladding and jacket diameters, respectively. The probe consists of the stripped optical fibre inserted in a micro-needle with an outer diameter of 260 μm [3]. Measurements will be made in the rat brain hippocampus. As commercial available probes have a 450 μm diameter, the use of only one optical fibre allows us to reduce the size of the probe to 58%.

2.2 Phantom and Simulation Models

Phantom Model. The phantom model was built with the purpose to evaluate, *in vitro*, the non-invasive flowmeter prototype response to moving fluid at different depths (as it can be found in skin) [6]. The phantom consists of a Teflon microtube rolled around an aluminium metal piece with a total of six layers. The inner and outer diameters of the microtube are 0.3 and 0.76 mm, respectively. Commercial skimmed milk has been chosen as a moving fluid because it has various components that act as scatterers, namely carbohydrates, fat, and protein. Moreover, it does not sediment like microspheres, and it has similar behaviour to intralipid solutions [7]. Finally, milk is easier for handling than blood and, besides, it is cheaper. However, as milk is unstable, we use the same milk solution for one day only. Milk is pumped in the microtubes with a motorized syringe with different velocities: 1.56, 3.12, 4.68, 6.25, 7.78, and 9.35 mm/s. Different milk solutions were used: 100% milk, and aqueous solutions with 50% and 25% of milk.

The first simulated model (presented in figure 1) consists of three main layers. The first layer is composed of a set of objects equivalent to the six microtube layers and it has a total depth of 5 mm. The two deeper layers mimic the aluminium plate and have a thickness of 0.1 mm each one: one acts as a scatterer with isotropic semi-spherical backscattering and the other is a totally reflecting layer. The laser light was considered as a pencil beam shape and it was positioned at the top of the most superficial tube. A parabolic profile was used for the milk flow simulations where the maximum velocity is twice the mean velocity.

The simulations were made only for 635 nm laser light wavelength due to the absence of information concerning milk and Teflon optical properties for 785 and 830 nm laser light. The milk optical properties used were published by Waterworth *et al.* [7], where the refractive index for milk is 1.346, absorption (μa) and scattering coefficients (μs) are 0.00052 and 52 mm^{-1}, respectively. The Teflon optical properties used were published by Li *et al.* [8] where the refractive index is 1.367, $\mu a = 0.001$ mm^{-1} and μs =167 mm^{-1}, respectively. Henyey-Greenstein phase function was used with g = 0.90 for both components [7] [8].

Simulations were made for six different velocities for milk speed, three different milk solutions and three different detection distances, which gives a total of 54 simulations, with 5,000,000 photons detected in each simulation.

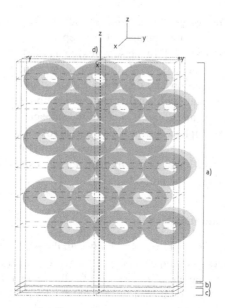

Fig. 1. Simulation phantom model: it consists of three main layers: a) the upper layer, composed of microtubes with skimmed milk as moving fluid; and the two deeper layers, b) and c) mimic the aluminium plate. d) represents the laser beam

Skin Model. The skin simulations were made for three wavelengths (635, 785 and 830 nm) and three different emitting-receiving fibre separations (0.14, 0.25 and 1.2 mm), the ones used in the non self mixing prototype.

The skin model was based on the model presented by Fredriksson *et al.* [9] [10]. It consists of 6 layers with different thicknesses and a given blood concentration at three different flow velocities with random direction (see table 2 from [10]). Oxygenated blood with hematocrit equal to 42% was considered (normal hematocrit values: 36-44% for females; 39-50% for males). A parabolic distribution was considered for the blood velocities.

The skin and blood optical properties for 635 and 785 nm were also based on Fredriksson *et al.* [10], whereas for 830 nm they were based on the results presented in [11] [12].

The skin and blood optical properties are summarized in table 1. Concerning the scattering functions, the blood was modelled with the Gegenbauer kernel scattering phase function, with $\alpha_{GK} = 1$ for all wavelengths and $g_{GK} = 0.95 \ mm^{-1}$ for 635 nm and $g_{GK} = 0.948 \ mm^{-1}$ for 785 and 830 nm. For static tissue the Henyey-Greenstein phase function was used with g = 0.85 mm^{-1}. The refractive index was set to 1.4 for all skin layers, 1.58 for the probe and 1 for the surrounding air.

The laser light was simulated as an external photon beam, with pencil beam shape with a perpendicular entrance in the tissue and the path tracking was recorded with $1/\mu'_s$ resolution (μ'_s refers to the reduced scattering coefficient). The numerical aperture (NA) of the fibres is 0.37. A total of 3 times 4 simulations were made where 10,000,000 photons were detected in each one.

Table 1. Optical properties for the six skin layers and oxygenated blood (hematocrit=42%) used in skin simulations for 635, 785 and 830 nm

	$\mu_a\ mm^{-1}$			$\mu_s'\ mm^{-1}$			g		
Wavelength (nm)	635	785	830	635	785	830	635	785	830
Epidermis	0.15	0.1	0.0122	4.8	3.5	1.81	0.85	0.85	0.9
Papillary dermis	0.15	0.1	0.0122	3	2	1.81	0.85	0.85	0.9
Superior blood net	0.15	0.1	0.0122	3	2	1.81	0.85	0.85	0.9
Reticular dermis	0.15	0.1	0.0122	3	2	1.81	0.85	0.85	0.9
Inferior blood net	0.15	0.1	0.0122	3	2	1.81	0.85	0.85	0.9
Subcutis	0.15	0.1	0.00856	2.4	2	1.12	0.85	0.85	0.9
Blood	0.34	0.5	0.52	2.13	2	2	0.991	0.991	0.991

The optical properties of deoxygenated and oxygenated blood are equal except for the absorption coefficient (see table 5 from [10]). However, this value is rather independent of the level of oxygenation of the blood, once the chosen wavelengths are close to the 800 nm isobestic point of oxygenated and deoxygenated haemoglobin.

Rat Brain Model. The rat hippocampus consists of several substances such as grey matter, white matter and blood vessels, among others. The blood percentage is nearly 4.5% and the white matter is up to 4% of the blood volume. As the percentage of white matter is very low we considered that the hippocampus has 95.5% of grey matter and 4.5% of blood (3.6% of oxygenated blood and 0.9% of deoxygenated blood) [10] [13].

The optical properties chosen in the simulations for the 785 nm wavelength were based on Fredriksson *et al.* [10]. The absorption coefficients used were 0.2, 0.5 and 0.64 mm^{-1} for grey matter, oxygenated and deoxygenated blood, respectively. The scattering coefficients were 0.78 mm^{-1} for the grey matter and 2 mm^{-1} for oxygenated and deoxygenated blood and the anisotropy factor was 0.900 for grey matter and 0.991 for oxygenated and deoxygenated blood [10]. Concerning the scattering functions, the blood (oxygenated and deoxygenated) was modeled with the Gegenbauer kernel scattering phase function, with $\alpha = 1$ and $g = 0.948\ mm^{-1}$ [10]. For grey matter the Henyey-Greenstein phase function was used with $g = 0.85\ mm^{-1}$ [10]. For blood a hematocrit equal to 42% was considered. The refractive index was set to 1.4 to all components and the laser light was simulated as a pencil beam with a perpendicular entrance in the tissue. The path tracking was recorded with $1/\mu_s'$ resolution, where μ_s' is the reduced scattering coefficient. The numerical aperture (NA) of the optical fibres is 0.11.

The simulations were performed only for the 785 nm laser light wavelength due to the absence of information concerning optical properties of grey matter, oxygenated and deoxygenated blood for the 1300 nm laser light beam.

Monte Carlo Simulations. For the simulations, Monte Carlo software MONTCARL from Frits de Mul was used [14] [15] (see also www.demul.net/frits).

2.3 Measurements

Phantom Measurements. Measurements were carried out with the non invasive prototype, with the 635 nm laser light, in the built phantom. The milk and two different aqueous milk solutions (50% and 25%) were pumped with a motorized syringe in the microtube with the simulated velocities during ten minutes with Perimed probe positioned perpendicular to the surface of the phantom. The mean perfusion, in perfusion units (PU), for each velocity and concentration was computed for three minutes of blood perfusion signal and a linear regression study was performed.

Skin Measurements. Measurements in the ventral side of the forearm, in 20 healthy non-smoking subjects, in the supine position, and the limb at heart level were performed (age 24.7 ± 4.2 years old, range 19-39 years old; 11 females; arterial pressure: min 7.1 ± 0.8 mmHg, max 11.4 ± 1; IMC: 22.1 ± 2.3). All subjects were Caucasian, except one female. The participants did not take any vasoactive medication and were asked to refrain from drinking coffee during the measurement tests. The tests were performed in a quiet room, where the temperature varied between 22 and 25 °C. Motility standard calibrations were performed before each measurement. The Ethics Committees of the Centro Cirúrgico de Coimbra (CCC) in Portugal, approved this study. Informed consent was obtained from the subjects before the recordings were performed.

Perfusion was recorded during thirty three minutes: baseline blood flux was recorded for 20 min. Then, an arterial occlusion test was performed with a pressure cuff placed around the upper limb, inflated for 3 min at 200 mm Hg in order to obtain the biological zero (BZ). The cuff was then released to obtain a post-occlusive reactive hyperemia and the signal was recorded during 10 min after the release of the occlusion. For each subject, the protocol was repeated using the three laser diodes existing in the prototype: 635, 785, and 830 nm.

Quantitative data were expressed as the median (first quartile, second quartile). The normal distribution of the data was evaluated (using Shapiro-Wilk statistics). The majority of the data of the different parameters were not normally distributed. Therefore, non parametric tests were used. The Wilcoxon test was used to test for significant differences between results obtained with different laser wavelengths, different emitting-receiving fibre distances and skin regions. Analysis was performed with SPSS v.17.0 for WINDOWS (SPSS Inc, Chicago, IL, USA). The results were considered statistically different for p-values lower than 0.05.

Rat Brain Measurements. *In vivo* tests were performed in male Wistar rats (9 weeks). Rats were anesthetized with urethane (1.25 g/kg, i.p.), and placed in a stereotaxic frame (Stoelting Co, USA). After having exposed the rat brain surface and removed the dura mater, one of the microprobes was stereotaxically inserted into the hippocampus. Together with our probe we inserted a commercial laser Doppler needle probe (Perimed reference: PF411; outer diameter, 450 μm; fibre separation, 150 μm; wavelength, 780 nm) connected to a laser Doppler flowmeter device (Periflux system 5000, Perimed, Sweden), in the opposite cerebral hemisphere. The simultaneous use of the probes allows the comparison of the signals acquired by our probes with the one of the commercial flowmeter probe. Measurements were made with the 785 nm micro-probe. After the probes implantation, a baseline was recorded during c.a. 5 minutes and then sodium nitrite (200 mg/kg in saline solution) was intraperitoneally injected, in order to induce metahemoglobinemia and finally cardiac arrest. The correlation between the results obtained with the Perimed probe and with both micro-probes was computed. The signals were normalized before the cross-correlation analysis.

Several signal processing methods applied to self-mixing signals were reported in the literature, namely, the Counting Method (CM), the Autocorrelation Method (AM), and the Power Spectral Method (PSM) [16]. In the PSM method the zero order moment (M0) and the first order moment (M1) were computed. These methods were evaluated thereafter.

3   Results and Discussion

3.1   Phantom

Phantom Simulations. The results obtained in the phantom model, namely the mean depth of the Doppler events per photon, the percentage of Doppler shifted photons detected and the mean of Doppler scattering events per photon are given in table 2 for 1.56 mm/s milk velocity.

Table 2. The mean depth of the Doppler events for each photon, the percentage of Doppler shifted photons detected and the mean of Doppler scattering events per photon for the phantom model, with milk pumped at 1.56 mm/s, for 635 nm laser light wavelength

Fibre distance mm	Mean Doppler depth (mm)			Detected Doppler (%)			Mean Doppler scattering		
	Milk concentration								
	25%	50%	100%	25%	50%	100%	25%	50%	100%
0	0.36	0.34	0.32	4.77	6.73	10.27	3.48	5.85	10.08
0.14	0.41	0.39	0.35	26	32.35	41.26	4.29	7.30	12.66
0.25	0.43	0.40	0.37	41.23	47.76	56.45	4.37	7.78	14.27
1.2	0.56	0.55	0.53	82	86	87.47	6.47	12.65	26.63

The emitting-receiving fibre separation influences the measurements, in such a way that for larger fibre separations, a larger sample volume is assumed to be probed probed. Therefore, increasing the fibre distance, photons travel through deeper objects leading to an increase in the mean depth of Doppler shifted photons. Our results are in agreement with previous studies based on light propagation in tissue using Monte Carlo computational simulations [9] [10]. The milk concentration also influences the mean depth, which decreases with the increase of milk concentration. This is due to a higher degree of multiple Doppler shifts registered for higher milk concentrations.

The percentage of Doppler shifted photons detected increases with the emitting-receiving separation for each velocity. This is expected because when the fibre distance increases, the measured sample volume increases, and the photons will encounter a larger amount of moving scatterers. When the concentration of milk increases the percentage of Doppler shifts detected also increases. The higher the milk concentration, higher scatterers are present, consequently, more scattering events occur.

Likewise, the mean Doppler scattering events per photon also increase with the fibre distance for each velocity. This is not surprising as we are considering a homogenous model (for the scatterers velocity and concentration). Therefore, an increase in the sampled volume will lead to more Doppler scattering events. Increasing the milk concentration, a higher degree of multiple Doppler scattering is reached, because the higher the concentration, the higher the scatterers.

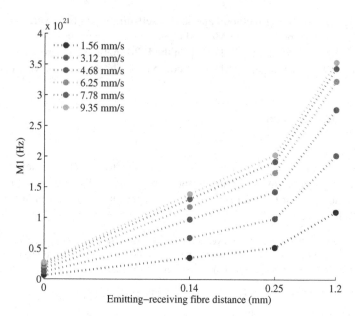

Fig. 2. M1 *vs.* emitting-receiving fibre distance, on the phantom model with 25% milk concentration for 635 nm laser light wavelength

Similar results were obtained for all velocities in what concerns the mean depth of the Doppler events per photon, the percentage of Doppler shifted photons detected and the mean of Doppler scattering events per photon, as these parameters are independent of the velocity of the moving fluid.

The first order moment of the Doppler power spectrum, M1, was also evaluated. Figure 2 shows the effect on M1 when the fiber distance increases, for each velocity and for a milk concentration of 25%. It can be seen that higher values of M1 come from larger fiber separations whereas the lower values of M1 are obtained for 0 mm fiber separation. Another observation is that, in general, M1 increases with the velocity and with the milk concentration. This is not surprising since M1 is proportional to perfusion (Perf), which in turn is proportional to the scatterers concentration times their average velocity. However, in some specific cases M1 does not increase with the velocity, especially for the two highest velocities for 1.2 mm fiber distance. This might be due to the phantom model that saturates in such extreme situations.

Phantom Measurements. In the measurements made, the perfusion increases with milk velocity and with the emitting-receiving fibre distance for the aqueous milk solution of 25%. For the other milk concentrations the perfusion saturates for the higher velocities. No perfusion tendency was obtained for the different scatterers concentration.

Positioning the probe in the top of the microtube phantom was difficult due to the microtube curvature. This, together with the small milk volume in the microtube, when compared with the tube volume, lead to the sub-estimation and uncertainties of the perfusion measurements. These factors could be the reason for the non-linearity obtained.

3.2   Skin

Skin Simulations. The mean depth of the Doppler events per photon, the percentage of Doppler shifted photons detected and the mean of Doppler scattering events per photon obtained in Monte Carlo simulations, for the skin model, are presented in table 3.

Table 3. The mean depth of the Doppler events for each photon, the percentage of Doppler events detected and the mean of Doppler scattering events per photon for the skin model

Fibre distance mm	Mean Doppler depth (mm)			Detected Doppler (%)			Mean Doppler scattering		
	Wavelength (nm)								
	635	785	830	635	785	830	635	785	830
0.00	0.24	0.25	0.27	1.94	1.97	3.54	1.19	1.16	1.15
0.14	0.27	0.29	0.31	9.62	10.71	14.54	1.23	1.22	1.23
0.25	0.28	0.30	0.33	15.45	16.29	20.39	1.26	1.23	1.25
1.20	0.37	0.38	0.41	47.70	41.89	43.45	1.49	1.41	1.46

The mean depth for the Doppler events, the percentage of Doppler shifted photons detected and the mean Doppler scattering per photon increase with fibre distance, as for the phantom model, because a larger emitting-receiving separation allows sampling a larger volume.

Furthermore, the mean measurement depth also increases with the wavelength. This is due to both skin absorption and scattering coefficients decrease with the wavelength, allowing the photons to travel a longer path. Similar results were obtained by Fredriksson et al. [10], with a measurement depth slightly smaller, but with the same order of magnitude. It can be noticed that the mean depth of the Doppler shifted photons never reaches the reticular dermis or the layers below this one, since reticular dermis lies at a depth of 1.175 mm and the mean depth predicted for the Doppler events is always lower than this value. In addition, detected photons reached the superior blood net dermis only when detected with the 1.2 mm fibre distance.

The percentage of Doppler shifted detected photons also increases with the wavelength excluding for the 1.2 mm emitting-receiving fibre distance. This may be related to the distribution of the Doppler events percentage in each layer (cf. table 4). It can be seen that the reticular dermis is the 2nd layer with the most detected Doppler photons for 1.2 mm fibre distance (for 785 and 830 nm laser light), whereas for the other fibre distances the 2nd layer with the most detected Doppler photons is the papillary dermis.

This proves that the photons detected at 1.2 mm from the emitting fibre cross a higher volume of blood. Besides, the Doppler events percentage in the inferior blood net (for 1.2 mm fibre distance) decreases going from 635 to 785 nm laser light and increases going from 785 to 830 nm laser light. This layer has the second higher blood concentration when compared with the other layers. The higher volume of blood crossed and the increasing of the blood absorption coefficient with the wavelength may cause this non-linearity.

In opposition, the mean Doppler scattering event does not follow a general trend when increasing the wavelength of the incident light but is smaller than 1.5, which means that there are few photons that suffer multiple Doppler shifts.

Table 4. Doppler events percentage in each layer for the skin model

Wavelength (nm)	Fibre distance (mm)	Skin layers					
		Epidermis	Papillary dermis	Superior blood net	Reticular dermis	Inferior blood net	Subcutis
635	0.00	0.00	41.07	51.19	7.24	0.50	0.00
	0.14	0.00	35.10	53.06	10.31	1.37	0.17
	0.25	0.00	31.98	55.23	11.10	1.49	0.21
	1.20	0.00	19.43	56.21	19.51	4.08	0.77
785	0.00	0.00	41.73	48.84	8.93	0.51	0.00
	0.14	0.00	30.85	54.86	12.97	1.33	0.00
	0.25	0.00	27.69	56.42	14.38	1.51	0.00
	1.20	0.00	18.81	55.44	22.24	3.51	0.00
830	0.00	0.00	38.76	49.36	10.35	1.54	0.00
	0.14	0.00	26.66	55.91	15.09	2.33	0.00
	0.25	0.00	23.09	57.05	17.27	2.59	0.00
	1.20	0.00	16.57	53.06	25.15	5.218	0.00

Table 5. First order moment of Doppler power spectrum (M1) for skin model

Wavelength (nm)	M1 (Hz)			
	Fibre distance (mm)			
	0.00	0.14	0.25	1.20
635	3.57×10^{18}	2.51×10^{19}	3.71×10^{19}	1.63×10^{20}
785	3.45×10^{18}	2.13×10^{19}	3.23×10^{19}	1.05×10^{20}
830	5.30×10^{18}	2.66×10^{19}	3.93×10^{19}	1.06×10^{20}

Simulation results also demonstrate that M1 increases with the emitting-receiving fibre distance (see table 5). Since M1 is proportional to the concentration of moving RBCs times its average velocity, and both parameters increase with the fibre distance, this was expected.

Regarding the wavelengths, M1 firstly decreases from 635 to 785 nm and then increases for 830 nm. This can be explained if we look at the Doppler photons percentage that exceeds the reticular dermis. This percentage is higher for 830 nm followed by 635 nm and lower for 785 nm, with the exception of the 0 mm fibre distance (table 4). As the inferior blood net has the highest concentration of the high velocity component of RBCs (30 mm/s), it results in higher Doppler shifts for 635 nm than for 785 nm photons. Therefore M1 will be higher for 635 nm than for 785 nm. Table 6 shows the results of the path tracking study for the skin model. It can be observed that the average path number travelled by each photon, the mean path depth and the average path length increase with emitting-receiving fibre distance. This occurs because increasing the fibre distance a greater tissue volume is probed, and so, more scattering events occur. The mean path number does not follow a general trend when increasing the wavelength of the incident light.

The path depth increases with the laser light wavelength due to both skin absorption and scattering coefficients decrease with the wavelength, allowing the photons to travel a longer path. This is in agreement with the mean depth of Doppler events results showed in table 3. The average path length increases with the wavelength, excluding for

Table 6. Mean path number, mean path depth and mean path length for photons, using the skin model

Wavelength (nm)	Fibre distance (mm)	\<Path number\> (mm)	\<Path depth\> (mm)	\<Path length\> (mm)
	0.00	3.16	0.02	0.23
635	0.14	6.20	0.08	0.78
	0.25	7.99	0.11	1.16
	1.20	19.77	0.27	4.11
	0.00	3.08	0.02	0.23
785	0.14	5.98	0.09	0.86
	0.25	7.46	0.12	1.26
	1.20	14.71	0.26	3.65
	0.00	3.77	0.04	0.34
830	0.14	6.93	0.12	1.12
	0.25	8.34	0.16	1.57
	1.20	14.86	0.28	4.01

Table 7. Perfusion median (first quartile - third quartile) values obtained in baseline, biological zero, and PORH in the ventral forearm

Wavelength (nm)	Fibre distance (mm)	Baseline (PU)	Biological Zero (PU)	PORH (PU)
	0.14	5.2 (3.5-11.5)	4.5 (2.9-8.8)	19.7 (13.6-52.7)
635	0.25	13.7 (9.2-23.1)	13.5 (8.6-20.3)	37.4 (23.3-57.2)
	1.20	55.8 (37.5-100.0)	49.0 (33.8-88.1)	116.8 (98.0-198.6)
	0.14	7.8 (5.3-10.3)	5.1 (4.2-7.7)	39.6 (29.4-66.0)
785	0.25	14.1 (10.6-17.9)	8.8 (6.8-12.4)	91.9 (48.0-140.0)
	1.20	38.2 (27.9-47.8)	22.6 (16.3-24.9)	227.5 (144.7-304.9)
	0.14	17.2 (3.0-58.3)	13.3 (2.0-31.8)	80.0 (12.7-246.2)
830	0.25	25.8 (3.2-62.3)	15.2 (2.0-30.7)	81.5 (20.4-303.8)
	1.20	31.7 (7.2-90.3)	13.5 (1.8-43.0)	187.5 (63.0-371.4)

the 1.2 mm emitting-receiving fibre distance. This may be related with the distribution of the Doppler events percentage in each layer (*cf.* table 4) as discussed later when the percentage of Doppler shifted detected photons is analyzed.

Skin Measurements. The median perfusion values measured in baseline, biological zero and PORH are presented in table 7. For all parameters, the perfusion increases with the fibre distance for each wavelength, except in the biological zero for 830 nm wavelength. This reflects the larger volume of tissue sampled by the higher fibre distances. A significant difference between different fibre distances for the same wavelength was obtained in baseline ($p < 0.03$) and in PORH ($p < 0.023$), for all wavelengths. In biological zero a significant difference was also obtained ($p < 0.003$), except for 830 nm wavelength.

When different wavelengths are compared at baseline, for the same fibre distance, the perfusion increases with the wavelength for F0.14 and F0.25 but decreases for F1.20. In PORH, the perfusion decreases between 785 and 830 nm for F0.25 and F1.20. These nonlinearities may be the result of the multiple scattering that occurs when the photon pathlength is long; this happens when the distance between emitting and receiving fibres is higher. The multiple scattering leads to an underestimation of the perfusion level. The multiple scattering underestimation effect could be avoided by using the correction factor presented by Nilsson *et al.* [2] but *in vitro* tests with blood are required to implement that.

In the simulation, results demonstrate that M1 increases with the emitting-receiving fibre distance. The same happens in the perfusion measurements, as it increases with the fibre distance. However, when wavelengths are compared, M1 firstly decreases from 635 to 785 nm and then increases for 830 nm. This may be due to the Doppler photons percentage that exceeds the reticular dermis. In the perfusion measurements, when different wavelengths are compared there was a trend to the decrease of the perfusion for the highest fibre distances and for the highest wavelengths, probably due too the multiple scattering.

3.3   Rat Brain

Rat Brain Simulations.   In the simulation made for the rat brain model, it can be seen that the photons Doppler shifted travelled a mean depth of 0.15 mm (cf. table 8). This value is in accordance with Frediksson *et al.* [10] which obtained 0.16 mm of measurement depth. Each photon suffers, in average, 2.23 scattering events. In a total of 5,000,000 photons detected 11.9% had suffered Doppler shift and M1 was predicted to be 3.51×10^{17} Hz. These results will help in the rat brain probe positioning as it shall be 0.15 mm above the mean measurement depth.

Table 8. The percentage of Doppler events detected, the mean of Doppler scattering, the mean depth of the Doppler events for each photon and M1 for the rat model

Mean depth Doppler (mm)	Detected Doppler (%)	Mean Doppler scattering	M1 (Hz)
0.15	11.9	2.23	3.51×10^{17}

Rat Brain Measurements.   The signals collected in the rat brain were processed and compared with those obtained with the Perimed probe. Signals were collected in one animal with the 785 nm micro-probe located in the rigth brain hemisphere and the Perimed probe located in the left rat brain hemisphere. The Periflux 5000, CM, AM, M0 and M1 results are presented in Fig. 3.

An initial baseline can be seen during the first 5 minutes in Perifllux 5000, CM, AM, M0 and M1 results. After the nitrite injection the mean amplitude decreases in Periflux 5000. This could be due to a probe displacement during the nitrite injection. Despite that, the perfusion increases in the next 19 minutes followed by an amplitude oscillation and finally the amplitude decreases after 27 minutes of acquisition during the cardiac arrest in Perimed, CM, AM, M0 and M1 results. A peak registered in Perimed at 28 min, caused by cardiac arrhythmias, can also be seen in the results obtained in our prototype.

In vivo results with the new probes show good agreement when compared with the signals obtained with the Perimed device. Blood perfusion variations promoted by nitrite intraperitoneal injection are clearly visible on the signals obtained with the tested

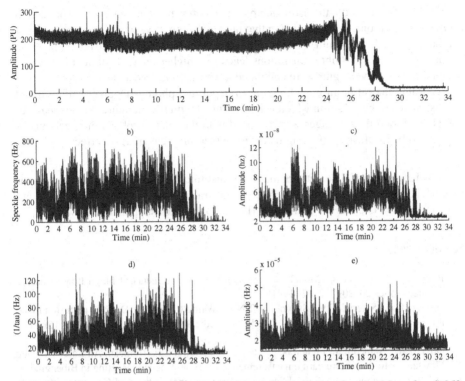

Fig. 3. *In vivo* validation results: a) Periflux 5000 results, b) CM results, c) AM results, d) M0 results and e) M1 results

microprobe: the slight increment verified after the nitrite injection and the abrupt decrease observed in consequence of the cardiac arrest. However, it should be remarked that some lacks of correlation may occur given that the commercial probe and the prototypes were located in different hemispheres, and blood flow changes can take place with different extent.

4 Conclusions

Monte Carlo simulations used for the two new LDF prototypes validation showed results in accordance with the literature. For the non invasive prototype, the phantom model presented here to evaluate the *in vitro* prototype response, has shown good agreement with theoretical expectations. M1 increases with the scatterers concentration and velocity, and with the fibre distances. The perfusion measured (estimated from M1) also increases with the fibre distance and with the scatterers velocity, but only for the aqueous milk solution of 25%. The mean depth increases with the fibre distance and decreases with the milk concentration. The non linearities obtained in the measurements leads us to built an acrylic phantom 1) with a plain surface easing the contact probe/phantom and 2) with a higher volume of milk.

For *in vivo* evaluation, the estimated parameters for the skin model corresponded to *a priori* expectations. We have shown that increasing the wavelength of incoming light (in the range of 635-830 nm) increases the mean depth probed. Moreover, an increase of the source-detection fibre separations leads to a higher mean depth and M1 value. When different wavelengths were compared it was not obtained a clear tendency.

Regarding the rat brain model, the mean depth that photons Doppler shifted travel was estimated to be 0.15 mm which is in agreement with the literature. In the measurements, the blood flow changes were detected with the micro-probe built which means that the probe localization (0.15 mm above the measurement area) is correct.

Acknowledgements. The authors thank the Instituto de Investigação Interdisciplinar (III) of the University of Coimbra, Acções Integradas Luso-Francesas (PAUILF) and Fundação para a Ciência e a Tecnologia (FCT), Lisbon, for supporting this work.

References

1. Bonner, R.F., Nossal, R.: Model for laser Doppler measurements of blood flow in tissue. Applied Optics 20, 2097–2107 (1981)
2. Nilsson, G.E., Salerud, E.G., Stromberg, T.N.O., Wardell, K.: Laser Doppler perfusion monitoring and imaging. In: Vo-Dinh, T. (ed.) Biomedical Photonics Handbook, ch. 15. CRC Press, Washington, D.C (2003)
3. Oliveira, R., Semedo, S., Figueiras, E., Requicha Ferreira, L.F., Humeau, A.: Laser Doppler Flowmeters for microcirculation measurements. In: 1st Portuguese Meeting in Bioengineering - Bioengineering and Medical Sciences - The Challenge of the XXI Century, Portuguese chapter of IEEE EMBS; Technical University of Portugal (2011)
4. Figueiras, E., Ferreira, L.F.R., De Mul, F.F.M., Humeau, A.: Monte Carlo Methods to Numerically Simulate Signals Reflecting the Microvascular Perfusion. In: Angermann, L. (ed.) Numerical Simulations - Applications, Examples and Theory, ch. 7. InTech, Rijeka (2011)
5. Braverman, I.M.: The Cutaneous Microcirculation. J. Investig. Dermatol. Symp. Proc. 5, 3–9 (2000)
6. Figueiras, E., Ferreira, L.F.R., Humeau, A.: Phantom validation for depth assessment in laser Doppler flowmetry technique. In: Proceedings of EOS, Topical Meeting on Diffractive Optics, 2413, Koli (Finland) (2010)
7. Waterworth, M.D., Tarte, B.J., Joblin, A.J., van Doorn, T., Niesler, H.E.: Optical transmission properties of homogenized milk used as a phantom material in visible wavelength imaging. Australasian Physical and Engineering Sciences in Medicine 18, 39–44 (1995)
8. Li, Q., Lee, B.J., Zhang, Z.M., Allen, D.W.: Light scattering of semitransparent sintered polytetrafluoroethylene films. Journal of Biomedical Optics 13(5), 054064 (2008)
9. Fredriksson, I., Larsson, M., Strömberg, T.: Optical microcirculatory skin model: assessed by Monte Carlo simulations paired with in vivo laser Doppler flowmetry. Journal of Biomedical Optics 13, 014015 (2008)
10. Fredriksson, I., Larsson, M., Strömberg, T.: Measurement depth and volume in laser Doppler flowmetry. Microvascular Research 78, 4–13 (2009)
11. Simpson, C.R., Kohl, M., Essenpreis, M., Cope, M.: Near infrared optical properties of ex-vivo human skin and subcutaneous tissues measured using the Monte Carlo inversion technique. Phys. Med. Biol. 43, 2465–2478 (1998)
12. Prahl, S.: Optical Absorption of Hemoglobin (1999),
   http://omlc.ogi.edu/spectra/hemoglobin/

13. Hamberg, L.M., Hunter, G.J., Kierstead, D., Lo, E.H., Gonzalez, R.G., Wolf, G.I.: Measurement of cerebral blood volume with substraction three-dimentional functional CT. Am. J. Neuroradiol. 17(10), 1861–1869 (1996)
14. De Mul, F.F.M., Koelink, M.H., Kok, M.L., Harmsma, P.J., Greve, J., Graaff, R., Aarnoudse, J.G.: Laser Doppler Velocimetry and Monte Carlo Simulations on Models for Blood Perfusion in Tissue. Applied Optics 34, 6595–6611 (1995)
15. De Mul, F.F.M.: Monte-Carlo simulation of Light transport in Turbid Media. In: Tuchin, V.V. (ed.) Handbook of Coherent Domain Optical Methods, Biomedical Diagnostics. Environment and Material Science, ch. 12. Kluwer Publishers, Dordrecht (2004)
16. Ozdemir, S.K., Ohno, I., Shinohara, S.: A comparative Study for the Assesment on Blood Flow Measurement Using Self-Mixing Laser Speckle Interferometer. IEEE Transactions on Instrumentation and Measurement 57, 33563 (2008)

Simulation of Prokaryotic Genome Evolution Subjected to Mutational Pressures Associated with DNA Replication

Paweł Błażej, Paweł Mackiewicz, and Stanisław Cebrat

Department of Genomics, Faculty of Biotechnology, University of Wrocław,
ul. Przybyszewskiego 63/77, 51-148 Wrocław, Poland
blazej@smorfland.uni.wroc.pl
http://www.smorfland.uni.wroc.pl

Abstract. Each of two differently replicated DNA strands (leading and lagging) is subjected to the distinct mutational pressure associated with its synthesis. To simulate the influence of these pressures on the gene and genome evolution we worked out a computer model in which protein coding sequences were mutated according to the direct pressure (of the strand on which they were located), the reverse pressure (of the opposite strand), and the changing pressure (when the latter pressures were applied alternately). Simulated genomes were eliminated by the occurrence of stop codons in the gene sequences and the loss of their coding properties. The selection against stop codons appeared more deleterious than for coding signal. The leading strand pressure eliminated more genes because of the coding signal loss whereas the lagging strand pressure generated more stop codons. Generally, the reverse and changing pressures destroyed the coding signal weaker than the direct pressure.

Keywords: Coding signal, DNA asymmetry, DNA replication, Genome evolution, Monte Carlo simulation, Mutational pressure, Selection, Stop codon.

1 Introduction

Because DNA is organized in two antiparallel strands, each showing 5' to 3' direction, and DNA polymerase can synthesize a new DNA strand only in the 5' to 3' direction, the replication of these strands proceeds differently. One of these strands named leading, is synthesized continuously toward the replication fork movement in contrast to the complementary lagging strand strand, that is synthesized of Okazaki fragments in the direction opposite to the movement of the fork. The distinct mode of replication causes that these strands are subjected to different pattern of nucleotide substitution [1–5]. These various directional mutational pressures lead to disparate nucleotide composition of the differently replicated DNA strands, which is very well pronounced in many bacterial genomes [6–13]. This compositional bias is called DNA asymmetry and is defined as a deviation from the equality between complementary nucleotides in a single DNA strand, i.e. [G]=[C] and [A]=[T]. The asymmetry is usually expressed as the normalized difference in the number of complementary nucleotides in one DNA strand, i.e. AT skew = [A-T]/[A+T] and GC skew = [G-C]/[G+C](Fig. 1).

J. Gabriel et al. (Eds.): BIOSTEC 2012, CCIS 357, pp. 150–161, 2013.

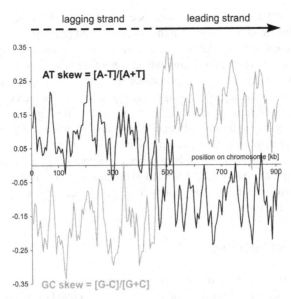

Fig. 1. AT skew and GC skew along the *Borrelia burgdorferi* chromosome calculated in sliding windows with the length of 10 kb and the shift of 5 kb. Organization of the chromosome into two differently replicated DNA strands (lagging and leading) is shown above the plot.

The strongest DNA asymmetry associated with replication is observed in the third codon position of protein coding genes, which indicates that the mutational pressure is the main reason of this bias [9, 11–13]. The leading strand is usually rich in guanine and thymine whereas the lagging strand shows excess of cytosine and adenine. It is assumed that the mutation C to T, which is the most common substitution type observed in the leading strand, is the main factor responsible for this DNA asymmetry [1] although analysis of several bacterial genomes revealed that similar compositional biases may result from different mutational patterns [5]. The effect of the mutational pressure is to some extent accepted by selection because the DNA asymmetry is also visible in codon usage of genes and amino acid composition of the coded proteins [14–17].

Interestingly, it was also found that the mutational pressure influences the evolution rate of genes in dependence of their location on the DNA strands. Genes coded in the leading strand generally accumulate less substitutions than the lagging strand genes whereas homologs located on the differently replicated strands in the compared genomes show the highest divergence [4, 18–20]. Computer simulation of gene evolution using different mutational pressures associated with replication confirmed these findings and revealed additionally that the best survival strategy for the majority of genes is switching their location between DNA strands to change the direction of the pressure from time to time [21–24]. It may explain the observed asymmetric translocations and non-conserved positions of many genes between differently replicated DNA strands in compared genomes during evolution [25, 26]. Exceptions from this tendency are genes coding for ribosomal proteins which do not profit very much from switching the directional pressure [21]. It is in agreement with their extremely conserved positions on the prokaryotic chromosomes because they, as also other essential genes for cell functioning, tend to be coded in the leading strand [26–28].

The previous computer simulation [21–24] included amino acid composition of the gene products as the selection constraint whereas here we have presented the results of other type of simulations in which selections against stop codon occurrence and for the preservation of the coding signal were applied. The selection for the coding signal was calculated by the algorithm for recognition of protein coding sequences [29, 30]. This algorithm exploites a specific way of genetic code degeneration and relations between mutational pressure and selection pressure shaping the amino acid usage in the proteomes. We used the algorithm to study how selection operating on the nucleotide level influences the elimination of genes subjected to the directional mutational pressure. We also analysed changes in the coding potential of genes under the mutational pressure during the simulation time.

2   Materials and Methods

We decided to perform our simulations on gene sequences of the *Borrelia burgdorferi* genome because it shows very strong compositional bias related to differently replicated DNA strands [14, 15, 31]. In addition to that, it does not show the selection for synonymous codon usage, and has the defined mutational pressure associated with replication for the both DNA strands [3]. Therefore, it appears to be very suitable for studies on DNA asymmetry and effects of the directional mutational pressure on protein coding sequences.

The simulation were applied to 475 protein coding sequences extracted from *B. burgdorferi* genome as annotated in the NCBI database [32]. This set of gene sequences will be further called 'genome' or 'individual'. The protein coding sequences that constituted the individual were divided into two subsets:

1. sequences lying on the leading DNA strand (including 333 genes of the total length 356, 034 nt),
2. sequences lying on the lagging strand (comprising 142 genes of the total length 173, 796 nt).

We considered in our simulation the population consisting of 72 individuals, which were eliminated and replaced by others during the simulations run.

A particular Monte Carlo step (MCS) of the simulation comprised two stages:

1. the mutation process of gene sequences,
2. the selection process of individuals.

In the mutation stage, a nucleotide from the genome sequence was substituted by another one. First, it was chosen for mutation using the Poisson process assuming one mutation per genome by average. Then, the selected nucleotide was substituted by another according to the probability given in one of two substitution matrices for the leading or the lagging strand, respectively (Tab. 1 and Tab. 2). The matrices describing the real mutational pressure for the differently replicated DNA strands in the *B. burgdorferi* genome were constructed empirically by the comparison of original gene sequences

Table 1. The substitution matrix describing mutational pressure in the leading DNA strand. A nucleotide in the first column is substituted by a nucleotide in the first row.

	A	T	G	C
A	0.81	0.10	0.07	0.02
T	0.07	0.87	0.03	0.03
G	0.16	0.12	0.71	0.01
C	0.07	0.26	0.05	0.62

Table 2. The substitution matrix describing mutational pressure in the lagging DNA strand. A nucleotide in the first column is substituted by a nucleotide in the first row.

	A	T	G	C
A	0.87	0.07	0.03	0.03
T	0.10	0.81	0.02	0.07
G	0.26	0.07	0.62	0.05
C	0.12	0.16	0.01	0.71

with their potential pseudogenes found in intergenic regions of the *B. burgdorferi* chromosome [3].

The mutated gene sequence was checked on account of two selectional assumptions:

1. appearance of stop translation codon inside the gene sequence,
2. strength of its coding signal.

We considered three standard stop codons (TAA, TAG, or TGA) whereas the coding signal was calculated according to the gene finding algorithm (called PMC) based on the theory of Markov chains [29, 30, 33]. This algorithm recognizes very efficiently protein coding sequences from prokaryotic genomes and uses three independent homogeneous Markov chains to describe occurrence of nucleotides for each of three codon positions in a given DNA sequence, separately. This method is based on specific correlations in the nucleotide composition observed in the first, the second, and the third codon positions of protein coding genes [34, 35]. In addition, this algorithm does not require learning sets of large sizes therefore the nucleotide transition matrices used by this method can be built on only a few coding sequences for its effective training [30, 33]. The individual that nested the mutated gene was eliminated if at least one stop codon was generated inside its sequence or if the sequence was not recognized by the PMC algorithm as a protein coding sequence in the first reading frame. The eliminated individual was replaced by another one chosen randomly from the population.

In our simulations, we have taken into account three different versions of the mutational pressure acting on gene sequences. In the first possibility (direct pressure), the genes from a given DNA strand (e.g. leading) were subjected to the matrix of the strand on which they were lying (i.e. leading). In the second variant, we considered genes that were under the pressure characteristic of the opposite strand (reverse pressure). Apart from these two constant pressures, we also applied the changing pressure. In this case the genes were subjected to the leading and lagging strand pressures that were switched every 0.5 million MCS. Such simulation mimics the inversion of the gene in chromosome, in which the gene is translocated from one DNA strand to the other. We have recorded several parameters during the simulations, which were finally presented as averages calculated over all genes and individuals in the population.

3   Results

The accumulated number of eliminated individuals during the simulation time for different mutational pressure scenarios is shown separately for these two types of selections in Fig. 2 and Fig. 3. Generally, the number of eliminated genomes increase in all cases during simulations but their elimination by stop codons generation occurs much more frequent than by the loss of the coding signal. It indicates that it is easier to introduce a stop codon to the gene sequence than to change its coding potential described by nucleotide composition in three codon positions. Moreover, the increase in the accumulated number of individuals eliminated because of the selection against stop codons begins already from the start of simulations whereas the number for the coding signal selection grows significantly only after 1 million MCS. Till this time, the elimination of individuals from the population by the coding signal loss is negligible.

However, the frequency of elimination of individuals depends clearly on the applied mutational pressure. The accumulated number of eliminated genomes because of the coding signal loss grows faster when the direct pressure is applied, i.e. when genes are subjected to the pressure from their own strand (Fig. 2). The number is lower in the case of the reverse pressure (i.e. when genes are under the pressure from the opposite strand) whereas the lowest accumulated number of eliminated individuals is observed for the changing pressures. If we consider the selection against the stop codon appearance, the accumulated number of eliminated genomes grows faster for the reverse and the changing pressure than for the direct pressure (Fig. 3).

Interestingly, the superposition of these two differently acting selections equalizes the potential differences between the applied mutational pressures according to the number of substitutions accumulated during simulations in gene sequences (Fig. 4). The numbers are almost identical for the direct and reverse mutational pressures but are slightly higher for the changing pressure.

The average protein coding signal (computed for all individuals who were still alive in a given simulation step) normalized by the average protein coding signal at the beginning of the simulation is another important feature which was considered during every simulation step (Fig. 5 and Fig. 6).

As one could expect, the average coding signal decreases faster in the simulation without selection constraints than in the simulation when the selection is applied (Fig. 5). However, the differences are not very large, which indicates that the applied selection is quite tolerant for the change in the coding signal. Some differences in the coding signal change are observed between simulations with selections when different mutational pressures are applied (Fig. 6). In the case of the sequences subjected to the reverse pressure, the coding signal is better preserved than for genes under the direct pressure. It well agrees with the results presented in (Fig. 2), which indicates that the direct pressure more often influences the coding signal and eliminates individuals because of the signal loss than the reverse pressure. Interestingly, the sequences exposed to the changing pressure keep their coding signal better than genes under other pressure to about 1.7 million MCS. After that the signal decreases but fluctuates.

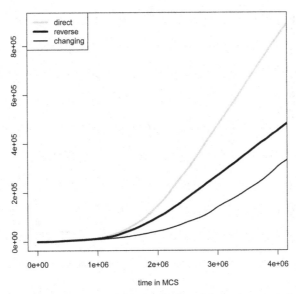

Fig. 2. The accumulated number of individuals eliminated from the population because of the coding potential loss by their gene sequences for simulations with three mutational pressures

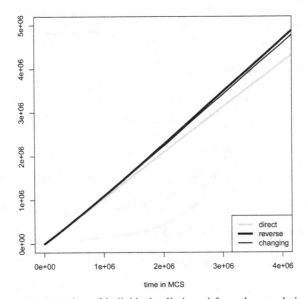

Fig. 3. The accumulated number of individuals eliminated from the population because of the stop translation codons appearance in their gene sequence for simulations with three mutational pressures

We also compared the accumulated number of damaged genes under different pressures for the leading and lagging DNA strand genes (Fig. 7 and Fig. 8). Because the leading strand genes are more than two times numerous than the lagging ones,

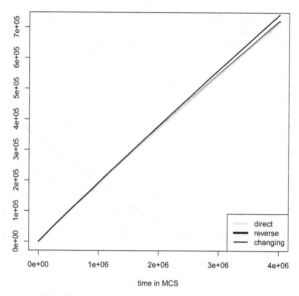

Fig. 4. The average number of mutations accumulated in gene sequences during simulations with three mutational pressures

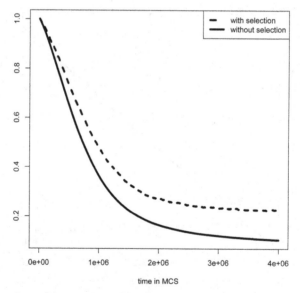

Fig. 5. The comparison of the average coding signal in the simulation run with selection (dashed line) and without selection (solid line). The change of the average coding signal was normalized by the average coding signal at the start of the simulation.

we normalized the numbers of damaged genes accordingly. The number of genes eliminated because of their coding signal loss are the highest when genes are under the mutational pressure typical of the leading strand (Fig. 7). It refers to the cases when the

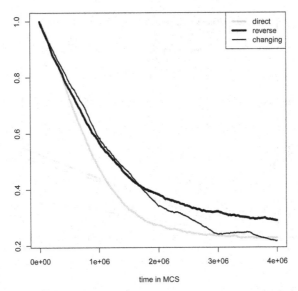

Fig. 6. The comparison of the average coding signal in simulations with selections run under different mutational pressures. The change of the average coding signal was normalized by the average coding signal at the start of the simulation.

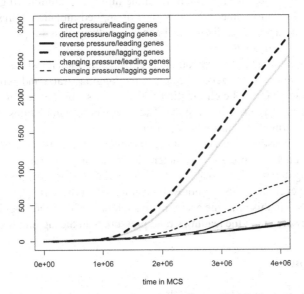

Fig. 7. The accumulated numbers of genes from the leading and lagging strands damaged because of the coding signal loss for different mutational scenarios

leading strand pressure is a direct pressure for the leading strand genes and reverse for the lagging strand genes. Significantly less genes from the both DNA strands are eliminated when their sequences are under the changing pressure whereas the least genes are damaged when they are under the influence of the lagging strand matrix. In the case

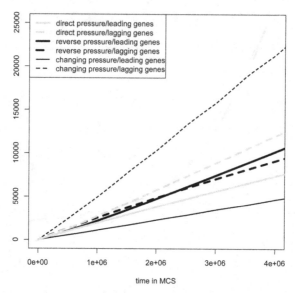

Fig. 8. The accumulated numbers of genes damaged because of the stop codon appearance in their gene sequence from the leading and lagging strands for different mutational scenarios

of the changing pressure, clear fluctuations in the number of eliminated genes are visible during the simulation time. For every type of applied mutational pressure (direct, reverse, and changing), the leading strand genes are always less frequently eliminated than the lagging strand genes.

When the selection against stop codons is considered, the largest number of eliminated genes are observed for genes located in the lagging strand and subjected to the changing pressure whereas the elimination is the weakest for the leading strand genes also under the changing pressure (Fig. 8). The numbers for simulations with constant pressures take intermediate values between these two extremes. Considering the constant mutational pressures, the accumulated numbers of eliminated genes are higher when genes are under the pressure characteristic of the lagging strand. It is in the case when the lagging pressure is reverse for the leading strand genes and direct for the lagging genes. However, the excess is not well pronounced in this case and is visible after 2.5 million MCS. Till this time, the accumulated numbers of damaged lagging genes are higher when they are under the leading strand pressure than the lagging strand pressure.

4   Discussion

In the paper, we have worked out a simulation model of bacterial genome evolution in which the algorithm for finding protein coding signal and stop codon occurrence played a role as selection criteria. In the simulations we used two mutational pressures associated with DNA replication, which were applied to mutating of protein coding genes from the leading and lagging DNA strands. The pressures were used as the constant pressure through the whole simulation or they were applied alternately as the changing pressure.

In general, the elimination of genomes from the population because of stop codon appearance is more frequent than their elimination by the loss of the coding signal. The elimination because of the coding signal is additionally delayed in the simulations, which indicates that gene sequences need some time to accumulate sufficient number of substitutions to change their coding properties. It seems to be important to notice, that the effect of selection pressure used in all simulations was exclusively negative. The higher robustness of coding sequences with enhanced coding signal after substitutions could be considered as some hidden no direct positive selection. Nevertheless, it would be interesting to introduce the direct positive selection effect of the increased coding signal.

The obtained results indicate that it is not indifferent to genes from the differently replicated DNA strands to which mutational pressure and selection they are subjected. The leading strand mutational pressure is more destructive for coding signal than the lagging strand pressure. Therefore it is more harmful for genes from the both DNA strands. On the other hand, the pressure typical of the lagging strand eliminates more genes because of stop codons occurrence than the leading strand pressure because the lagging strand substitutions generate such codons with higher frequency. The results are in agreement with analysis of nucleotide usage biases in four-fold degenerated sites in codons from bacterial genes [36].

Interestingly, the reverse and changing pressures destroy the coding signal weaker than the direct pressure. As a result of this, the number of eliminated genes subjected to these pressures is smaller than for the direct pressure and the average number of accumulated substitutions shows some excess for the simulations with the changing pressure. These findings agree with other simulations of gene evolution using selection constraints on amino acid composition [21–24]. They also showed that genes are less frequently eliminated when they change their mutational pressure between the differently replicated DNA strands periodically. These results of computer simulations are in agreement with the comparative genome analysis showing that genes very often change their position between the different strands [26] and homologs located on these strands evolve faster than homologs located in the same type of DNA strand [4, 18–20].

In contrast to that, generally more genes are eliminated under the reverse and changing pressures when the selection against stop codons is considered although the difference to the direct pressure is not very big. However, if the location of the genes on the DNA strands is taking into account, the lagging strand genes under the changing pressure are the most often eliminated whereas the leading strand genes are the least frequently removed by this type of selection of all possible scenarios. The applied period of changing the mutational matrix equal to 0.5 million MCS was chosen arbitrarily and probably there exist more optimal conditions for different genes as they were found in other type of simulations [21].

We expect that the obtained results of simulations should be very similar to those using other selection algorithms predicting protein coding sequences based on other coding measures, e.g. codon or dicodon usage, because the measures used by the algorithm applied here are strongly correlated with the others. The presented model of bacterial genome evolution, which was shown in the example of *B. burgdorferi*, should give similar general results for other bacterial genomes because their DNA asymmetry resembles that from the species analysed here.

References

1. Frank, A., Lobry, J.: Asymmetric substitution patterns: a review of possible underlying mutational or selective mechanisms. Gene 238, 65–77 (1999)
2. Kowalczuk, M., Mackiewicz, P., Mackiewicz, D., Nowicka, A., Dudkiewicz, M., Dudek, M.R., Cebrat, S.: DNA asymmetry and the replicational mutational pressure. J. Appl. Genet. 42, 553–557 (2001)
3. Kowalczuk, M., Mackiewicz, P., Mackiewicz, D., Nowicka, A., Dudkiewicz, M., Dudek, M.R., Cebrat, S.: High correlation between the turnover of nucleotides under mutational pressure and the DNA composition. BMC Evol. Biol. 1, 1–13 (2001)
4. Rocha, E., Danchin, A.: Ongoing evolution of strand composition in bacterial genomes. Mol. Biol. Evol. 18, 1789–1799 (2001)
5. Rocha, E., Touchon, M., Feil, E.: Similar compositional biases are caused by very different mutational effects. Genome Res. 16, 1537–1547 (2006)
6. Lobry, J.: Asymmetric substitution patterns in the two DNA strands of bacteria. Mol. Biol. Evol. 13, 660–665 (1996)
7. Freeman, J., Plasterer, T., Smith, T., Mohr, S.: Patterns of genome organization in bacteria. Science 279, 1827 (1998)
8. Grigoriev, A.: Analysing genomes with cumulative skew diagrams. Nucleic Acids Res. 26, 2286–2290 (1998)
9. McLean, M., Wolfe, K., Devine, K.: Base composition skews, replication orientation, and gene orientation in 12 prokaryote genomes. J. Mol. Evol. 47, 691–696 (1998)
10. Mrazek, J., Karlin, S.: Strand compositional asymmetry in bacterial and large viral genomes. Proc. Natl. Acad. Sci. 95, 3720–3725 (1998)
11. Mackiewicz, P., Gierlik, A., Kowalczuk, M., Dudek, M.R., Cebrat, S.: Asymmetry of nucleotide composition of prokaryotic chromosomes. J. Appl. Genet. 40, 1–14 (1999)
12. Tillier, E., Collins, R.: The contributions of replication orientation, gene direction, and signal sequences to base composition asymmetries in bacterial genomes. J. Mol. Evol. 51, 459–463 (2000)
13. Lobry, J., Sueoka, N.: Asymmetric directional mutation pressures in bacteria. Genome Biol. 3, 58 (2002)
14. McInerney, J.: Replicational and transcriptional selection on codon usage in *Borrelia burgdorferi*. Proc. Natl. Acad. Sci. 95, 10698–10703 (1998)
15. Lafay, B., Lloyd, A., McLean, M., Devine, K., Sharp, P., Wolfe, K.: Proteome composition and codon usage in spirochaetes: species-specific and DNA strand-specific mutational biases. Nucleic Acids Res. 27, 1642–1649 (1999)
16. Mackiewicz, P., Gierlik, A., Kowalczuk, M., Dudek, M.R., Cebrat, S.: How does replication-associated mutational pressure influence amino acid composition of proteins? Genome Res. 9, 409–416 (1999)
17. Rocha, E., Danchin, A., Viari, A.: Universal replication biases in bacteria. Mol. Microbiol. 32, 11–16 (1999)
18. Tillier, E., Collins, R.: Replication orientation affects the rate and direction of bacterial gene evolution. J. Mol. Evol. 51, 459–463 (2000)
19. Szczepanik, D., Mackiewicz, P., Kowalczuk, M., Gierlik, A., Nowicka, A., Dudek, M.R., Cebrat, S.: Evolution rates of genes on leading and lagging DNA strands. J. Mol. Evol. 52, 426–433 (2001)
20. Mackiewicz, P., Mackiewicz, D., Kowalczuk, M., Dudkiewicz, M., Dudek, M.R., Cebrat, S.: High divergence rate of sequences located on different DNA strands in closely related bacterial genomes. J. Appl. Genet. 44, 561–584 (2003)

21. Mackiewicz, P., Dudkiewicz, M., Kowalczuk, M., Mackiewicz, D., Banaszak, J., Polak, N., Smolarczyk, K., Nowicka, A., Dudek, M.R., Cebrat, S.: Differential Gene Survival under Asymmetric Directional Mutational Pressure. In: Bubak, M., van Albada, G.D., Sloot, P.M.A., Dongarra, J. (eds.) ICCS 2004. LNCS, vol. 3039, pp. 687–693. Springer, Heidelberg (2004)
22. Dudkiewicz, M., Mackiewicz, P., Kowalczuk, M., Mackiewicz, D., Nowicka, A., Polak, N., Smolarczyk, K., Kriaga, J., Dudek, M.R., Cebrat, S.: Simulation of gene evolution under directional mutational pressure. Physica A 336, 63–73 (2004)
23. Dudkiewicz, M., Mackiewicz, P., Mackiewicz, D., Kowalczuk, M., Nowicka, A., Polak, N., Smolarczyk, K., Kiraga, J., Dudek, M.R., Cebrat, S.: Higher mutation rate helps to rescue genes from the elimination by selection. Biosystems 80, 192–199 (2005)
24. Mackiewicz, D., Cebrat, S.: To understand nature computer modelling between genetics and evolution. In: Miekisz, J., Lachowicz, M. (eds.) From Genetics to Mathematics. Series on Advances in Mathematics for Applied Sciences, vol. 79, pp. 1–33. World Scientific (2009)
25. Mackiewicz, P., Szczepanik, D., Gierlik, A., Kowalczuk, M., Nowicka, A., Dudkiewicz, M., Dudek, M.R., Cebrat, S.: The differential killing of genes by inversions in prokaryotic genomes. J. Mol. Evol. 53, 615–621 (2001)
26. Mackiewicz, D., Mackiewicz, P., Kowalczuk, M., Dudkiewicz, M., Dudek, M.R., Cebrat, S.: Rearrangements between differently replicating DNA strands in asymmetric bacterial genomes. Acta Microbiologica Polonica 52, 245–261 (2003)
27. Rocha, E., Danchin, A.: Gene essentiality determines chromosome organisation in bacteria. Nucleic Acids Res. 31, 5202–5211 (2003)
28. Rocha, E., Danchin, A.: Essentiality, not expressiveness, drives gene strand bias in bacteria. Nature Genetics 34, 377–378 (2003)
29. Błażej, P., Mackiewicz, P., Cebrat, S.: Using the genetic code wisdom for recognizing protein coding sequences. In: Proceedings of the 2010 International Conference on Bioinformatics & Computational Biology, BIOCOMP 2010, pp. 302–305. CSREA Press, Las Vegas (2010)
30. Błażej, P., Mackiewicz, P., Cebrat, S.: Algorithm for finding coding signal using homogeneous markov chains independently for three codon positions. In: Proceedings of the 2011 International Conference on Bioinformatics and Computational Biology, ICBCB 2011, pp. 20–24. IEEE, Chengdu (2011)
31. Mackiewicz, P., Gierlik, A., Kowalczuk, M., Szczepanik, D., Dudek, M.R., Cebrat, S.: Mechanisms generating long-range correlation in nucleotide composition of the Borrelia burgdorferi genome. Physica A 273, 103–115 (1999)
32. National Center for Biotechnology Information, http://www.ncbi.nlm.nih.gov
33. Wańczyk, M., Błażej, P., Mackiewicz, P.: Comparison of two algorithms based on markov chains applied in recognition of protein coding sequences in prokaryotes. In: Proceedings of the Seventeeth National Conference on Applications of Mathematics in Biology and Medicine, pp. 118–123. QPrint, Warsaw (2011)
34. Cebrat, S., Dudek, M.R., Mackiewicz, P., Kowalczuk, M., Fita, M.: Asymmetry of coding versus non-coding strand in coding sequences of different genomes. Microbial and Comparative Genomics 2, 259–268 (1997)
35. Cebrat, S., Dudek, M.R., Mackiewicz, P.: Sequence asymmetry as a parameter indicating coding sequence in Saccharomyces cerevisiae genome. Theory in Biosciences 117, 78–89 (1998)
36. Khrustalev, V., Barkovsky, E.: The probability of nonsense mutation caused by replication-associated mutational pressure is much higher for bacterial genes from lagging than from leading strands. Genomics 96, 173–180 (2010)

Single Tandem Halving by Block Interchange

Antoine Thomas, Aïda Ouangraoua, and Jean-Stéphane Varré

LIFL, UMR 8022 CNRS, Université Lille 1
INRIA Lille, Villeneuve d'Ascq, France

Abstract. We address the problem of finding the minimal number of block interchanges required to transform a duplicated unilinear genome into a single tandem duplicated unilinear genome. We provide a formula for the distance as well as a polynomial time algorithm for the sorting problem. This is the extended version of [1].

1 Introduction

Genomic rearrangements are known to play a central role in the evolutionary history of the species. Several operations act on the genome, shaping the sequence of genes. A number of rearrangement operations to sort a genome into another, and evaluate the evolutionary distance between genomes, have been studied: reversals, transpositions, translocations, block interchanges, fusions, fissions, and more recently Double-Cut-and-Join (DCJ). In this paper, we focus on the *block interchange* operation, that consists in exchanging two intervals of a genome.

Block interchanges scenarios have been studied for the first time by Christie [2]. He proposed a $O(n^2)$ time algorithm for computing the minimum number of block interchanges for transforming a linear chromosome with unique gene content into another one. Lin *et al.* [3] proposed later the best algorithm to date in $O(\gamma n)$ where γ is the minimum number of block interchanges required for the transformation. Yancopoulos *et al.* [4] introduced the DCJ operation which consist in cutting the genomes in two points and joining the four resulting extremities in a different way. Interestingly, they noticed that a block interchange can be simulated by two consecutive DCJ operations.

Another very important feature in genome evolution is that genomes often undergo genome duplication events, both segmental and whole-genome duplications. For instance, a *tandem duplication event* is a segmental duplication that duplicates a genomic sequence and results in a segment made of two consecutive occurrences of the genomic sequence, called a *tandem duplicated segment*, in the genome. Genome duplication events are followed by other rearrangements events which result in a scrambled genome. The *Genome Halving problem* introduced by El-Mabrouk *et al.* [5] consists in finding the sequence of rearrangement events that allow one to go back from the scrambled genome to the original duplicated one, when the duplication event is a *whole genome duplication* event. The *Single Tandem Halving problem* we introduce consists in finding the original single tandem duplicated genome, considering the duplication event was a tandem duplication event. In this case however, as we are about to see, it can be seen as a particular restriction of the Genome Halving problem, implied by the block interchange rearrangement model.

J. Gabriel et al. (Eds.): BIOSTEC 2012, CCIS 357, pp. 162–174, 2013.
© Springer-Verlag Berlin Heidelberg 2013

Genome Halving has been studied under several models: reversals [5], transloca-
tion/reversals [6], breakpoints [7]. Most of the results led to polynomial time algorithms. Particularly, the Genome Halving by DCJ was studied in [8,9], and in [9] some
useful data structures were presented leading to a linear time algorithm for the Genome
Halving by DCJ. Following these results on the Genome Halving by DCJ, a natural
problem to consider is the Genome Halving by block interchange. However, as block
interchanges cannot split chromosomes, the original duplicated genome in this case
would rather be a *single tandem duplicated genome*, ie. we consider that the cause of
duplicated content was a tandem duplication event on the whole genome.

In this paper, we study the Single Tandem Halving by block interchange on a duplicated genomic segment resulting from a tandem-duplication event, followed by block
interchange events that have scrambled the gene content of the segment. This duplicated genomic segment is represented as a linear chromosome with duplicated gene
content w.l.o.g, and we search for a parsimonous scenario of block interchange operations transforming the linear chromosome into a linear tandem duplicated chromosome.
We answer *yes* to the question: *Does there exist a parsimonious sequence of block interchange operations, such that, replacing each block interchange by two consecutive
DCJ operations yields a parsimonious sequence of DCJ operations ?*. Based on the
adequate data structure to represent potential DCJ operations and their overlapping relations, we derive a quadratic time algorithm for the Single Tandem Halving by block
interchanges. Very recently, Kováč *et al.* [10] addressed the problem of reincorporating
the temporary circular chromosomes induced by DCJs immediately after their creation
considering the Genome Halving. This problem is obviously related to the problem addressed in the present paper, but the aim and results are different. We are interested
in linear genomes, not in multilinear ones, and we focus on pure block interchange
scenarios that can be simulated by particular types of DCJ operations called *excisions*
and *integrations*, whereas Kováč *et al.* focused on general DCJ scenarios simulating
reversals, translocation, fusion, fissions along with excisions, integrations, and block
interchanges.

Section 2 gives definitions. In Section 3, we first give a lower bound on the distance
with helpful properties for the rest of the paper. In Section 4, we prove the analytical
formula for the distance. We conclude in Section 5 with a quadratic time and space
algorithm to obtain a parsimonious scenario.

2   Preliminaries: Duplicated Genomes, Rearrangement, Genome Halving Problems

In this section we give the main definitions and notations used in the paper.

2.1   Duplicated Genomes

A genome is composed of genomic markers organized in linear or circular chromosomes.
A linear chromosome is represented by an ordered sequence of unsigned integers, each
standing for a marker, surrounded by two abstract markers ∘ at each end indicating the
telomeres. A circular chromosome is represented by a circularly ordered sequence of

unsigned integers representing markers. For example, (1 2 3) (○ 4 5 6 7 ○) is a genome constituted of one circular and one linear chromosome. Note that all genomes are considered *unsigned* in this paper w.l.o.g, because block interchange operations do not modify the signs of markers.

Definition 1. *A* totally duplicated genome *is a genome in which each marker appears twice.*

In a totally duplicated genome, two copies of a same marker are called paralogs. We distinguish paralogs by denoting one marker by x and its paralog by $\bar{x}$. By convention $\bar{\bar{x}} = x$. For example, the following genome is a totally duplicated genome: (○ 1 $\bar{1}$ 3 2 4 5 6 $\bar{6}$ 7 $\bar{3}$ 8 $\bar{2}$ $\bar{4}$ $\bar{5}$ 9 $\bar{8}$ $\bar{7}$ $\bar{9}$ ○).

An *adjacency* in a genome is a pair of consecutive markers. For example, the genome (○ 1 2 ○) (3 4 5) has six adjacencies, (○ 1), (1 2), (2 ○), and (3 4), (4 5), (5 3). The linear or circular order of the markers in a chromosome naturally induces an order on the adjacencies that we denote by $<$. For example in the previous genome the order induced on the adjacencies is: (○ 1) < (1 2) < (2 ○), and (3 4) < (4 5) < (5 3) < (3 4).

A *double-adjacency* in a genome G is an adjacency $(a\ b)$ such that $(\bar{a}\ \bar{b})$ is an adjacency of G as well. Note that a genome always has an even number of double-adjacencies. For example, the four double-adjacencies in the following genome are indicated by dots :

$$G = (○ \ 1 \ \bar{1} \ 3 \ 2 \ \cdot \ 4 \ \cdot \ 5 \ 6 \ \bar{6} \ 7 \ \bar{3} \ 8 \ \bar{2} \ \cdot \ \bar{4} \ \cdot \ \bar{5} \ 9 \ \bar{8} \ \bar{7} \ \bar{9} \ ○)$$

A consecutive sequence of double-adjacencies can be rewritten as a single marker; this process is called *reduction*. For example, genome G can be reduced by rewriting $2 \cdot 4 \cdot 5$ and $\bar{2} \cdot \bar{4} \cdot \bar{5}$ as 10 and $\overline{10}$, yielding the following genome:

$$G^r = (○ \ 1 \ \bar{1} \ 3 \ 10 \ 6 \ \bar{6} \ 7 \ \bar{3} \ 8 \ \overline{10} \ 9 \ \bar{8} \ \bar{7} \ \bar{9} \ ○)$$

Definition 2. *A* single tandem duplicated genome *is a totally duplicated genome composed of a single linear chromosome which can be reduced to a chromosome of the form* (○ x $\bar{x}$ ○).

In other words, a single tandem duplicated genome is composed of a single linear chromosome where all adjacencies, except the two containing the marker ○ and the central adjacency, are double-adjacencies. For example, the genome (○ $1 \cdot 2 \cdot 3 \cdot 4$ $\bar{1} \cdot \bar{2} \cdot \bar{3} \cdot \bar{4}$ ○) is a single tandem duplicated genome that can be reduced to (○ 5 $\bar{5}$ ○) by rewritting $1 \cdot 2 \cdot 3 \cdot 4$ and $\bar{1} \cdot \bar{2} \cdot \bar{3} \cdot \bar{4}$ as 5 and $\bar{5}$.

Definition 3. *A* perfectly duplicated genome *is a totally duplicated genome such that each adjacency is a double-adjacency.*

For example, the genome (1 2 $\bar{1}$ $\bar{2}$) (○ 3 4 ○) (○ $\bar{3}$ $\bar{4}$ ○) is a perfectly duplicated genome composed of one single circular chromosome and two linear chromosomes.

In other words, a single tandem duplicated genome is the representation of a duplicated segment resulting from a tandem-duplication of a genomic sequence, and a perfectly duplicated genome represents the result of a whole-genome duplication event that has duplicated all chromosomes.

2.2 Rearrangements

A rearrangement operation on a given genome cuts a set of adjacencies of the genome called *breakpoints* and forms new adjacencies with the exposed extremities, while altering no other adjacency. In the sequel, the adjacencies cut by a rearrangement operation are indicated in the genome by the symbol ▲.

An *interval* in a genome is a set of markers that appear consecutively in the genome. Given two different adjacencies $(a\ b)$ and $(c\ d)$ in a genome G such that $(a\ b) < (c\ d)$, $[b\ ;\ c]$ denotes the interval of G beginning with marker b and ending with marker c.

In this paper, we consider two types of rearrangement operations called *block interchange (BI)* and *double-cut-and-join (DCJ)*.

A *block interchange* (BI) on a genome G is a rearrangement operation that acts on four adjacencies in G, $(a\ b) < (c\ d) \le (u\ v) < (x\ y)$ such that the intervals $[b\ ;\ c]$ and $[v\ ;\ x]$ do not overlap, swapping the intervals $[b\ ;\ c]$ and $[v\ ;\ x]$. For example, the following block interchange acting on adjacencies $(\bar{1}\ 2) < (6\ \bar{6}) < (\bar{3}\ 8) < (\bar{8}\ \bar{7})$ consists in swapping the intervals $[2, 6]$ and $[8, \bar{8}]$.

$$(\circ\ 1\ \bar{1}\ ▲\ 2\ 3\ \bar{2}\ 4\ 5\ 6\ ▲\ \bar{6}\ 7\ \bar{3}\ ▲\ 8\ \bar{4}\ 9\ \bar{5}\ \bar{8}\ ▲\ \bar{7}\ \bar{9}\ \circ)$$
$$\downarrow$$
$$(\circ\ 1\ \bar{1}\ 8\ \bar{4}\ 9\ \bar{5}\ \bar{8}\ \bar{6}\ 7\ \bar{3}\ 2\ 3\ \bar{2}\ 4\ 5\ 6\ \bar{7}\ \bar{9}\ \circ)$$

A *double-cut-and-join* (DCJ) operation on a genome G cuts two different adjacencies in G and glues pairs of the four exposed extremities to form two new adjacencies. Here, we focus on two types of DCJ operations called *excision* and *integration*.

An *excision* is a DCJ operation acting on a single chromosome by extracting an interval from it, making this interval a circular chromosome, and making the remainder a single chromosome. For example, the following excision extracts the circular chromosome $(2\ 3\ 4)$:

$$(\circ\ 1\ ▲\ 2\ 3\ 4\ ▲\ 5\ 6\ \circ) \to (2\ 3\ 4)(\circ 1\ 5\ 6\ \circ)$$

An *integration* is the inverse of an excision; it is a DCJ operation that acts on two chromosomes, one being a circular chromosome, to produce a single chromosome. For example, the following operation is an integration of the circular chromosome $(2\ 3\ 4)$:

$$(2\ ▲\ 3\ 4)(\circ 1\ 5\ 6\ ▲\ \circ) \to (\circ\ 1\ 5\ 6\ 3\ 4\ 2\ \circ)$$

We now give an obvious, but very useful, property linking BI operations to DCJ operations.

Property 1. A single BI operation on a linear chromosome is equivalent to two DCJ operations: an excision followed by an integration.

Proof. Let $(\circ\ 1\ U\ 2\ V\ 3\ \circ)$ be a genome, U and V the two intervals that are to be swapped by a block interchange operation, 1 2 and 3 the intervals constituting the rest of the genome (note that each of them may be empty).

The first DCJ operation is the excision that produces the adjacency $(1\ V)$ by extracting and circularizing the interval $[U\ ;\ 2]$:

$$(\circ\ 1\ _{\blacktriangle}\ U\ 2\ _{\blacktriangle}\ V\ 3\ \circ) \rightarrow (\circ\ 1\ V\ 3\ \circ)(U\ 2\ )$$

The second DCJ operation is the integration that produces the adjacency $(U\ 3)$ by reintegrating the circular chromosome $(U\ 2)$ in the appropriate way:

$$(\circ\ 1\ V\ _{\blacktriangle}\ 3\ \circ)(U\ 2\ _{\blacktriangle}) \rightarrow (\circ\ 1\ V\ 2\ U\ 3\ \circ).$$

A *rearrangement scenario* between two genomes A and B is a sequence of rearrangement operations allowing one to transform A into B.

Definition 4. *A BI (resp. DCJ) scenario is a rearrangement scenario composed of BI (resp. DCJ) operations.*

The length of a rearrangement scenario is the number of rearrangement operations composing the scenario.

Definition 5. *The BI (resp. DCJ) distance between two genomes A and B, denoted by $d_{BI}(A, B)$ (resp. $d_{DCJ}(A, B)$), is the minimal length of a BI (resp. DCJ) scenario between A and B.*

2.3   Single Tandem Halving

We now state the single tandem halving problem considered in this paper.

Definition 6. *Given a totally duplicated genome G composed of a single linear chromosome, the BI single tandem halving problem consists in finding a single tandem duplicated genome H such that the BI distance between G and H is minimal.*

In order to solve the BI single tandem halving problem, we use some results on the *DCJ genome halving problem* that were stated in [9] as a starting point. However, unlike the single tandem halving problem, the aim of the genome halving problem is to find a perfectly duplicated genome instead of a single tandem duplicated genome.

Definition 7. *Given a totally duplicated genome G, the DCJ genome halving problem consists in finding a perfectly duplicated genome H such that the DCJ distance between G and H is minimal.*

The BI and DCJ genome halving problems lead to two definitions of *halving distances*: the *BI single tandem halving distance* (resp. *DCJ genome halving distance*) of a totally duplicated genome G is the minimum BI (resp. DCJ) distance between G and any single tandem duplicated genome (resp. any perfectly duplicated genome) ; we denote it by $d^t_{BI}(G)$ (resp. $d^p_{DCJ}(G)$).

3   Lowerbound for the BI Single Tandem Halving Distance

In this section we give a lowerbound on the BI single tandem halving distance of a totally duplicated genome. We use a data structure representing the genome called the *natural graph* introduced in [9].

Definition 8. [9] *The natural graph of a totally duplicated genome G, denoted by* NG(G), *is the graph whose vertices are the* adjacencies *of G, and for any marker u there is one edge between* $(u\ v)$ *and* $(\overline{u}\ w)$, *and one edge between* $(x\ u)$ *and* $(y\ \overline{u})$.

Note that the number of edges in the natural graph of a genome G containing n distinct markers, each one present in two copies, is always $2n$. Moreover, since every vertex has degree one or two, then the natural graph consists only of cycles and paths. For example, the natural graph of genome $G = (\circ\ 1\ \overline{2}\ \overline{1}\ \overline{4}\ 3\ 4\ \overline{3}\ 2\ \circ)$ is depicted in figure 1.

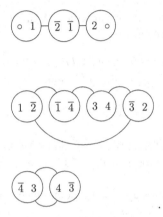

Fig. 1. The natural graph of genome $G = (\circ\ 1\ \overline{2}\ \overline{1}\ \underline{\overline{4}\ 3\ 4\ \overline{3}}\ 2\ \circ)$; it is composed of one path and two cycles

Definition 9. *Given an integer k, a k−cycle (resp. k−path) in the natural graph of a totally duplicated genome is a cycle (resp. path) that contains k edges. If k is even, the cycle (resp. path) is called* even, *and* odd *otherwise.*

Based on the natural graph, a formula for the DCJ halving distance was given in [9]. Given a totally duplicated genome G such that the number of even cycles and the number of odd paths in NG(G) are respectively denoted by EC and OP, the DCJ halving distance of G is:

$$d^p_{DCJ}(G) = n - \text{EC} - \left\lfloor \frac{\text{OP}}{2} \right\rfloor$$

In the case of the BI single tandem halving distance, some peculiar properties of the natural graph need to be stated, allowing one to simplify the formula of the DCJ halving distance, and leading to a lowerbound on the BI single tandem halving distance.

In the following properties, we assume that G is a genome composed of a single linear chromosome containing n distinct markers, each one present in two copies in G.

Property 2. The natural graph NG(G) contains only even cycles and paths:

1. All cycles in the natural graph NG(G) are even.
2. The natural graph NG(G) contains only one path, and this path is even.

Proof. First, if $(a \ x)$ is a vertex of the graph that belongs to a cycle C, then there exists an edge between $(a \ x)$ and a vertex $(\bar{a} \ y)$. These two adjacencies are the only two containing a copy of the marker a at the first position. So, if we consider the set of all the first markers in all adjacencies contained in the cycle C, then each marker in this set is present exactly twice. Therefore, the cycle C is an even cycle.

Secondly, the graph contains exactly two vertices (adjacencies) containing the marker $\circ$ which are both necessarily ends of a path in $\mathrm{NG}(G)$. Thus there can be only one path in the graph. Since the number of edges in the graph is even and all cycles are even, then the single path is also even.

We now give a lowerbound on the minimum length of DCJ scenario transforming G into a single tandem duplicated genome.

Lemma 1. *Let $d^t_{DCJ}(G)$ be the minimum DCJ distance between G and any single tandem duplicated genome. If $\mathrm{NG}(G)$ contains C cycles then a lowerbound on $d^t_{DCJ}(G)$ is given by:*

$$d^t_{DCJ}(G) \geq n - C - 1$$

Proof. First, since all cycles of $\mathrm{NG}(G)$ are even and $\mathrm{NG}(G)$ contains no odd path, then, from the DCJ halving distance formula, the DCJ halving distance of G is $d^p_{DCJ}(G) = n - C$.

Now, since any single tandem duplicated genome can be transformed into a perfectly duplicated genome with one DCJ, then $d^t_{DCJ} + 1 \geq d^p_{DCJ}$. Therefore, we have $d^t_{DCJ} \geq d^p_{DCJ} - 1 \geq n - C - 1$.

We are now ready to state a lowerbound on the BI single tandem halving distance of a totally duplicated genome G.

Theorem 1. *If $\mathrm{NG}(G)$ contains C cycles, then a lowerbound on the BI single tandem halving distance is given by:*

$$d^t_{BI}(G) \geq \left\lfloor \frac{n - C}{2} \right\rfloor$$

Proof. We denote by $\ell(S)$ the length of a rearrangement scenario S. Let S_{BI} be a BI scenario transforming G into a single tandem duplicated genome. From property 1, we have that S_{BI} is equivalent to a DCJ scenario S_{DCJ} such that $\ell(S_{DCJ}) = 2 * \ell(S_{BI})$. Now, suppose that $\ell(S_{BI}) < \lfloor \frac{n-C}{2} \rfloor$, then $\ell(S_{BI}) \leq \lfloor \frac{n-C}{2} \rfloor - 1 \leq \lceil \frac{n-C-1}{2} \rceil - 1$.

This implies $\ell(S_{DCJ}) \leq 2\lceil \frac{n-C-1}{2} \rceil - 2 \leq n - C - 2 < n - C - 1$. Thus, from Lemma 1 we have $\ell(S_{DCJ}) < d^t_{DCJ}$ which contradicts the fact that d^t_{DCJ} is the minimal number of DCJ operations required to transform G into a single tandem duplicated genome.

In conclusion, we always have $d^t_{BI}(G) \geq \lfloor \frac{n-C}{2} \rfloor$.

4   Formula for the BI Single Tandem Halving Distance

In this section, we show that the BI single tandem halving distance of a totally dupli-cated genome G with n distinct markers such that $\text{NG}(G)$ contains C cycles is exactly:

$$d^t_{BI}(G) = \left\lfloor \frac{n - C}{2} \right\rfloor$$

In other words, we show that enforcing the constraint that successive couples of con-secutive DCJ operations have to be equivalent to BI operations does not change the distance even though it obviously restricts the DCJ that can be performed at each step of the scenario.

In the following, G denotes a totally duplicated genome G constisting in a single linear chromosome with n distinct markers after the reduction process, and such that $\text{NG}(G)$ contains C cycles. We begin by recalling some useful definitions and properties of the DCJ operations that allow one to decrease the DCJ halving distance by 1 in the resulting genome.

Definition 10. *A DCJ operation on G producing genome G' is sorting if it decreases the DCJ halving distance by 1:* $d^p_{DCJ}(G') = d^p_{DCJ}(G) - 1 = n - C - 1$.

Since the number of distinct markers in G' is n and $d^p_{DCJ}(G') = n - C - 1$, then $\text{NG}(G')$ contains $C + 1$ cycles. In other words, a DCJ operation is sorting if it increases the number of cycles in $\text{NG}(G)$ by 1.

Given $(u \ v)$ an adjacency of G that is not a double-adjacency, we denote by $DCJ(u \ v)$ the DCJ operation that cuts adjacencies $(\overline{u} \ x)$ and $(y \ \overline{v})$ to form adjacencies $(\overline{u} \ \overline{v})$ and $(y \ x)$, making $(u \ v)$ a double-adjacency.

Property 3. *Let $(u \ v)$ be an adjacency of G that is not a double-adjacency, $DCJ(u \ v)$ is a sorting DCJ operation.*

Proof. $DCJ(u \ v)$ increases the number of cycles in $\text{NG}(G)$ by 1, by creating a new cycle composed of adjacencies $(u \ v)$ and $(\overline{u} \ \overline{v})$.

Definition 11. *Let $(u \ v)$, $(\overline{u} \ x)$, and $(y \ \overline{v})$ be adjacencies of G. The* interval *of the adjacency $(u \ v)$, denoted by $I(u \ v)$ is either:*

- *the interval $[x \ ; \ y]$ if $(\overline{u} \ x) < (y \ \overline{v})$. In this case, we denote it by $]\overline{u} \ ; \ \overline{v}[$, or*
- *the interval $[\overline{v} \ ; \ \overline{u}]$ if $(y \ \overline{v}) < (\overline{u} \ x)$.*

For example, the intervals of the adjacencies in genome $(\circ \ 2 \ 1 \ \overline{2} \ 3 \ \overline{1} \ 3 \ \circ)$ are depicted in figure 2. Note that, given an adjacency $(u \ v)$ of G, if $(u \ v)$ is a double-adjacency then the interval $I(u \ v)$ is empty, otherwise $DCJ(u \ v)$ is the excision operation that extracts the interval $I(u \ v)$ to make it circular, thus producing the adjacency $(\overline{u} \ \overline{v})$.

Two intervals $I(a \ b)$ and $I(x \ y)$ are said to be *overlapping* if their intersection is non-empty, and none of the intervals is included in the other. It is easy to see, following Property 1, that given two adjacencies $(a \ b)$ and $(x \ y)$ of G such that $I(a \ b)$ and $I(x \ y)$ are non-empty intervals, the successive application of $DCJ(a \ b)$ and $DCJ(x \ y)$

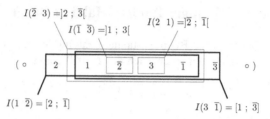

Fig. 2. $\mathcal{I}(G) = \{]\overline{2};\overline{1}[,[\mathbf{2};\overline{1}],]2;\overline{3}[,[1;\overline{3}],]1;3[\}$, the set of intervals of $G = (\circ\ 2\ 1\ \overline{2}\ 3\ \overline{1}\ \overline{3}\ \circ)$ depicted as boxes. The two boxes with thick lines represent two overlapping intervals of $\mathcal{I}(G)$ inducing a BI which exchanges 2 and $\overline{3}$

is equivalent to a BI operation if and only if $I(a\ b)$ and $I(x\ y)$ are overlapping. Note that in this case neither $(a\ b)$, nor $(x\ y)$ can be double-adjacencies in G since their intervals are non-empty. Figure 2 shows an example of two overlapping intervals.

The following property states precisely in which case the successive application of $DCJ(a\ b)$ and $DCJ(x\ y)$ decreases the DCJ halving distance by 2, meaning that both DCJ operations are sorting.

Property 4. Given two adjacencies $(a\ b)$ and $(x\ y)$ of G, such that $I(a\ b)$ and $I(x\ y)$ are overlapping, the successive application of $DCJ(a\ b)$ and $DCJ(x\ y)$ decreases the DCJ halving distance by 2 if and only if $x \neq \overline{a}$ and $y \neq \overline{b}$.

Proof. If $x \neq \overline{a}$ and $y \neq \overline{b}$, then the successive application of $DCJ(a\ b)$ and $DCJ(x\ y)$ increases the number of cycles in $\mathrm{NG}(G)$ by 2, by creating two new 2-cycles. Otherwise, $DCJ(a\ b)$ first creates a new cycle that is then destroyed by $DCJ(x\ y)$.

We denote by $\mathcal{I}(G)$, the set of intervals of all the adjacencies of G that do not contain marker $\circ$.

Remark 1. Note that, if G contains n distinct markers, then there are $2n-1$ adjacencies in G that do not contain marker $\circ$, defining $2n-1$ intervals in $\mathcal{I}(G)$.

Definition 12. *Two intervals $I(a\ b)$ and $I(x\ y)$ of $\mathcal{I}(G)$ are said to be* compatible *if they are overlapping and $x \neq \overline{a}$ and $y \neq \overline{b}$.*

In the following, we prove the BI single tandem halving distance formula by showing that if genome G contains more than three distinct markers, $n > 3$, then there exist two compatible intervals in $\mathcal{I}(G)$, and if $n = 2$ or $n = 3$ then $d^t_{BI}(G) = 1$ and $2 \leq d^p_{DCJ}(G) \leq 3$. This means that there exists a BI halving scenario S such that all BI operations in S, possibly excluding the last one, are equivalent to two successive sorting DCJ operations.

From now on, until the end of the section, $(a\ b)$ is an adjacency of G that is not a double-adjacency, A is a genome consisting in a linear chromosome $\mathfrak{L}$ and a circular chromosome $\mathfrak{C}$, obtained by applying the *sorting DCJ*, $DCJ(a\ b)$, on G.

If there exists an interval $I(x\ y)$ in $\mathcal{I}(G)$ compatible with $I(a\ b)$, then applying $DCJ(x\ y)$ on A consists in the integration of the circular chromosome $\mathfrak{C}$ into the linear chromosome $\mathfrak{L}$ such that the adjacency $(\overline{x}\ \overline{y})$ is formed. Such an *integration* can only be performed by cutting an adjacency $(\overline{x}\ u)$ in $\mathfrak{C}$ and an adjacency $(v\ \overline{y})$ in $\mathfrak{L}$ (or inversely) to produce adjacencies $(\overline{x}\ \overline{y})$ and $(v\ u)$. This means that there must be an adjacency $(x\ y)$ in either $\mathfrak{C}$ or $\mathfrak{L}$ such that $\overline{x}$ is in $\mathfrak{C}$ and $\overline{y}$ in $\mathfrak{L}$ or inversely. Hence, we have the following property :

Property 5. $\mathfrak{C}$ *cannot* be reintegrated into $\mathfrak{L}$ by applying a sorting DCJ, $DCJ(x\ y)$, on A if and only if either:

(1) for any adjacency $(x\ y)$ in $\mathfrak{C}$ (resp. $\mathfrak{L}$), markers $\overline{x}$ and $\overline{y}$ are in $\mathfrak{L}$ (resp. $\mathfrak{C}$), or
(2) for any adjacency $(x\ y)$ in $\mathfrak{C}$ (resp. $\mathfrak{L}$), markers $\overline{x}$ and $\overline{y}$ are also in $\mathfrak{C}$ (resp. $\mathfrak{L}$).

Proof. If there exists no adjacency $(x\ y)$ in A such that $\overline{x}$ is in $\mathfrak{C}$ and $\overline{y}$ in $\mathfrak{L}$ or inversely, then A necessarily satisfies either (1), or (2).

Definition 13. *An interval* $I(a\ b)$ *in* $\mathcal{I}(G)$ *is called* interval of type 1 *(resp.* interval of type 2*) if* $DCJ(a\ b)$ *produces a genome* A *satisfying configuration* (1) *(resp. configuration* (2)*) described in Property 5.*

For example, in genome $(\circ\ 2\ 1\ \overline{1}\ 3\ \overline{2}\ 3\ \circ)$, $I(\overline{1}\ 3)$ is of type 1 as $DCJ(\overline{1}\ 3)$ produces genome $(\circ\ 2\ 1\ \overline{3}\ \circ)(\overline{1}\ 3\ \overline{2})$; $I(\overline{2}\ 3)$ is of type 2 as $DCJ(\overline{2}\ 3)$ produces genome $(\circ\ 2\ 3\ \overline{2}\ \overline{3}\ \circ)(1\ \overline{1})$.

Now we give the maximum numbers of intervals of type 1 and type 2 that can be contained in genome G.

Lemma 2. *The maximum number of intervals of type 1 in* $\mathcal{I}(G)$ *is 2.*

Proof. First, note that there cannot be two intervals I and J of $\mathcal{I}(G)$ such that $I \neq J$, and both I and J are of type 1. Now, if I is an interval of type 1, there can be at most two different adjacencies $(x\ y)$ and $(u\ v)$ such that $I(x\ y) = I(u\ v) = I$. In this case G necessarily has a chromosome of the form $(\ldots\ \overline{x}\ \overline{v}\ \ldots\ \overline{u}\ \overline{y}\ \ldots)$ or $(\ldots\ \overline{u}\ \overline{y}\ \ldots\ \overline{x}\ \overline{v}\ \ldots)$. Therefore, there are at most two intervals of type 1 in $\mathcal{I}(G)$.

Lemma 3. *The maximum number of intervals of type 2 in* $\mathcal{I}(G)$ *is n.*

Proof. First, note that for two adjacencies $(x\ y)$ and $(\overline{x}\ z)$ in G that do not contain marker $\circ$, if $(x\ y)$ is of type 2 then $(\overline{x}\ z)$ cannot be of type 2. Now, there is only one marker u such that $(\overline{u}\ \circ)$ is an adjacency of G. Let $(u\ v)$ be the adjacency of G having u as first marker, then at most half of the intervals in $\mathcal{I}(G) - \{I(u\ v)\}$ can be of type 2. Therefore, there are at most n intervals of type 2 in $\mathcal{I}(G)$.

Theorem 2. *If* $NG(G)$ *contains* C *cycles, then the BI single tandem halving distance of* G *is given by:*

$$d^t_{BI}(G) = \left\lfloor \frac{n-C}{2} \right\rfloor$$

Proof. Since there are $2n - 1$ intervals in $\mathcal{I}(G)$, and at most $n + 2$ are of type 1 or 2, then if G is a genome containing more than three distinct markers $n > 3$, then $2n - 1 > n + 2$ and there exist two compatible intervals in $\mathcal{I}(G)$ inducing a BI operation that decreases the DCJ distance by 2.

Next, we show that if $n = 2$ or $n = 3$, then $d^t_{BI}(G) = 1$ and $2 \le d^p_{DCJ}(G) \le 3$.

If $n = 2$, then the genome can be written, either as $(\circ \, a \, b \, \bar{b} \, \bar{a} \, \circ)$, in which case a BI can swap a and b to produce a single tandem duplicated genome, or as $(\circ \, a \, \bar{a} \, b \, \bar{b} \, \circ)$, in which case a BI can swap a and $\bar{a} \, b$ to produce a single tandem duplicated genome.

If $n = 3$, then the genome has two double-adjacencies to be constructed, of the form $(\bar{a} \, \bar{b})$, $(\bar{x} \, \bar{y})$, with $(a \, b)$ and $(x \, y)$ being two adjacencies already present in the genome such that $b = x$ or $b = \bar{x}$ and a and y are distinct markers. One can rewrite $(a \, b)$ and $(x \, y)$ as single markers since they will not be splitted, which makes a genome with 4 markers such that at most 2 are misplaced. Then, a single BI can produce a single tandem duplicated genome.

Now, it is easy to see to see that if $n = 2$ or $n = 3$, then $d^p_{DCJ}(G) = n - C \le 3$. Finally, if $n = 2$ or $n = 3$, then $d^p_{DCJ}(G) \ge 2$, otherwise we would have $d^p_{DCJ}(G) = 1$ which would imply, as G consists in a single linear chromosome, $d^t_{BI}(G) = 0$. In conclusion, if $n > 3$ then there exist two compatible intervals in $\mathcal{I}(G)$, otherwise if $n = 2$ or $n = 3$, then $d^t_{BI}(G) = 1$ and $2 \le d^p_{DCJ}(G) \le 3$. Therefore $d^t_{BI} = \lfloor \frac{d^p_{DCJ}}{2} \rfloor = \lfloor \frac{n-C}{2} \rfloor$.

5 Sorting Algorithm

In Section 4, we showed that if a genome G contains more than three distinct markers after reduction then there exist two compatible intervals in $\mathcal{I}(G)$ inducing a BI to perform. If G contains two or three distinct markers then the BI to perform can be trivially computed. Thus the main concern of this section is to describe an efficient algorithm for finding compatible intervals when $n > 3$.

As in Section 4, in the following, G denotes a genome consisting of n distinct markers after reduction. It is easy to show that the set of intervals $\mathcal{I}(G)$ can be built in $O(n)$ time and space complexity.

We now show that finding 2 compatible intervals in $\mathcal{I}(G)$ can be done in $O(n)$ time and space complexity.

Property 6. If $n > 3$, then all the smallest intervals in $\mathcal{I}(G)$ that are not of type 2 admit compatible intervals.

Proof. Let J be a smallest interval that is not of type 2 in $\mathcal{I}(G)$. As J is not of type 2, then J has compatible intervals if J is not of type 1.

Let us suppose that J is of type 1, then for any adjacency $(a \, b)$ such that markers a and b are not in J, $\bar{a}$ and $\bar{b}$ are in J, and then $I(a \, b)$ is strictly included in J and $I(a \, b)$ can't be of type 2. Such adjacency does exist as there are $n > 3$ markers not included in J. Therefore J cannot be a smallest interval that is not of type 2.

We are now ready to give the algorithm for sorting a duplicated genome G into a single tandem duplicated genome with $\lfloor \frac{n-C}{2} \rfloor$ BI operations.

Theorem 3. *Algorithm 1 reconstructs a single tandem duplicated genome with a BI scenario of length $\lfloor \frac{n-C}{2} \rfloor$ in $O(n^2)$ time and space complexity, by computing pairs of sorting DCJ operations.*

Algorithm 1. Reconstruction of a single tandem duplicated genome

---

1: **while** G contains more than 3 markers **do**
2:    Construct $\mathcal{I}(G)$
3:    Pick a smallest interval $I(a\ b)$ that is not of type 2 in $\mathcal{I}(G)$
4:    Find an interval $I(x\ y)$ in $\mathcal{I}(G)$ compatible with $I(a\ b)$
5:    Perform the BI equivalent to $DCJ(a\ b)$ followed by $DCJ(x\ y)$
6:    Reduce G
7: **end while**
8: **if** G contains 2 or 3 markers **then**
9:    Find the last BI operation and perform it
10: **end if**

Proof. Building $\mathcal{I}(G)$ and finding two compatible intervals can be done in $O(n)$ time and space complexity. It follows that the while loop in the algorithm can be computed in $O(n^2)$ time and space complexity.

Finding and performing the last BI operation when $2 \leq n \leq 3$ can be done in constant time and space complexity.

Moreover, all BI operations, possibly excluding the last one, are computed as pairs of compatible intervals, ie. pairs of sorting DCJ operations, which ensures that the length of the scenario is $\lfloor \frac{n-C}{2} \rfloor$. □

Corollary 1. *Any BI scenario computed by Algorithm 1 is also a most parsimonious DCJ scenario, twice as long since a BI is equivalent to 2 DCJ.*

An example of scenario is shown in figure 3.

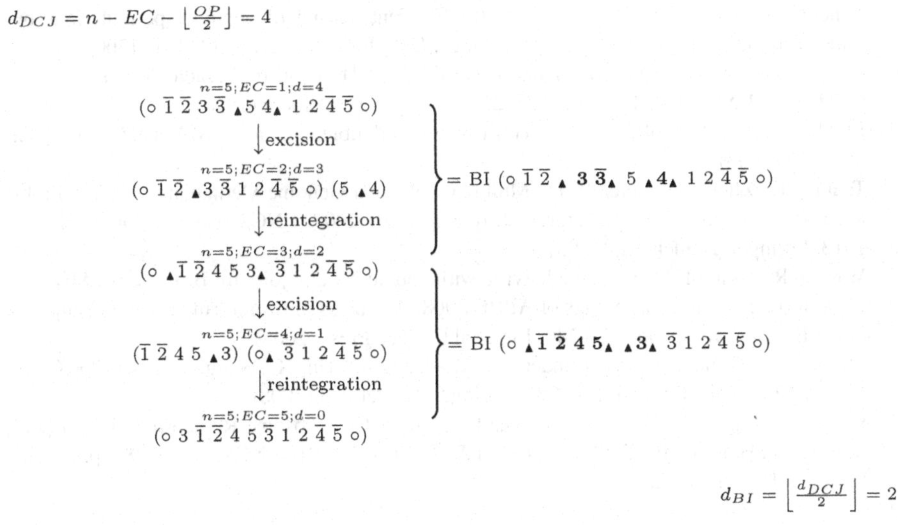

$$d_{DCJ} = n - EC - \lfloor \tfrac{OP}{2} \rfloor = 4$$

Fig. 3. A BI scenario computed by algorithm 1

6   Conclusions

In this paper, we introduced the single tandem halving problem. We used the DCJ model to simulate BI operations and we showed that it is always possible to choose two consecutive sorting DCJ operations such that they are equivalent to a BI operation. This is an interesting result as it shows that restricting the scope of allowed DCJ operations under the constraint of performing only BI doesn't affect our halving distance. This also means our BI scenarios are in fact optimal DCJ scenarios. We thus provided a quadratic time and space algorithm to obtain a most parsimonious scenario for the BI single tandem halving problem, but also a most parsimonious scenario for the DCJ single tandem halving problem for genomes that present the orientation constraint that all markers must have the same orientation as their paralogs. A further extension of these results will be a generalization to the more general DCJ model, without the orientation constraints on the genome that the BI model implies, as well as ways of reconstructing more than a single tandem. Another direction for further studies of variants of the BI single tandem halving problem is to consider multichromosomal genomes.

References

1. Thomas, A., Ouangraoua, A., Varré, J.S.: Genome halving by block interchange. In: BIOIN-FORMATICS 2012, Third International Conference on Bioinformatics Models, Methods and Algorithms, Vilamoura, Portugal, February 1-4 (2012)
2. Christie, D.A.: Sorting permutations by block-interchanges. Inf. Process. Lett. 60, 165–169 (1996)
3. Lin, Y.C., Lu, C.L., Chang, H.Y., Tang, C.Y.: An efficient algorithm for sorting by block-interchanges and its application to the evolution of vibrio species. Journal of Computational Biology 12, 102–112 (2005)
4. Yancopoulos, S., Attie, O., Friedberg, R.: Efficient sorting of genomic permutations by translocation, inversion and block interchange. Bioinformatics 21, 3340–3346 (2005)
5. El-Mabrouk, N., Nadeau, J.H., Sankoff, D.: Genome Halving. In: Farach-Colton, M. (ed.) CPM 1998. LNCS, vol. 1448, pp. 235–250. Springer, Heidelberg (1998)
6. El-Mabrouk, N., Sankoff, D.: The reconstruction of doubled genomes. SIAM J. Comput. 32, 754–792 (2003)
7. Tannier, E., Zheng, C., Sankoff, D.: Multichromosomal Genome Median and Halving Problems. In: Crandall, K.A., Lagergren, J. (eds.) WABI 2008. LNCS (LNBI), vol. 5251, pp. 1–13. Springer, Heidelberg (2008)
8. Warren, R., Sankoff, D.: Genome halving with double cut and join. In: Brazma, A., Miyano, S., Akutsu, T. (eds.) Proceedings of APBC 2008. Advances in Bioinformatics and Computational Biology, vol. 6, pp. 231–240. Imperial College Press (2008)
9. Mixtacki, J.: Genome Halving under DCJ Revisited. In: Hu, X., Wang, J. (eds.) COCOON 2008. LNCS, vol. 5092, pp. 276–286. Springer, Heidelberg (2008)
10. Kováč, J., Braga, M.D.V., Stoye, J.: The Problem of Chromosome Reincorporation in DCJ Sorting and Halving. In: Tannier, E. (ed.) RECOMB-CG 2010. LNCS, vol. 6398, pp. 13–24. Springer, Heidelberg (2010)

Fast RNA Secondary Structure Prediction Using Fuzzy Stochastic Models

Markus E. Nebel and Anika Scheid*

Department of Computer Science, University of Kaiserslautern,
P.O. Box 3049, D-67653 Kaiserslautern, Germany
{nebel,a_scheid}@cs.uni-kl.de

Abstract. Computational prediction of RNA secondary structures has been an active area of research over the past decades and since become of great relevance for practical applications in structural biology. To date, many popular state-of-the-art prediction tools have the same worst-case time and space requirements of $\mathcal{O}(n^3)$ and $\mathcal{O}(n^2)$ for sequence length n, limiting their applicability for practical purposes. Accordingly, biologists are interested in getting results faster, where a moderate loss of accuracy would willingly be tolerated in favor of saving a significant amount of computation time. Motivated by these facts, we invented a novel algorithm for predicting the secondary structure of RNA molecules that manages to reduce the worst-case time complexity by a linear factor to $\mathcal{O}(n^2)$, while on the other hand it is still capable of producing highly accurate results. Basically, the presented method relies on a probabilistic statistical sampling approach which is actually based on an appropriate *stochastic context-free grammar (SCFG)*: for any given input sequence, it generates a random set of candidate structures (from the ensemble of all feasible foldings) according to a "noisy" distribution (obtained by heuristically approximating the inside-outside values for the input sequence), such that finally a corresponding prediction can be efficiently derived. Notably, this method may be employed with different sampling strategies. Therefore, we not only consider a popular common strategy but also introduce a novel one that is supposed to fit especially well in connection with fuzzy stochastic models. A major advantage of the proposed prediction approach is that sampling can easily be parallelized on modern multi-core architectures or grids. Furthermore, it can be done in-place, that is only the best (here most probable) candidate structure(s) generated so far need(s) to be stored and finally collected. The combination of these two benefits immediately allows for an efficient handling of the increased sample sizes that are often necessary to achieve competitive prediction accuracy in connection with the noisy distribution.

1 Introduction

Over the past years, several new approaches towards the prediction of RNA secondary structures from a single sequence have been invented which are based on generating statistically representative and reproducible samples of the entire ensemble of feasible

---

* Corresponding author. The research of this author is supported by Carl Zeiss Foundation, Germany.

J. Gabriel et al. (Eds.): BIOSTEC 2012, CCIS 357, pp. 175–194, 2013.

structures for the given sequence. For example, the popular Sfold software [1,2] employs a sampling extension of the partition function (PF) approach [3] to produce statistically representative subsets of the Boltzmann-weighted ensemble. More recently, a corresponding probabilistic method has been studied [4] which actually samples the possible foldings from a distribution implied by a sophisticated stochastic context-free grammar (SCFG).

Notably, both sampling methods imply the same worst-case time and space requirements of $\mathcal{O}(n^3)$ and $\mathcal{O}(n^2)$, respectively, for generating a fix-size sample of random secondary structures for a given input sequence of length n and can easily be extended to structure prediction. In fact, a corresponding prediction can be derived from any representative statistical sample without increasing the overall time or space complexity. Thus, the worst-case time and storage requirements for computing structure predictions via statistical sampling are equal to those of modern state-of-the-art tools like for instance the commonly used minimum free energy (MFE) based Mfold [5,6] and Vienna RNA [7,8] packages or the popular SCFG based Pfold software [9,10].

Furthermore, applications to structure prediction showed that neither of the two competing sampling approaches (SCFG and PF based method) generally outperforms the other and consequently, it is not obvious which one should rather be preferred in practice. This somehow contradicts the fact that the best physics-based prediction methods still generally perform significantly better than the best probabilistic approaches. In principle, only if the computational effort of one particular variant could be improved without significant losses in quality (that is if one of them requires considerably less time than the others while it sacrifices only little predictive accuracy), then the corresponding method would be undoubtably the number one choice for practical applications, indeed outperforming all other modern computational tools for predicting the secondary structure of RNA sequences. This, by the way, due to the often quite large sizes of native RNA molecules considered in practice, meets exactly the demands imposed by biologists on computational prediction procedures: rather getting moderately less accurate (but still good quality) results in less time than needing significantly more time for obtaining results that are expectedly not considerably more accurate.

Note that recently, there already have been several practical heuristic speedups [11,12]. Particularly, the approach of [11] for folding single RNA sequences manages to speed up the standard dynamic programming algorithms without sacrificing the optimality of the results, yielding an expected time complexity of $\mathcal{O}(n^2 \cdot \psi(n))$, where $\psi(n)$ is shown to be constant *on average* under standard polymer folding models. In [12], it is shown how to reduce those average-case time and space complexities in the sparse case. Furthermore, the practical technique from [13] achieves an improved worst-case time complexity of $\mathcal{O}(n^3 / \log(n))$, and with the (MFE and SCFG based) algorithms from [14], a slight worst-case speedup of $\mathcal{O}(n^3 \cdot \log(\log(n))^{1/2} / \log(n)^{1/2})$ time can be reached (whose practicality is unlikely and unestablished).

In this article, we present a new way to reduce the worst-case time complexity of SCFG based statistical sampling by a linear factor, making it possible to predict for instance the most probable (MP) structure among all sampled foldings for a given input sequence of length n (in direct analogy to conventional structure prediction via

SCFGs) with only $\mathcal{O}(n^2)$ time and space requirements. This complexity improvement is basically realized by employing an appropriate heuristic instead of the corresponding exact algorithm for preprocessing the input sequence, i.e. for deriving a "noisy" distribution (induced by heuristic approximations of the corresponding inside and outside probabilities) on the entire structure ensemble for the input sequence. From this distribution candidate structures can be efficiently sampled[1]. Moreover, we will consider two different sampling strategies: (a slight modification of) the widely known sampling procedure from [1,4] which basically generates a random structure from outside to inside, and a novel alternative strategy that obeys to contrary principles and employs a reverse course of action (from inside to outside) and manages to take more advantage of the approximative preprocessing.

As we will see, even building on our new heuristic preprocessing step, both sampling strategies can be applied to obtain MP structure predictions of respectable accuracy. In principle, for sufficiently large sample sizes we obtain a similar high predictive accuracy as in the case of exact calculations[2]. The seemingly sole pitfall is that due to the noisy ensemble distribution resulting from approximative computations, the produced samples are no longer guaranteed to primarily contain rather likely structures (with respect to the *exact* distribution of feasible foldings for a given input sequence), such that we usually have to generate more candidate structures (i.e., consider larger sample sizes) in oder to ensure reproducible structure predictions. However, this is quite unproblematic in practice: firstly, we can generate the candidate structures in-place (only the so far most probable structure(s) need(s) to be stored), such that large sample sizes give no rise to increased memory consumption and secondly, generating samples can easily be parallelized on modern multi-core architectures or grids.

The rest of this paper is organized as follows: Sect. 2 briefly recaps the principles of probabilistic statistical sampling and provides the needed formal framework. Sect. 2.3 contains a (short and exemplary) analysis on how different types and levels of disturbances in the underlying ensemble distribution affect the resulting sampling quality. This actually yields an impression on the required precision of an adequate heuristic approximation scheme for inside-outside calculations. Section 3 describes all facts concerning the approximative preprocessing step that needs to be applied for decreasing the worst-case time requirements. A (slightly modified) common sampling strategy and an alternative novel strategy (intended to match well with our heuristic method) are introduced and opposed in Sect. 4. In Sect. 5, the overall quality of generated sample sets and their applicability to RNA structure prediction are investigated. We present experiments which show how the prediction accuracy grows with the sample size together with considerations on how an efficient implementation can deal with large sample sets. Finally, Sect. 6 concludes the paper.

---

[1] With purposive proof-of-concept implementations (in Wolfram Mathematica 7.0), for instance the overall preprocessing time for *E.coli* tRNA^{Ala} (of length $n = 76$) could be reduced from 49.0 (traditional cubic algorithm) to only 3.7 (new quadratic strategy) seconds.

[2] For *E.coli* tRNA^{Ala}, we for instance observed the same sensitivity and PPV values of 1.0 and 0.91, respectively, with a particular application of our heuristic method and the corresponding exact variant.

2 Preliminaries

In the sequel, given an RNA molecule r consisting of n nucleotides, we denote the corresponding sequence fragment from position i to position j, $1 \leq i \leq j \leq n$, by $R_{i,j} = r_i r_{i+1} \ldots r_{j-1} r_j$. Accordingly, $S_{i,j}$ denotes a feasible secondary structure on $R_{i,j}$.

2.1 Sampling Based on SCFG Models

Briefly, probabilistic sampling based on a suitable SCFG $\mathcal{G}_s$ with sets $\mathcal{I}_{\mathcal{G}_s}$ and $\mathcal{R}_{\mathcal{G}_s}$ of intermediate symbols and productions, respectively, and axiom $S \in \mathcal{I}_{\mathcal{G}_s}$ (that models the class of all feasible secondary structures) has two basic steps: In the first step (preprocessing), all inside probabilities

$$\alpha_X(i,j) := \Pr(X \Rightarrow_{lm}^* r_i \ldots r_j) \tag{1}$$

and all outside probabilities

$$\beta_X(i,j) := \Pr(S \Rightarrow_{lm}^* r_1 \ldots r_{i-1} X r_{j+1} \ldots r_n) \tag{2}$$

for a sequence r of size n, $X \in \mathcal{I}_{\mathcal{G}_s}$ and $1 \leq i, j \leq n$, are computed. According to [4], this can be done with a special variant of an Earley-style parser (such that the considered grammar does not need to be in *Chomsky normal form (CNF)*), where the grammar parameters (trained beforehand on a suitable RNA structure database) are splitted into a set of *transition probabilities* $\Pr_{tr}^0(rule)$ for $rule \in \mathcal{R}_{\mathcal{G}_s}$ and two sets of *emission probabilities* $\Pr_{em}^1(\cdot)$ for the 4 unpaired bases and the 16 possible base pairings. For any such SCFG $\mathcal{G}_s$, there results $\mathcal{O}(n^3)$ time complexity and $\mathcal{O}(n^2)$ memory requirement for this preprocessing step. Note that in this work, we will use the sophisticated grammar from [4] which has been parameterized to impose two relevant restrictions on the class of all feasible structures, namely a minimum length of $\min_{HL}$ for hairpin loops and a minimum number of $\min_{hel}$ consecutive base pairs for helices.

The second step takes the form of a recursive sampling algorithm to randomly draw a complete secondary structure by consecutively sampling substructures (defined by base pairs and unpaired bases). Notably, different sampling strategies may be employed for realizing this step; two contrary variants that will be considered within this work are described in detail in Sect. 4. In general, for any sampling decision (for example choice of a new base pair), the strategy considers the respective set of all possible choices that might actually be formed on the currently considered fragment of the input sequence. Any of these sets contains exactly the mutually and exclusive cases as defined by the alternative productions (of a particular intermediate symbol) of the underlying grammar. The corresponding random choice is then drawn according to the resulting conditional sampling distribution (for the considered sequence fragment). This means that the respective sampling distributions are defined by the inside and outside values derived in step one (providing information on the distribution of all possible choices according to the actual input sequence) and the grammar parameters (transition probabilities).

Since every of the before mentioned conditional distributions needed for randomly drawing one of the respective possible choices can be derived in linear time (during the sampling process), any *valid*[3] base pair can be sampled in time $\mathcal{O}(n)$. Thus, since any structure of size n can have at most $\lfloor \frac{n-\min_{HL}}{2} \rfloor$ base pairs, a random candidate structure for the given input sequence can be generated in $\mathcal{O}(n^2)$ time.

Thus, one straightforward approach for improving the performance of the overall sampling algorithm in the worst-case is to reduce the $\mathcal{O}(n^3)$ time complexity required for the preprocessing step at least to the quadratic time of the sampling strategy. To us, this means we might be able to save a significant amount of time by replacing the exact inside-outside calculations with a corresponding heuristic method yielding only approximative inside-outside values for a given input sequence. To see if this might actually be successful, we next want to determine to which extend the sampling strategy reacts to different types and degrees of disturbances in the inside and outside probabilities in order the get evidence if it could actually be possible to find an appropriate heuristic.

2.2 Considered Disturbance Types and Levels

We decided to disturb the exact inside and outside probabilities for a given input sequence r of length n in the following ways: For each $X \in \mathcal{I}_{\mathcal{G}_s}$ and $1 \leq i, j \leq n$, redefine the corresponding inside value according to

$$\alpha_X(i,j) := \max(\min(\alpha_X(i,j) + \alpha_{err}, 1), 0), \tag{3}$$

where α_{err} is randomly chosen from the following interval or set:

$$[-\text{max}_{\text{ErrPerc}}\alpha_A(i,j), +\text{max}_{\text{ErrPerc}}\alpha_A(i,j)] \text{ or}$$
$$\{-\text{fix}_{\text{ErrPerc}}\alpha_A(i,j), +\text{fix}_{\text{ErrPerc}}\alpha_A(i,j)\}$$

(relative errors), with $\text{max}_{\text{ErrPerc}}, \text{fix}_{\text{ErrPerc}} \in (0, 1]$ defining percentages, or else,

$$[-\text{max}_{\text{ErrVal}}, +\text{max}_{\text{ErrVal}}] \text{ or } \{-\text{fix}_{\text{ErrVal}}, +\text{fix}_{\text{ErrVal}}\}$$

(absolute errors), with $\text{max}_{\text{ErrVal}}, \text{fix}_{\text{ErrVal}} \in (0, 1]$ being fixed values. Random errors on all outside values $\beta_X(i,j)$, $X \in \mathcal{I}_{\mathcal{G}_s}$ and $1 \leq i, j \leq n$, can be generated in the same way.

The needed conditional sampling distributions (as considered by a particular strategy) are then derived from the exact grammar parameters and the disturbed inside-outside probabilities for the input sequence. This might create the need to (slightly) modify a particularly employed sampling strategy for being capable of dealing with these skewed distributions, as we will see in Sect. 4.1.

---

[3] One may for example consider only the 6 different most stable *canonical* pairs as valid ones (like usually done in physics-based approaches due to missing thermodynamics parameters for *non-canonical* pairs). However, we decided to drop this restriction, considering all possible non-crossing base pairings to be valid.

2.3 Influence of Disturbances

To get a first impression on the influence of disturbances (in the ensemble distribution for a given input sequence) on the quality of generated sample sets, we opted for the potentially most intuitive application in this context, namely *probability profiling* for unpaired bases within particular loop types (see, e.g., [1]). In principle, for each nucleotide position i, $1 \le i \le n$, of a given sequence of length n, one computes the probabilities that i is an unpaired base within a specific loop type. These probabilities are given by the observed frequencies in a representative statistical sample of the complete ensemble (of all possible secondary structures) for the given input sequence.

Furthermore, in order to investigate to what extend the accuracy of predicted foldings changes when different dimensions of relative disturbances are incorporated into the needed sampling probabilities, we will additionally derive the most probable (MP) structure in the generated samples, respectively, as prediction.

Note that for our examinations, we will exemplarily consider a well-known trusted tRNA structure, *Escherichia coli* tRNA^{Ala}, since this molecule folds into the typical cloverleaf structure, making it very easy to judge the accuracy of the resulting profiles and predictions.

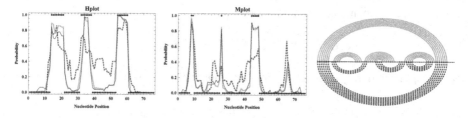

Fig. 1. Loop profiles and MP predictions obtained for *E.coli* tRNA^{Ala}. Hplot and Mplot display the probability that an unpaired base lies in a hairpin and multibranched loop, respectively. All results have been derived from samples of size 1,000, generated with $\min_{hel} = 2$ and $\min_{HL} = 3$. Errors were produced with $\max_{ErrPerc} = 0.99$ (thick gray lines) and $\mathrm{fix}_{ErrPerc} = 0.99$ (thick dotted darker gray lines). The profiles also display the respective exact results (thin black lines) and the native folding of *E.coli* tRNA^{Ala} (black points).

Figure 1 indicates that even in the case of large relative errors, the sampled structures still exhibit the typical cloverleaf structure of tRNAs, especially for the extenuated disturbance variant according to $\max_{ErrPerc}$ which seems to have practically no effect on the resulting sampling quality and prediction accuracy. However, Fig. 2 perfectly demonstrates that if the disturbances have been created by generating absolute errors on all inside values, then – even for rather small values – the resulting samples (and corresponding predictions as well) seem to be useless. Nevertheless, it seems reasonable to believe that the inside and outside probabilities do not necessarily have to be computed in an exact way, but it may probably suffice to only (adequately) approximate them.

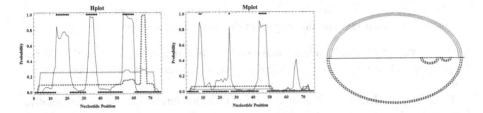

Fig. 2. Sampling results for *E.coli* tRNA^{Ala} corresponding to those presented in Fig. 1, where $\max_{\text{ErrVal}} = 10^{-9}$ (thick gray lines) and $\text{fix}_{\text{ErrVal}} = 10^{-9}$ (thick dotted darker gray lines) have been chosen for generating the disturbances

3 Reducing the Preprocessing Time

According to the previous discussion, the proclaimed aim of this section is to lower the $\mathcal{O}(n^3)$ time complexity for preliminary inside-outside calculations to $\mathcal{O}(n^2)$, such that the preprocessing has the same worst-case time requirements as the subsequent sampling process (for constructing a constant number of random secondary structure of size n).

3.1 Basic Idea

The main idea for reaching this time complexity reduction by a factor n in the worst-case is actually quite simple: Instead of deriving the inside values $\alpha_X(i, j)$ (and the corresponding outside probabilities $\beta_X(i, j)$), $X \in \mathcal{I}_{\mathcal{G}_s}$, for any combination of start position i and end position j, $1 \leq i, j \leq n$, we abstract from the actual position of subword $R_{i,j} = r_i \ldots r_j$ in the input sequence and consider only its length $d = |r_i \ldots r_j|$. Thus, for any $X \in \mathcal{I}_{\mathcal{G}_s}$, we do not need to calculate $\mathcal{O}(n^2)$ values $\alpha_X(i, j)$ (and $\beta_X(i, j)$) for $1 \leq i, j \leq n$, but only $\mathcal{O}(n)$ values $\alpha_X(d)$ (and $\beta_X(d)$) for $0 \leq d \leq n$. However, the problem with this approach is that distance d alone may be associated with any of the strings in $\{r_i \ldots r_j \mid j - i + 1 = d\}$, i.e. without using positions i and j we are inevitably forced to additionally abstract from the actual input sequence r.

Note that it is also possible to combine both alternatives, that is we can first use the traditional algorithms to calculate exact values $\alpha_X(i, j)$ (and $\beta_X(i, j)$) within a window of fixed size W_{exact}, i.e. for $j - i + 1 \leq \text{W}_{exact}$ (and $j - i + 1 \geq n - \text{W}_{exact}$), and afterwards derive the remaining values for $\text{W}_{exact} < d \leq n$ (and $0 \leq d < n - \text{W}_{exact}$) in an approximate fashion by employing the time-reduced variant for obtaining $\alpha_X(d)$ (and $\beta_X(d)$) for each $X \in \mathcal{I}_{\mathcal{G}_s}$. Since W_{exact} is constant, this effectively yields an improvement in the time complexity of the corresponding complete inside computation, which is then given by $\mathcal{O}(n^2 \cdot \text{W}_{exact})$. However, even for fix W_{exact} the time requirements for such a mixed outside computation are $\mathcal{O}(n^3)$.

3.2 Approximation of Emission Probabilities

Due to the unavoidable abstraction from sequence, we have to determine some approximated terms for the emissions of unpaired bases and base pairs, respectively, that

- do not depend on the positions of subwords within the overall input word, but
- should at least depend on the lengths of the corresponding subwords,

where it is strongly recommended to make sure that as much information on the composition of the actual input sequence as possible is incorporated into these approximated terms.

Therefore, we decided to use the following emission terms that incorporate relative frequencies $\mathrm{rf}^1_{em}(r_i, i - i + 1)$ and $\mathrm{rf}^2_{em}(r_i r_j, j - i + 1)$ for unpaired bases and base pairs, respectively, that can be efficiently derived from the actual input sequence:

$$\widehat{\mathrm{Pr}}^1_{em}(1) := \sum_{u \in \Sigma_{\mathcal{G}_r}} \mathrm{Pr}^1_{em}(u) \cdot \mathrm{rf}^1_{em}(u, 1), \tag{4}$$

$$\widehat{\mathrm{Pr}}^2_{em}(d) := \sum_{p_1 p_2 \in \Sigma^2_{\mathcal{G}_r}} \mathrm{Pr}^2_{em}(p_1 p_2) \cdot \mathrm{rf}^2_{em}(p_1 p_2, d). \tag{5}$$

3.3   (Improved) Approximated Sampling Probabilities

Fortunately, during the complete sampling process, not only the start and end positions of the currently considered sequence fragment $R_{i,j}, 1 \le i, j \le n$, but also the actual input sequence r are always known. Thus, we can in certain cases easily remove some approximate factors in the corresponding approximated inside and outside probabilities and replace them with the respective correct terms (depending on i, j and r) in order to obtain more reliable values.

Therefore, for any sampling strategy, the sampling probabilities from which the respective (conditional) distributions for possible choices are inferred should be defined by using such improved inside and outside probabilities (instead of the corresponding uncorrected precomputed ones). For example, if $X \in \mathcal{I}_{\mathcal{G}_s}$ generates hairpin loops, we should use

$$\widehat{\alpha}_X(i, j) := \begin{cases} \alpha_X(i, j), & \text{if } (j - i + 1) \le W_{exact}, \\ \alpha_X(j - i + 1) \cdot c^1_{em}(i, j), & \text{else,} \end{cases} \tag{6}$$

and

$$\widehat{\beta}_X(i, j) := \begin{cases} \beta_X(i, j), & \text{if } (j - i + 1) \ge n - W_{exact}, \\ \beta_X(j - i + 1) \times \\ \quad c^2_{em}(i - \min_{\mathrm{hel}}, j + \min_{\mathrm{hel}}, \min_{\mathrm{hel}}), & \text{else,} \end{cases} \tag{7}$$

where

$$c^1_{em}(s, e) := \frac{\prod_{k=s}^{e} \mathrm{Pr}^1_{em}(r_k)}{\widehat{\mathrm{Pr}}^1_{em}(1)^{e-s+1}} \tag{8}$$

and

$$c^2_{em}(i, j, l) := \frac{\prod_{k=0}^{l-1} \mathrm{Pr}^2_{em}(r_{i+k} r_{j-k})}{\prod_{k=0}^{l-1} \widehat{\mathrm{Pr}}^2_{em}((j - k) - (i + k) + 1)}. \tag{9}$$

4 Considered Sampling Strategies

For the subsequent examinations, we will employ two different sampling strategies, which are introduced now.

4.1 Well–Established Strategy

Let us first consider a slightly modified variant of the rather simple and widely known sampling strategy from [1,4]. Briefly, this well-established strategy samples a complete secondary structure $S_{1,n}$ for a given input sequence r of length n in the following recursive way: Start with the entire RNA sequence $R_{1,n}$ and consecutively compute the adjacent substructures (single-stranded regions and paired substructures) of the exterior loop (from left to right). Any (paired) substructure on fragment $R_{i,j}$, $1 \leq i < j \leq n$, is folded by recursively constructing substructures (hairpins, stacked pairs, bulges, interior and multibranched loops) on smaller fragments $R_{l,h}$, $i \leq l < h \leq j$. That is, fragments are sampled in an *outside-to-inside* fashion.

Notably, without disturbances of the underlying probabilistic model, it is guaranteed that any sampled loop type for a considered sequence fragment can be successfully generated (otherwise its probability would have been 0). As this must not hold in disturbed cases (like e.g. those of Sect. 2.3), the most straightforward modification to solve this problem is that in any such case where the chosen substructure type can not be successfully generated, the strategy returns the partially formed substructure. Figure 3 gives a schematic overview on this inherently controlled sampling strategy; a simple example is presented in Fig. 4.

As regards this particular sampling strategy, the outside values can easily be omitted from the corresponding formulæ for defining the needed sampling probabilities, since in any case they contribute the same multiplicative factor to the distinct sampling probabilities for mutually exclusive and exhaustive cases, such that they finally do not influence the sampling decision at all.

The correctness of this simplification can easily be verified by considering a particular set $ac_X(i,j)$ of all choices for (valid) derivations of intermediate symbol $X \in \mathcal{I}_{\mathcal{G}_s}$ on sequence fragment $R_{i,j}$, $1 \leq i < j \leq n$, which actually correspond to possible substructures on $R_{i,j}$. Under the assumption that the alternatives for intermediate symbol X are $X \to Y$ and $X \to VW$, the (valid) mutually exclusive and exhaustive cases are defined by:

$$acX(i,j) := acX_Y(i,j) \cup acX_{VW}(i,j), \tag{10}$$

where

$$acX_Y(i,j) := \{(0,p) \mid p \neq 0 \text{ for}$$
$$p = \beta_X(i,j) \cdot \alpha_Y(i,j) \cdot \mathrm{Pr}_{tr}^0(X \to Y)\} \tag{11}$$

and

$$acX_{VW}(i,j) := \{(k,p) \mid i \leq k \leq j \text{ and } p \neq 0 \text{ for}$$
$$p = \beta_X(i,j) \cdot \alpha_V(i,k)\alpha_W(k+1,j) \cdot \mathrm{Pr}_{tr}^0(X \to VW)\}. \tag{12}$$

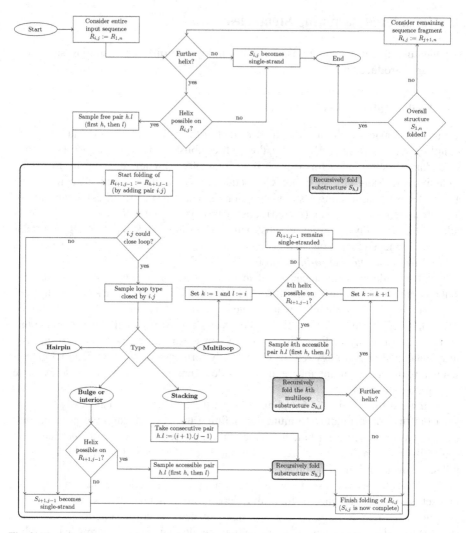

Fig. 3. Flowchart for recursive sampling of an RNA secondary structure $S_{1,n}$ for a given input sequence r of length n according to an inherently controlled strategy with predetermined order, similar to that of [1,4])

We then sample from the probability distribution induced by $acX(i,j)$ (conditioned on fragment $R_{i,j}$), which implies

$$1 = \sum_{(k,p) \in acX(i,j)} \frac{p}{z} = \frac{\beta_X(i,j)}{z} \sum_{(k,p) \in acX(i,j)} \frac{p}{\beta_X(i,j)}, \tag{13}$$

with

$$z := \sum_{(k,p) \in acX(i,j)} p, \tag{14}$$

since $\beta_X(i,j) \neq 0$ due to the definition of $acX(i,j)$ (multiplicative term).

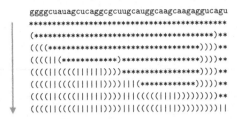

Fig. 4. Illustration of the recursive process of sampling an RNA secondary structure for an exemplary input sequence according to the common sampling strategy. Note that base pairs are represented by pairs of corresponding brackets ( ), unpaired bases by symbols | and bases which have not been solved yet (i.e., which have not been determined to be paired or unpaired so far) are depicted by symbols ∗.

4.2 Alternative Strategy

Unfortunately, the common sampling strategy from Sect. 4.1 lacks the ability to take full advantage of the exact inside values $\widehat{\alpha}_X(i,j) = \alpha_X(i,j)$, for $X \in \mathcal{I}_{\mathcal{G}_s}$ and $j - i + 1 \leq W_{exact}$, obtained by employing a particular mixed preprocessing variant according to $0 \leq W_{exact} < n$. Particularly, the strategy in general inevitably has to sample the first base pairs from corresponding conditional probability distributions for rather large fragments $R_{i,j}$ with $j - i + 1 > W_{exact}$, which are indeed induced by approximated sampling probabilities rather than exact ones. Therefore, we designed an alternative to this well-established sampling strategy that obeys to contrary principles, resulting in a reverse sampling direction.

Basically, a complete secondary structure $S_{1,n}$ for a given input sequence r of length n can alternatively albeit unconventionally be sampled in the following (deliberately less controlled) way: Start with the entire RNA sequence $R_{start,end} = R_{1,n}$ and randomly construct adjacent substructures (paired substructures preceeded by potentially empty single-stranded regions) of the exterior loop on the considered sequence fragment $R_{start,end}$ (where the construction does not follow a particular order, e.g. does not sample from left to right), as long as no further paired substructure can be folded. Any (paired) substructure on fragment $R_{start,end}$, $1 \leq start \leq end \leq n$, is created by sampling a random hairpin loop (with closing base pair $i.j$, for $start < i < j < end$) – here we can take advantage of exact inside values from a mixed preprocessing since most likely i and j are close – and extending it (towards the ends of $R_{start,end}$) by successively drawing closing base pairs. During this extension, basically all known substructures (stacked pairs, bulges, interior and multibranched loops, that obey to certain restrictions which will be discussed later) may be folded, where each substructure (e.g. multiloop) has to be completed before its closing base pair is added and the corresponding helix can actually be further extended[4]. The process of folding a particular paired substructure ends with a complete and valid paired structure (of the currently folding multiloop or of the exterior loop), either with or without a directly preceeding unpaired region, both on the

---

[4] Note that the sampling strategy proposed here is only heuristic in that it does not sample structures precisely according to the distribution implied by the underlying SCFG. See [18] for a discussion why this is unavoidable when sampling inside-out.

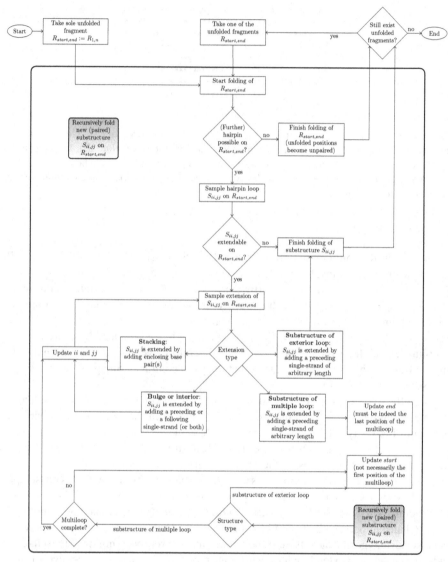

Fig. 5. Flowchart for recursive sampling of an RNA secondary structure $S_{1,n}$ for a given input sequence r of length n according to a less restrictive strategy with extensively more freedom (that requires dynamic validation of possible random choices during the sampling process)

considered fragment $R_{start,end}$. Figure 5 gives a schematic overview on this *inside-out* fashion sampling strategy; a simple example is presented in Fig. 6.

Note that in order to ensure that all sampled substructures can be successfully folded, especially in the case of multiloops, we have to take care that at any point, the strategy may only draw such random choices that do not make it impossible to successfully finish the currently running construction process (of a particular loop). As this strongly depends on the actual positions and types of all previously folded paired substructures,

```
ggggcuauagcucaggcgcuugcauggcaagcaagaggucagu
*******************************************
*********(||||||)**************************
***(||(((((||||||))))****************)*****
***(||(((((||||||))))*******(||||)****)*****
***(||(((((||||||))))|||(((((||||))))))*****
***(||(((((||||||))))|||(((((||||))))))*****
(((((||(((((||||||))))|||(((((||||)))))))))**
(((((||(((((||||||))))|||(((((||||)))))))))||
```

Fig. 6. Illustration of the recursive process of sampling an RNA secondary structure for an exemplary input sequence according to our alternative sampling strategy

the algorithm obviously needs to dynamically determine the respective set of all *valid* choices (during the sampling process itself) before a corresponding probability distribution (needed for drawing a particular random choice) can be derived.

This, however, may cause severe problems as regards the time complexity for randomly sampling the next extension (or base pair). Nevertheless, in order to guarantee that the worst-case time complexity for drawing any random choice remains in $\mathcal{O}(n)$, we only need to impose a few restrictions concerning the lengths of single-stranded regions in some types of loops. In detail, we have to consider a maximum allowed number of nucleotides in unpaired regions of hairpin loops ($\max_{hairpin}$), bulge or interior loops ($\max_{bulge}$), and multiloops ($\max_{strand}$)[5].

For example, if $X \in \mathcal{I}_{\mathcal{G}_s}$ generates hairpin loops, then the set of all possible hairpin loops that can be validly folded on sequence fragment $R_{start,end}$ is given by

$$\text{pcHL}(start, end) := \big\{(i, j, p) \mid$$
$$start + \min_{\text{hel}} \leq i \leq j \leq end - \min_{\text{hel}} \text{ and}$$
$$i + \min_{HL} - 1 \leq j \leq i + \max_{hairpin} - 1 \text{ and}$$
$$R_{i-\min_{\text{hel}}, j+\min_{\text{hel}}} \text{ not folded and}$$
$$p = \widehat{\beta}_X(i, j) \cdot \widehat{\alpha}_X(i, j) \neq 0 \big\}. \tag{15}$$

Obviously, $\max_{hairpin}$ indeed ensures that $\text{pcHL}(start, end)$ can be computed in $\mathcal{O}(n)$ time.

Finally, it should be noted that this sampling strategy needs to additionally consider outside probabilities, for two reasons: First, for "normalizing" the resulting sampling probabilities. This is due to the fact that the different possible choices (i, j, p) usually imply substructures $S_{i,j}$ of different lengths $j - i + 1$, such that only $p = \widehat{\alpha}_X(i, j) \cdot \widehat{\beta}_X(i, j)$ ensures that the probabilities of all possible choices are of the same order of magnitude and hence imply a reasonable probability distribution for drawing a random choice. Second, the outside values are required for guaranteeing that sampled substructures can be validly extended. This means that only such hairpin loops and ex-

---

[5] Note that these restrictions are not as severe as it may seem, since for example choosing the constant value 30 for all three parameters can be expected to hardly have a negative impact on the resulting sampling quality. In fact, many MFE based prediction algorithms also restrict the lengths of particular single-stranded regions, at least for long bulge and interior loops (where the proposed constant value of 30 is considered a common choice).

Table 1. Comparison of the considered sampling strategies (for an arbitrary input sequence of length n)

Aspect	Conventional Strategy	Alternative Strategy
Preprocessing time	$\mathcal{O}(n^3)$ for exact calculations, $\mathcal{O}(n^2)$ for approximate variant, $\mathcal{O}(n^2)$ with constant $W_{exact} \geq 0$	$\mathcal{O}(n^3)$ for exact calculations, $\mathcal{O}(n^2)$ for approximate variant, $\mathcal{O}(n^3)$ with constant $W_{exact} \geq 0$
Constraints	None	Constant $\max_{hairpin}$, $\max_{bulge}$ and $\max_{strand}$
Characteristics and course of action	Inherently controlled, ordered: - helices from left to right, - sampling proceeds "inwards": construction of substructure $S_{i,j}$ starts by considering $R_{i,j}$ and ends by generating an unpaired region (usually a hairpin loop)	Extensively more freedom, less restrictive: - helices in arbitrary order, - sampling proceeds "outwards": construction of new substructure on unfolded fragment $R_{start,end}$ starts with random hairpin loop which is extended to a complete and valid (paired) substructure $S_{i,j}$ on $R_{start,end}$
Benefits of sampling direction	(Sub)structures are folded in accordance with the generation of the corresponding (unique leftmost) derivation (sub)tree by the underlying SCFG	Takes more advantage of inside probabilities for shorter fragments containing less approximated terms and thus less inaccuracies (although this potential is narrowed by the outside values for which the contrary holds)
Function of outside values	Not considered (do not influence sampling distributions)	1) "Normalize" sampling probabilities 2) Ensure valid extensions
Identification of valid choices	Not required (all possible choices are principally valid)	Dynamic checking required (due to dependence on previously folded substructures)
Folding time	$\mathcal{O}(n^2)$	$\mathcal{O}(n^2)$ with larger constants
Overall time complexity for MP predictions	$\mathcal{O}(n^3)$ with exact variant, $\mathcal{O}(n^2)$ with constant $W_{exact} \geq 0$ or in completely approximated case	$\mathcal{O}(n^3)$ in case of exact computations or mixed variants using $W_{exact} \geq 0$, $\mathcal{O}(n^2)$ only in completely approximated case

tensions (implying a surrounding base pair $i.j$) may be sampled that can actually lead to the generation of a corresponding valid helix.

We conclude this section by referring to Table 1 that summarizes the main differences of both sampling strategies. Note that despite the significant algorithmic differences between both sampling variants (our new strategy is computationally more complex), there exists a fundamental difference when it comes to producing identical outcomes: When applying our dynamic strategy, a complete secondary structure (e.g., the one shown in Figs. 4 and 6) can generally be sampled in more than one way (due to the higher degree of freedom), whereas with the common strategy, there is always only one unique way for sampling a particular structure.

5 Applications

Attempting to quantify the decline in accuracy that results from approximative preprocessing, we first considered the *E.coli* tRNA^{Ala} sequence for probability profiling. The

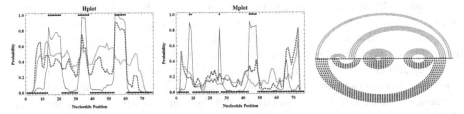

Fig. 7. Sampling results for *E.coli* tRNA^{Ala}, derived with the common strategy (under the assumption of min$_{\text{hel}}$ = 2 and min$_{HL}$ = 3), where we used sample size 100,000, 10,000 and 1,000 for W$_{exact}$ = −1 (no window, thick gray lines), W$_{exact}$ = 30 (moderate window, thick dotted darker gray lines) and W$_{exact}$ = +∞ (complete window, thin black lines), respectively

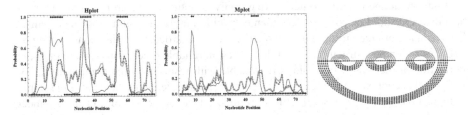

Fig. 8. Sampling results corresponding to those of Fig. 7, obtained by employing the alternative sampling strategy

sampling results shown in Fig. 7 indicate that for the common sampling strategy, considering a window of constant size W$_{exact}$ (chosen to cover the size of hairpin-loops) with a mixed preprocessing variant, actually yields a slight improvement of the resulting sampling quality, where the same time requirements are needed for generating the respective sample sets.

Contrary to this observation, Fig. 8 demonstrates that when employing our alternative sampling strategy, the corresponding results are not significantly different for the completely approximate preprocessing variant and for a mixed version on the basis of a constant value for W$_{exact}$. Thus, to our surprise it does not matter if we consider a constant window for exact calculations or simply approximate all inside and outside values, which is not only an interesting observation itself, but also fortunately prevents us from having to deal with an undesirable trade-off between reducing the worst-case time complexity (by a linear factor) and sacrificing less of the resulting sampling quality. In fact, this means we may (without resulting significant quality losses) always use the more efficient approximative preprocessing variant in order to reduce the worst-case time complexity of the overall sampling algorithm.

However, all profiles perfectly demonstrate that due to the noisy ensemble distribution caused by approximating the highly relevant sequence-dependent emission probabilities, the resulting sample sets usually contain many foldings that are rather unlikely according to the exact distribution for the considered input sequence. For this reason, it can not be recommended to employ one of the following otherwise reasonable construction schemes for deriving predictions according to the entire sample set: we should rather neither predict γ-MEA nor γ-centroid structures of the generated sample set as defined in [4], since those effectively reflect the overall behavior of the sample. Those predictions must anyway be considered inappropriate choices in our case, since their computation requires $\mathcal{O}(n^3)$ time, which would inevitably undo the time reduction reached by approximating (expect for in the case of γ-centroid structures for the default choice $\gamma = 1$, i.e. unique centroids, as those can be derived in $\mathcal{O}(n^2)$ time). Nevertheless, we can without significant losses in performance (without increasing the worst-case time complexity of the overall algorithm) identify the MP structure(s) of the generated sample[6], in strong analogy to traditional SCFG approaches. Since for this selection principle, we can actually rely on the exact distribution of feasible structures[7], this seems to be the right choice indeed.

It should be mentioned that one could, however, alternatively consider a particular subset of the overall sample that contains only those candidate foldings satisfying a preliminary defined quality criterion (e.g., only structures with probabilities above a specified threshold or with not less than a specified minimum number of base pairs). This means candidate folding of low quality are disregarded, such that constructing γ-MEA or γ-centroid structures might then result in reliable predictions (where only the derivation of the unique centroid is reasonable with respect to time complexity). Notably, in this context, one should apply a corresponding rejection scheme during sample composition (i.e., generated structures are only added to the sample set if they meet the preliminary specified requirements, otherwise they are rejected), since this is obviously more efficient than collecting all generated structures and afterwards employing a corresponding filtering process in order to identify the subset to be considered. Utilizing a reasonable rejection criterion, it actually becomes possible to generate any desired number of candidate foldings that obey to the imposed restrictions: new structures are generated until the resulting sample set is large enough. In connection with noisy ensemble distributions, choosing only moderate sizes for such filtered samples might suffice under certain circumstances, which might then result in a reduced runtime compared to the generation of large unfiltered sample sets. Anyway, in this work we will exclusively consider unfiltered samples and MP predictions.

On the basis of a series of experiments, we observed that stability in resulting predictions and a competitive prediction accuracy can only be reached by increasing the sample size, especially in the case of complete approximation for the preprocessing step and sampling according to the alternative strategy introduced in Sect. 4.2. That is, more candidate structures ought to be generated for guaranteeing that the resulting MP predictions are reproducible (by independent runs for the same input sequence) and of hight quality. This negative effect is considerably lowered by using (larger) constant values of $W_{exact} \geq 0$, and is actually less recognizable when employing the

---

[6] The probability of each structure can either be determined on the fly while sampling, multiplying the probabilities of the production rules which correspond to the respective sampling decisions, and otherwise – since the underlying SCFG from [4] is unambiguous – are computable in $\mathcal{O}(n^2)$ time making use, e.g., of an Earley-style parser.

[7] Note that the probability for a particular folding of a given RNA sequence is equal to a product of (different powers of the diverse) transition and emission probabilities (according to the corresponding derivation tree), which means it depends only on the exact trained parameter values of the underlying SCFG.

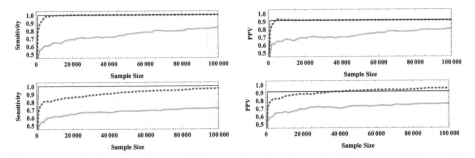

Fig. 9. Sensitivity and PPV of prediction as a function of sample size derived for *E.coli* tRNA^{Ala}. Top (bottom) line shows the common (alternative) sampling strategy for $W_{exact} = -1$ (no window, thick gray lines), $W_{exact} = 30$ (moderate window, thick dotted darker gray lines) and $W_{exact} = +\infty$ (complete window, thin black lines), respectively

conventional sampling strategy recapped in Sect. 4.1. Figure 9 shows the averaged sensitivities and PPVs obtained for 50 independent runs of continuously sampling secondary structures taking the so far most probable one as the actual prediction (which determines sensitivity and PPV for the actual sample size).

We observe that when making use of approximate probabilities sample sizes about 40 to 50 times as large as for a precise preprocessing are needed to generate competitive predictions. Thus, for a naive implementation the speedup gained by approximation may partly be lost. However, unlike prediction algorithms using dynamic programming, sampling can easily be parallelized. Making use of a grid environment where today one may assume a processor to have about 8 cores, a grid of size 5 or 6 computers is sufficient to compensate the increased sample size. Furthermore, since we only make use of the most probable sampled structure for our prediction, sampling can be done in-place, storing in each core only the best structure(s) seen so far. This reduces the memory requirements and keeps the communication costs rather moderate since it is finally only necessary to gather m structures from m cores and select the best. Notably, we performed a series of experiments, making use of Mathematica's parallel computation features, which proved that the overall process scales linearly in the number of cores used with a non-measurable communication overhead.

This initially proves the applicability of our approach providing a factor n speedup compared to established prediction tools but still maintaining the limits implied by a quadratic memory consumption (in our case used to store parameter values).

Anyway, for more reliable evaluation results, we performed a corresponding empirical study on the basis of comprehensive tRNA data (specifically, on 100 randomly chosen tRNA sequences from [15]), where for any sequence we picked the most probable structure in a particular sample as prediction. The plots presented in Fig. 10 show the resulting sensitivities and PPVs (averaged over all sequences) as functions of the considered sample size. Similar to the observations made above, we find that our heuristic variants can still yield highly accurate predictions, but may require the consideration of larger sample sizes.

Furthermore, the best accuracies as observed for one of the considered sample sizes (as recorded in Table 2(b)) are in any case superior to the accuracies obtained with

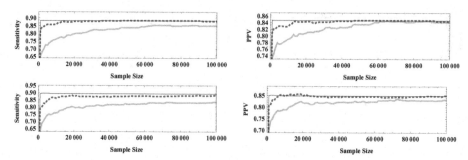

Fig. 10. Sensitivity and PPV of prediction as a function of sample size obtained by considering 100 distinct tRNA sequences from [15]. Plots correspond to those of Fig. 9.

Table 2. Sensitivity and PPV of predicted foldings for 100 distinct tRNA sequences from [15]

(a) Accuracies derived with leading prediction tools.

Tool	Sens.	PPV
CONTRAfold	80	71
PPfold	59	84
RNAshapes	66	60
Sfold, mfe	64	57
Sfold, ec	61	64
UNAFold	64	59
ViennaRNA	64	57

(b) Accuracies obtained with heuristic sampling. Tabulated are the best accuracies (maximizing the sum of sensitivity and PPV) as observed for one of the considered sample sizes.

Strategy	W_{exact}	Time	Sens.	PPV	S. size
Conventional	∞	$\mathcal{O}(n^3)$	88	84	1000
	30	$\mathcal{O}(n^2)$	88	84	60000
	-1	$\mathcal{O}(n^2)$	85	84	86000
Alternative	∞	$\mathcal{O}(n^3)$	89	85	2000
	30	$\mathcal{O}(n^3)$	87	85	19000
	-1	$\mathcal{O}(n^2)$	83	83	73000

several leading RNA secondary structure prediction tools (on the basis of the same tRNA sequences, as collected in Table 2(a)). Despite the rather simple structure of tRNA molecules (short sequences with comparably low structural variety), these first results anyhow indicate the validity of the proposed heuristic sampling approach. Notably, our novel sampling strategy seems to produce more accurate results than the common one for particular settings, for instance in case of exact preprocessing, although it was originally designed to fit especially well with our approximative variants.

In summary, the presented results are quite encouraging, but undoubtably more reliable empirical studies in connection with other classes of RNAs (e.g., the ones considered in [4] for exact preprocessing in combination with the common sampling strategy) and further prediction selection schemes (e.g., centroids) are required for evaluating our heuristic method(s) and for identifying potential sites for improvement.

6   Conclusions

The major advantage of the presented approximative method is that it is more efficient than all other modern prediction algorithms (implemented in popular tools like Mfold [6], ViennaRNA [8], Pfold [10], Sfold [2] or CONTRAfold [16]), reducing the worst-case time complexity by a linear factor, such that the time and space requirements are both bounded by $\mathcal{O}(n^2)$. However, a potential drawback lies in the observation that the overall quality of generated samples decreases (as indicated by probability profiling for specific loop types), which is due to the approximated ensemble distribution. As a consequence, we usually need to use larger sample sizes for obtaining a competitive prediction accuracy and stable predictions, i.e., more candidate structures for a given input sequence have to be generated to ensure that the approximation method outputs rather identical predictions in independent runs for that sequence. According to our experiments, an efficient implementation that really takes advantage of the accelerated preprocessing (3.7 compared to 49 seconds for our proof-of-concept implementation in Wolfram Mathematica) but handles large sample sizes can be obtained by parallelization.

Note that all results presented in this article have been derived with a purposive proof-of-concept implementation of the described methods. A more sophisticated tool will be realized in the future, hoping that the proposed prediction approach proves capable of yielding acceptable accuracies even for such types of RNAs whose molecules imply a great variety of structural features (due to large sequence lengths). In fact, we here only considered exemplary applications for simple tRNA sequences (specifically, for one particular tRNA molecule and a collection of 100 distinct tRNAs, respectively) in order to get positive feedback that (at least) the MP predictions obtained via approximated SCFG based sampling can be of high quality. Accordingly, more general experiments are needed, e.g., in connection with RNA molecules of sizes $n = 3000 - 30000$ (for which the memory constraints of our approach are not restrictive assuming 1GB of memory for each core) and where long distance base pairs in a global folding are of interest. In such a scenario, the proposed algorithm could be the method of choice – provided it performs similarly well.

This line of research is work in progress, but we found the first impressions presented within this note so motivating that we wanted to share them with the scientific community already at this point, primarily because this work leaves a number of open questions that may be inspiration for further research of other groups. For instance, recall that we used a sophisticated SCFG (representing a formal language counterpart to the thermodynamic model applied in the Sfold program) as probabilistic basis for the considered sampling strategies. However, it would also be possible to employ other SCFG designs, for example one of the commonly known *lightweight* grammars from [17]. This might of course yield at least noticeable if not significant changes in the resulting sampling quality, which could be an interesting subject to be explored.

It should also be noted that a similar approximative approach could potentially be considered when attempting to reduce the worst-case time complexity of the sampling extension of the PF approach. In fact, since sequence information is incorporated into the used (equilibrium) PFs and corresponding sampling probabilities only in the form of particular sequence-dependent free energy contributions, it seems reasonable to believe that the time complexity for the forward step (preprocessing) could possibly be reduced by a linear factor to $\mathcal{O}(n^2)$ when using some sort of approximated (averaged) free energy contributions that do not depend on the actual sequence (but contain as much sequence information as possible), in analogy to the approximated preprocessing step (inside and outside calculations) considered in this work, where we only had to use averaged emission terms instead of the exact emission probabilities in order to save time.

References

1. Ding, Y., Lawrence, C.E.: A statistical sampling algorithm for RNA secondary structure prediction. Nucleic Acids Research 31, 7280–7301 (2003)
2. Ding, Y., Chan, C.Y., Lawrence, C.E.: Sfold web server for statistical folding and rational design of nucleic acids. Nucleic Acids Research 32, W135–W141 (2004)
3. McCaskill, J.S.: The equilibrium partition function and base pair binding probabilities for RNA secondary structure. Biopolymers 29, 1105–1119 (1990)
4. Nebel, M.E., Scheid, A.: Evaluation of a sophisticated SCFG design for RNA secondary structure prediction. Theory in Biosciences 130, 313–336 (2011)
5. Zuker, M.: On finding all suboptimal foldings of an RNA molecule. Science 244, 48–52 (1989)
6. Zuker, M.: Mfold web server for nucleic acid folding and hybridization prediction. Nucleic Acids Res. 31, 3406–3415 (2003)
7. Hofacker, I., Fontana, W., Stadler, P., Bonhoeffer, S., Tacker, M., Schuster, P.: Fast folding and comparison of rna secondary structures (the Vienna RNA package). Monatsh Chem. 125, 167–188 (1994)
8. Hofacker, I.L.: The vienna RNA secondary structure server. Nucleic Acids Research 31, 3429–3431 (2003)
9. Knudsen, B., Hein, J.: RNA secondary structure prediction using stochastic context-free grammars and evolutionary history. Bioinformatics 15, 446–454 (1999)
10. Knudsen, B., Hein, J.: Pfold: RNA secondary structure prediction using stochastic context-free grammars. Nucleic Acids Research 31, 3423–3428 (2003)
11. Wexler, Y., Zilberstein, C., Ziv-Ukelson, M.: A study of accessible motifs and RNA folding complexity. Journal of Computational Biology 14, 856–872 (2007)
12. Backofen, R., Tsur, D., Zakov, S., Ziv-Ukelson, M.: Sparse RNA folding: Time and space efficient algorithms. Journal of Discrete Algorithms 9, 12–31 (2011)
13. Frid, Y., Gusfield, D.: A simple, practical and complete $\mathcal{O}(n^3/\log(n))$-time algorithm for RNA folding using the Four-Russians speedup. Algorithms for Molecular Biology 5, 5–13 (2010)
14. Akutsu, T.: Approximation and exact algorithms for RNA secondary structure prediction and recognition of stochastic context-free languages. J. Comb. Optim. 3, 321–336 (1999)
15. Sprinzl, M., Horn, C., Brown, M., Ioudovitch, A., Steinberg, S.: Compilation of tRNA sequences and sequences of tRNA genes. Nucleic Acids Res. 26, 148–153 (1998)
16. Do, C.B., Woods, D.A., Batzoglou, S.: CONTRAfold: RNA secondary structure prediction without physics-based models. Bioinformatics 22, e90–e98 (2006)
17. Dowell, R.D., Eddy, S.R.: Evaluation of several lightweight stochastic context-free grammars for RNA secondary structure prediction. BMC Bioinformatics 5, 71 (2004)
18. Scheid, A.: Sampling and Approximation in the Context of RNA Secondary Structure Prediction, PhD-Thesis Kaiserslautern University, Germany (2012)

A Vaccination Strategy Based on a State Feedback Control Law for Linearizing SEIR Epidemic Models

S. Alonso-Quesada[1], M. De la Sen[1], and A. Ibeas[2]

[1] Department of Electricity and Electronics, Faculty of Science and Technology, University of the Basque Country, UPV/EHU, Campus of Leioa, 48940-Leioa, Bizkaia, Spain
{santiago.alonso,manuel.delasen}@ehu.es
[2] Departamento de Telecomunicación e Ingeniería de Sistemas, Escuela Técnica Superior de Ingeniería, Universitat Autònoma, Barcelona, Spain
Asier.Ibeas@uab.es

Abstract. A vaccination strategy for fighting against the propagation of epidemic diseases within a host population is purposed. A SEIR epidemic model is used to describe the propagation of the illness. This compartmental model divides the population in four classes by taking into account their status related to the infection. In this way, susceptible, exposed, infectious and recovered populations are included in the model. The vaccination strategy is based on a continuous-time nonlinear control law synthesized via an exact feedback input-output linearization approach. The asymptotic eradication of the infection from the host population under such a vaccination is proved. Moreover, the positivity and stability properties of the controlled system are investigated.

Keywords: SEIR epidemic models, Vaccination, State feedback control, Stability, Positivity.

1   Introduction

A relevant area in the mathematical theory of epidemiology is the development of models for studying the propagation of epidemic diseases within a host population. The epidemic mathematical models include the most basic ones [1-5], namely: (i) SI models where only susceptible and infected populations are assumed to be present in the model, (ii) SIR models which include susceptible plus infected plus removed-by-immunity populations and (iii) SEIR models where the infected population is split into two ones, namely, the "infected" (or "exposed") which incubate the disease but they do not still have any disease symptoms and the "infectious" (or "infective") which do have the external disease symptoms. There are many variants of the above models as, for instance, the SVEIR epidemic models which incorporate the dynamics of a vaccinated population [6], [7], the SEIQR-SIS model which adds a quarantine population [8] and the model proposed in [9] which incorporates vaccinated, quarantine and hospitalized populations. Other variant consists of the generalization of such models by incorporating point and/or distributed delays [10], [11]. Another one is concerned with the inclusion of a saturated disease transmission incidence rate for

J. Gabriel et al. (Eds.): BIOSTEC 2012, CCIS 357, pp. 195–209, 2013.
© Springer-Verlag Berlin Heidelberg 2013

taking into account the inhibition effect from the behavioral change of susceptible individuals when the infectious individual number increases [12].

The analysis of the existence of equilibrium points, relative to either the persistence or extinction of the epidemics, the conditions for the existence of a backward bifurcation where both equilibrium points co-exist and the constraints for guaranteeing the positivity and the boundedness of the solutions of such models have been some of the main objectives in the aforementioned papers. Also, the conditions that generate an oscillatory behavior in such solutions have been dealt with in the literature about epidemic models [13]. Other important aim is that relative to the design of control strategies in order to eradicate the persistence of the infection within the host population [1], [4], [6], [9]. In this context, an explicit vaccination function of many different kinds may be considered, namely: constant, continuous-time, impulsive, mixed constant/impulsive, mixed continuous-time/impulsive, discrete-time and so on.

In this paper, a SEIR epidemic model is considered. The dynamics of susceptible (S) and immune (R) populations are directly affected by a vaccination function $V(t)$, which also has indirectly influence in the time evolution of infected or exposed (E) and infectious (I) populations. In fact, such a vaccination function has to be suitably designed in order to eradicate the infection from the population. This model has been already studied in [1] from the viewpoint of equilibrium points in the controlled and free-vaccination cases. A vaccination auxiliary control law being proportional to the susceptible population was proposed in order to achieve the whole population be asymptotically immune. Such an approach assumed that the parameters of the model were known and the illness transmission was not critical. Moreover, some important issues of positivity, stability and tracking of the SEIR model were discussed. *The present paper proposes an alternative method to obtain the vaccination control law to asymptotically eradicate the epidemic disease. Concretely, the vaccination function is synthesized by means of an input-output exact feedback linearization technique. Such a linearization control strategy constitutes the main contribution of the paper. Moreover, mathematical proofs about the disease eradication based on such a controlled SEIR epidemic model while maintaining the non-negativity of all the partial populations for all time are issued.* The exact feedback linearization can be implemented by using a proper nonlinear coordinate transformation and a static-state feedback control. The use of such a linearization strategy is motivated by three main facts, namely: (i) it is a power tool for controlling nonlinear systems which is based on well-established technical principles [14], (ii) the given SEIR model is highly nonlinear and (iii) such a control strategy has not been yet applied in epidemic models.

2   SEIR Epidemic Model

Let $S(t)$, $E(t)$, $I(t)$ and $R(t)$ be, respectively, the susceptible, infected (or exposed), infectious and removed-by-immunity populations at time t. Consider a time-invariant true-mass action type SEIR epidemic model given by:

$$\dot{S}(t) = -\mu S(t) + \omega R(t) - \beta \frac{S(t)I(t)}{N} + \mu N[1 - V(t)] \qquad (1)$$

$$\dot{E}(t) = -(\mu+\sigma)E(t)+\beta\frac{S(t)I(t)}{N} \tag{2}$$

$$\dot{I}(t) = -(\mu+\gamma)I(t)+\sigma E(t) \tag{3}$$

$$\dot{R}(t) = -(\mu+\omega)R(t)+\gamma I(t)+\mu NV(t) \tag{4}$$

subject to initial conditions $S(0) \geq 0$, $E(0) \geq 0$, $I(0) \geq 0$ and $R(0) \geq 0$ under a vaccination function $V : \mathbb{R}_{0+} \to \mathbb{R}_{0+}$, with $\mathbb{R}_{0+} \triangleq [0, \infty) \cap \mathbb{R}$. In the above SEIR model, $N > 0$ is the total population at any time instant $t \in \mathbb{R}_{0+}$, $\mu > 0$ is the rate of deaths and births from causes unrelated to the infection, $\omega \geq 0$ is the rate of losing immunity, $\beta > 0$ is the transmission constant (with the total number of infections per unity of time at time t being $\beta S(t)I(t)/N$) and, $\sigma^{-1} > 0$ and $\gamma^{-1} > 0$ are, respectively, the average durations of the latent and infective periods. The total population dynamics can be obtained by summing-up (1)-(4) yielding:

$$\dot{N}(t) = \dot{S}(t)+\dot{E}(t)+\dot{I}(t)+\dot{R}(t) = 0 \tag{5}$$

so that the total population $N(t) = N(0) = N$ is constant $\forall t \in \mathbb{R}_{0+}$. Then, this model is suitable for epidemic diseases with very small mortality incidence caused by infection and for populations with equal birth and death rates so that the total population may be considered constant for all time.

3   Vaccination Strategy

An ideal control objective is that the removed-by-immunity population asymptotically tracks the whole population. In this way, the joint infected plus infectious population asymptotically tends to zero as $t \to \infty$, so the infection is eradicated from the population. A vaccination control law based on a static-state feedback linearization strategy is developed for achieving such a control objective. This technique requires a nonlinear coordinate transformation, based on the Lie derivatives Theory [14], in the system representation.

The dynamics equations (1)-(3) of the SEIR model can be equivalently written as the following nonlinear control affine system:

$$\dot{x}(t) = f(x(t))+g(x(t))u(t) \quad ; \quad y(t) = h(x(t)) \tag{6}$$

where $y(t) = I(t) \in \mathbb{R}_{0+}$, $u(t) = V(t) \in \mathbb{R}_{0+}$ and $x(t) = [I(t) \ E(t) \ S(t)]^T \in \mathbb{R}_{0+}^3$ are, respectively, considered as the output signal, the input signal and the state vector of the system $\forall t \in \mathbb{R}_{0+}$ and $R(t) = N-S(t)-E(t)-I(t)$ has been used, with:

$$f(x(t)) = \begin{bmatrix} -(\mu+\gamma)I(t)+\sigma E(t) \\ -(\mu+\sigma)E(t)+\beta_1 I(t)S(t) \\ -\omega(I(t)+E(t))+(\mu+\omega)(N-S(t))-\beta_1 I(t)S(t) \end{bmatrix} \in \mathbb{R}^3 \tag{7}$$

$$g(x(t)) = [0 \ 0 \ -\mu N]^T \in \mathbb{R}_{0-}^3 \quad ; \quad h(x(t)) = I(t) \in \mathbb{R}_{0+}$$

where $\beta_1 = \beta/N$ and $\mathbb{R}_{0-} \triangleq (-\infty, 0] \cap \mathbb{R}$. The first step to apply a coordinate transformation based on the Lie derivation is to determine the relative degree of the system. For such a purpose, the following definitions are taken into account: (i) $L_f^k h(x(t)) \triangleq \dfrac{\partial \left(L_f^{k-1} h(x(t)) \right)}{\partial x} f(x(t))$ is the kth-order Lie derivative of $h(x(t))$ along $f(x(t))$ with $L_f^0 h(x(t)) \triangleq h(x(t))$ and (ii) the relative degree r of the system is the number of times that the output must be differentiated to obtain the input explicitly, i.e., the number r so that $L_g L_f^k h(x(t)) = 0$ for $k < r-1$ and $L_g L_f^{r-1} h(x(t)) \neq 0$.

From (7), $L_g h(x(t)) = L_g L_f h(x(t)) = 0$ while $L_g L_f^2 h(x(t)) = -\mu\sigma\beta I(t)$, so the relative degree of the system is 3 in $D \triangleq \left\{ x = [I \ E \ S]^T \in \mathbb{R}_{0+}^3 \mid I \neq 0 \right\}$, i.e., $\forall x \in \mathbb{R}_{0+}^3$ except in the singular surface $I = 0$ of the state space where the relative degree is not well-defined. Since the relative degree of the system is exactly equal to the dimension of the state space for any $x \in D$, the nonlinear coordinate change

$$\begin{aligned}
\bar{I}(t) &= L_f^0 h(x(t)) = I(t) \\
\bar{E}(t) &= L_f h(x(t)) = [1 \ 0 \ 0] f(x(t)) = -(\mu+\gamma)I(t) + \sigma E(t) \\
\bar{S}(t) &= L_f^2 h(x(t)) = [-(\mu+\gamma) \ \sigma \ 0] f(x(t)) = (\mu+\gamma)^2 I(t) - \sigma(2\mu+\sigma+\gamma)E(t) + \sigma\beta_1 I(t)S(t)
\end{aligned} \tag{8}$$

allows to represent the model in the called normal form in a neighborhood of any $x \in D$. Namely:

$$\dot{\bar{x}}(t) = \bar{f}(\bar{x}(t)) + \bar{g}(\bar{x}(t))u(t) \ ; \ y(t) = h(\bar{x}(t)) \tag{9}$$

where $\bar{x}(t) = \left[\bar{I}(t) \ \bar{E}(t) \ \bar{S}(t) \right]^T$ and:

$$\bar{f}(\bar{x}(t)) = \left[\bar{E}(t) \ \bar{S}(t) \ \varphi(\bar{x}(t)) \right]^T ; \ \bar{g}(\bar{x}(t)) = \left[0 \ 0 \ -\mu\sigma\beta\bar{I}(t) \right]^T ; \ h(\bar{x}(t)) = \bar{I}(t) \tag{10}$$

with:

$$\begin{aligned}
\varphi(\bar{x}(t)) = &(\mu+\omega)\left[\sigma\beta - (\mu+\sigma)(\mu+\gamma)\right]\bar{I}(t) - (\mu+\omega)(2\mu+\sigma+\gamma)\bar{E}(t) \\
&- (3\mu+\sigma+\gamma+\omega)\bar{S}(t) - \beta_1\left[\omega(\mu+\sigma+\gamma) + (\mu+\sigma)(\mu+\gamma)\right]\bar{I}^2(t) \\
&- \beta_1(2\mu+\sigma+\gamma+\omega)\bar{I}(t)\bar{E}(t) - \beta_1\bar{I}(t)\bar{S}(t) + \frac{\bar{E}(t)\bar{S}(t)}{\bar{I}(t)} + (2\mu+\sigma+\gamma)\frac{\bar{E}^2(t)}{\bar{I}(t)}
\end{aligned} \tag{11}$$

The following result about the system input-output linearization is established.

Theorem 1. The state feedback control law

$$u(t) = \frac{-L_f^3 h(x(t)) - \lambda_0 h(x(t)) - \lambda_1 L_f h(x(t)) - \lambda_2 L_f^2 h(x(t))}{L_g L_f^2 h(x(t))} \tag{12}$$

where λ_i, for $i \in \{0, 1, 2\}$, are the controller tuning parameters, induces the linear closed-loop dynamics

$$\dddot{y}(t) + \lambda_2 \ddot{y}(t) + \lambda_1 \dot{y}(t) + \lambda_0 y(t) = 0 \tag{13}$$

around any point $x \in D$.

Proof. The state equation for the closed-loop system

$$\begin{bmatrix} \dot{\overline{I}}(t) \\ \dot{\overline{E}}(t) \\ \dot{\overline{S}}(t) \end{bmatrix} = \begin{bmatrix} \overline{E}(t) \\ \overline{S}(t) \\ \varphi(\overline{x}(t)) - L_f^3 h(x(t)) - \lambda_0 \overline{I}(t) - \lambda_1 \overline{E}(t) - \lambda_2 \overline{S}(t) \end{bmatrix} \tag{14}$$

is obtained by introducing the control law (12) in (9) and taking into account the fact that $L_g L_f^2 h(x(t)) = -\mu\sigma\beta I(t) = -\mu\sigma\beta\overline{I}(t) \neq 0 \ \forall x \in D$ and the coordinate transformation (8). Moreover, it follows by direct calculations that:

$$L_f^3 h(x(t)) = \left[\sigma\beta(\mu+\omega) - (\mu+\gamma)^3\right] I(t) + \sigma\left[(\mu+\gamma)^2 + (2\mu+\sigma+\gamma)(\mu+\sigma)\right] E(t)$$
$$- \sigma\beta_1\omega I(t)[I(t) + E(t)] + \sigma^2\beta_1 E(t)S(t) - \sigma\beta_1(4\mu+\sigma+2\gamma+\omega)I(t)S(t) - \sigma\beta_1^2 I^2(t)S(t) \tag{15}$$

One may express $L_f^3 h(x(t))$ in the state space defined by $\overline{x}(t)$ via the application of the reverse coordinate transformation to that in (8). Then, it follows directly that $L_f^3 h(x(t)) = \varphi(\overline{x}(t))$. Thus, the state equation of the closed-loop system in the state space defined by $\overline{x}(t)$ can be written as:

$$\dot{\overline{x}}(t) = A\overline{x}(t) \tag{16}$$

with

$$A = \begin{bmatrix} 0 & 1 & 0 \\ 0 & 0 & 1 \\ -\lambda_0 & -\lambda_1 & -\lambda_2 \end{bmatrix}. \tag{17}$$

Furthermore, the output equation of the closed-loop system is $y(t) = C\overline{x}(t)$ with $C = [1 \ 0 \ 0]$ since $y(t) = I(t) = \overline{I}(t)$. From (16)-(17) and the closed-loop output equation, it follows that:

$$y^{(\ell)}(t) = CA^\ell e^{At}\overline{x}(0) \tag{18}$$

for $\ell \in \{0, 1, 2, 3\}$ denoting the order of the differentiation of $y(t)$. Finally, the dynamics of the closed-loop system (13) is directly obtained from (18).   ***

Remark 1. The controller parameters λ_i, for $i \in \{0, 1, 2\}$, will be adjusted such that the roots of the closed-loop system characteristic polynomial $P(s) = \text{Det}(sI_3 - A) = (s+r_1)(s+r_2)(s+r_3)$, with $I_3 \in \mathbb{R}^{3\times 3}$ denoting the identity matrix, be located at prescribed positions. i.e., $\lambda_i = \lambda_i(-r_j)$ for $i \in \{0, 1, 2\}$ and $j \in \{1, 2, 3\}$, with $(-r_j)$ denoting the desired roots of $P(s)$. If one of the control objectives is to guarantee the exponential stability of the closed-loop system then $\text{Re}\{r_j\} > 0$ for all $j \in \{1, 2, 3\}$. Then, the values $\lambda_0 = r_1 r_2 r_3 > 0$, $\lambda_1 = r_1 r_2 + r_1 r_3 + r_2 r_3 > 0$ and $\lambda_2 = r_1 + r_2 + r_3 > 0$ for the controller parameters have to be chosen in order to achieve such a stability result. It implies that the strictly positivity of the controller parameters is a necessary condition for the exponential stability of the closed-loop system.   ***

Remark 2. The control (12) may be rewritten as:

$$
\begin{aligned}
u(t) = {} & \frac{(\mu+\omega)\sigma\beta - (\mu+\gamma)^3 + \lambda_0 - \lambda_1(\mu+\gamma) + \lambda_2(\mu+\gamma)^2}{\mu\sigma\beta} - \frac{\omega}{\mu N}[I(t)+E(t)] \\
& - \frac{(3\mu+\sigma+2\gamma-\lambda_2)}{\mu N}S(t) + \frac{\sigma}{\mu N}\frac{E(t)S(t)}{I(t)} - \frac{\beta}{\mu N^2}I(t)S(t) \\
& + \frac{(\mu+\gamma)^2 + (2\mu+\sigma+\gamma)(\mu+\sigma) + \lambda_1 - \lambda_2(2\mu+\sigma+\gamma)}{\mu\beta}\frac{E(t)}{I(t)}
\end{aligned}
\tag{19}
$$

by using (8) and (15). ***

Remark 3. The control law (12) is well-defined for all $x \in \mathbb{R}_{0+}^3$ except in the surface $I = 0$. However, the infection may be considered eradicated from the population once the infectious population strictly exceeds zero while it is smaller than one individual, so the vaccination strategy may be switched off when $0 < I(t) \le \delta < 1$. This fact implies that the singularity in the control law is not reached. i.e., such a control law is well-defined by the nature of the system. In this sense, the control law

$$
u_p(t) = \begin{cases} u(t) & \text{for } 0 \le t \le t_f \\ 0 & \text{for } t > t_f \end{cases}
\tag{20}
$$

may be used instead of (12) in a practical situation. The signal $u(t)$ in (20) is given by the linearizing control law (12) while t_f denotes the eventual time instant after which the infection propagation may be assumed ended. Formally, such a time instant is defined as:

$$
t_f \triangleq \mathrm{Min}\{t \in \mathbb{R}_0^+ \mid I(t_f) < \delta \text{ for some } 0 < \delta < 1\} .
\tag{21}
$$

Then, the control action is maintained active while the infection persists within the population and it is switched off once the epidemics is eradicated. ***

3.1   Control Parameters Choice

The application of the control law (12), obtained from the exact input-output linearization strategy, makes the closed-loop dynamics of the infectious population be given by (13). Such a dynamics depends on the control parameters λ_i, for $i \in \{0, 1, 2\}$. Such parameters have to be appropriately chosen in order to guarantee the following suitable properties: (i) the stability and positivity of the controlled SEIR model and (ii) the eradication of the infection, i.e., the asymptotic convergence of $I(t)$ and $E(t)$ to zero as time tends to infinity. The following theorems related to the choice of the controller tuning parameter in order to meet such properties are proven.

Theorem 2. Assume that the initial condition $x(0) = [I(0)\ E(0)\ S(0)]^T \in \mathbb{R}_{0+}^3$ is bounded and all roots $(-r_j)$ for $j \in \{1, 2, 3\}$ of the characteristic polynomial $P(s)$ associated with the closed-loop dynamics (13) are of strictly negative real part via an appropriate choice of the free-design controller parameters $\lambda_i > 0$, for $i \in \{0, 1, 2\}$. Then, the control law (12) guarantees the exponential stability of the controlled SEIR model (6)-(11) while achieving the eradication of the infection from the population as $t \to \infty$. Moreover, the SEIR model (1)-(4) has the properties: $E(t)$, $I(t)$, $S(t)I(t)$ and $S(t) + R(t) = N - [E(t) + I(t)]$ are bounded for all time, $E(t) \to 0$, $I(t) \to 0$, $S(t) + R(t) \to N$ and $S(t)I(t) \to 0$ exponentially as $t \to \infty$, and $I(t) = o(1/S(t))$.

Proof. The dynamics of the controlled SEIR model (13) can be equivalently rewritten with the state equation (16)-(17) and the output equation $y(t) = C\bar{x}(t)$, where $C = [1 \; 0 \; 0]$, by taking into account that $y(t) = \bar{I}(t)$, $\dot{y}(t) = \bar{E}(t)$ and $\ddot{y}(t) = \bar{S}(t)$. The initial condition $\bar{x}(0) = [\bar{I}(0) \; \bar{E}(0) \; \bar{S}(0)]^T$ in such a realization is bounded since it is related to $x(0)$ via the coordinate transformation (8) and $x(0)$ is assumed to be bounded. The controlled SEIR model is exponentially stable since the eigenvalues of the matrix A are the roots $-r_j < 0$ for $j \in \{1, 2, 3\}$ of $P(s)$ which are assumed to be in the open left-half plane. Then, the state vector $\bar{x}(t)$ exponentially converges to zero as $t \to \infty$ while being bounded for all time. Moreover, $I(t)$ and $E(t)$ are also bounded and converge exponentially to zero as $t \to \infty$ from the boundedness and exponential convergence to zero of $\bar{x}(t)$ as $t \to \infty$ according to the coordinate transformation (8). Then, the infection is eradicated from the host population. Furthermore, the boundedness of $S(t) + R(t)$ follows from that of $E(t)$ and $I(t)$, and the fact that the total population is constant for all time. Also, the exponentially convergence of $S(t) + R(t)$ to the total population as $t \to \infty$ is derived from the fact that $S(t) + E(t) + I(t) + R(t) = N \;\; \forall t \in \mathbb{R}_{0+}$ and the exponential convergence to zero of $I(t)$ and $E(t)$ as $t \to \infty$. Finally, from the third equation of (8), it follows that $S(t)I(t)$ is bounded and it converges exponentially to zero as $t \to \infty$ from the boundedness and convergence to zero of $I(t)$, $E(t)$ and $\bar{x}(t)$ as $t \to \infty$. The facts that $I(t) \to 0$ and $S(t)I(t) \to 0$ as $t \to \infty$ imply directly that $I(t) = o(1/S(t))$.   ***

Remark 4. Theorem 2 implies the existence of a finite time instant t_f after which the infectious disease is eradicated if the vaccination control law (20) is used instead of (12). Concretely, such an existence derives from the fact that $I(t) \to 0$ as $t \to \infty$ via the application of the control law (12).   ***

Theorem 3. Assume an initial condition for the SEIR model satisfying $R(0) \geq 0$, $x(0) \in \mathbb{R}_{0+}^3$, i.e., $I(0) \geq 0$, $E(0) \geq 0$ and $S(0) \geq 0$, and the constraint $S(0) + E(0) + I(0) + R(0) = N$. Assume also that some strictly positive real numbers r_j for $j \in \{1, 2, 3\}$ are chosen such that:

(a) $0 < r_1 < \mu + \text{Min}\{\sigma, \gamma\}$, $r_2 = \mu + \gamma$ and $r_3 > \mu + \text{Max}\{\sigma, \gamma\}$, and

(b) r_1 and r_3 satisfy the inequalities:

$$r_1 + r_3 \geq 2\mu + \sigma + \gamma + \beta - \omega \; ; \quad r_1 r_3 \geq (\mu + \sigma)(r_1 + r_3) + (\gamma - \sigma)(2\mu + \sigma + \gamma) - (\mu + \gamma)^2 \; ;$$
$$(r_3 - r_1)(r_3 - \mu - \gamma) \geq \sigma \beta$$

Then:

(i) the application of the control law (12) to the SEIR model guarantees that the epidemics is asymptotically eradicated from the population while $I(t) \geq 0$, $E(t) \geq 0$ and $S(t) \geq 0$ $\forall t \in \mathbb{R}_{0+}$, and

(ii) the application of the control law (20) guarantees the epidemics eradication after a finite time t_f, the positivity of the controlled SEIR epidemic model $\forall t \in [0, t_f)$ and that $u(t) = V(t) \geq 1$ $\forall t \in [0, t_f)$ so that $u(t) \geq 0$ $\forall t \in \mathbb{R}_{0+}$,

provided that the controller tuning parameters λ_i, $i \in \{0, 1, 2\}$, are chosen so that $(-r_j)$, $j \in \{1, 2, 3\}$, be the roots of the characteristic polynomial $P(s)$ associated with the closed loop dynamics (13).

Proof.

(i) On one hand, the epidemics asymptotic eradication is proved by following the same reasoning that in Theorem 2. On the other hand, the dynamics (13) of the controlled SEIR model can be written in the state space defined by $\bar{x}(t) = [\bar{I}(t) \; \bar{E}(t) \; \bar{S}(t)]^T$ as in (16)-(17). From such a realization and taking into account the first equation in (8) and that $(-r_j)$ for $j \in \{1, 2, 3\}$ are the eigenvalues of A, it follows that:

$$I(t) = \bar{I}(t) = y(t) = c_1 e^{-r_1 t} + c_2 e^{-r_2 t} + c_3 e^{-r_3 t} \tag{22}$$

$\forall t \in \mathbb{R}_{0+}$ for some constants c_j for $j \in \{1, 2, 3\}$ being dependent on the initial conditions $y(0)$, $\dot{y}(0)$ and $\ddot{y}(0)$. In turn, such initial conditions are related to the initial conditions of the SEIR model in its original realization, i.e., in the state space defined by $x(t) = [I(t) \; E(t) \; S(t)]^T$ via (8). The constants c_j for $j \in \{1, 2, 3\}$ can be obtained by solving the following set of linear equations:

$$\begin{aligned}
\bar{I}(0) &= y(0) = c_1 + c_2 + c_3 = I(0) \\
\bar{E}(0) &= \dot{y}(0) = -(c_1 r_1 + c_2 r_2 + c_3 r_3) = -(\mu + \gamma)I(0) + \sigma E(0) \\
\bar{S}(0) &= \ddot{y}(0) = c_1 r_1^2 + c_2 r_2^2 + c_3 r_3^2 = (\mu + \gamma)^2 I(0) - \sigma(2\mu + \sigma + \gamma)E(0) + \sigma \beta_1 I(0) S(0)
\end{aligned} \tag{23}$$

where (8) and (22) have been used. Such equations can be compactly written as $R_p \cdot K = M$ where:

$$R_p = \begin{bmatrix} 1 & 1 & 1 \\ r_1 & r_2 & r_3 \\ r_1^2 & r_2^2 & r_3^2 \end{bmatrix} \; ; \quad K = \begin{bmatrix} c_1 \\ c_2 \\ c_3 \end{bmatrix} \; ; \quad M = \begin{bmatrix} I(0) \\ (\mu + \gamma)I(0) - \sigma E(0) \\ (\mu + \gamma)^2 I(0) - \sigma(2\mu + \sigma + \gamma)E(0) + \sigma \beta_1 I(0) S(0) \end{bmatrix} \tag{24}$$

Then, once the desired roots of the characteristic equation of the closed-loop dynamics have been prefixed the constants c_j for $j \in \{1, 2, 3\}$ of the time-evolution of $I(t)$ are obtained from $K = R_p^{-1}M$ since R_p is a non-singular matrix, i.e., an invertible matrix. In this sense, note that $\text{Det}(R_p) = (r_2 - r_1)(r_3 - r_1)(r_3 - r_2) \neq 0$ since R_p is a Vandermonde matrix [15] and the roots $(-r_j)$ for $j \in \{1, 2, 3\}$ have been chosen different among them. Namely:

$$\begin{bmatrix} c_1 \\ c_2 \\ c_3 \end{bmatrix} = \begin{bmatrix} \dfrac{F(r_2, r_3) I(0) + \sigma G(r_2, r_3) E(0) + \sigma \beta_1 I(0) S(0)}{(r_2 - r_1)(r_3 - r_1)} \\ -\dfrac{F(r_1, r_3) I(0) + \sigma G(r_1, r_3) E(0) + \sigma \beta_1 I(0) S(0)}{(r_2 - r_1)(r_3 - r_2)} \\ \dfrac{F(r_1, r_2) I(0) + \sigma G(r_1, r_2) E(0) + \sigma \beta_1 I(0) S(0)}{(r_3 - r_1)(r_3 - r_2)} \end{bmatrix} \tag{25}$$

where $F: \mathbb{R}_+^2 \to \mathbb{R}$ and $G: \mathbb{R}_+^2 \to \mathbb{R}$ are defined as:

$$F(v, w) \triangleq vw - (\mu + \gamma)(v + w) + (\mu + \gamma)^2 \quad ; \quad G(v, w) \triangleq v + w - (2\mu + \sigma + \gamma) \tag{26}$$

Note that $c_1 = \dfrac{\sigma(r_3 - \mu - \gamma)E(0) + \sigma\beta_1 I(0)S(0)}{(\mu + \gamma - r_1)(r_3 - r_1)} > 0$ since $I(0) \geq 0$, $E(0) \geq 0$, $S(0) \geq 0$, $F(r_2, r_3) = 0$, $G(r_2, r_3) = r_3 - \mu - \gamma > 0$, $\mu + \gamma - r_1 > 0$ and $r_3 - r_1 > 0$ by taking into account the constraints in (a). On one hand, $I(t) \geq 0$ $\forall t \in \mathbb{R}_{0+}$ is proved directly from (22) as follows. One 'a priori' knows that $c_1 > 0$. However, the sign of both c_2 and c_3 may not be 'a priori' determined from the initial conditions and constraints in (a). The following four cases may be possible: (i) $c_2 \geq 0$ and $c_3 \geq 0$, (ii) $c_2 \geq 0$ and $c_3 < 0$, (iii) $c_2 < 0$ and $c_3 \geq 0$, and (iv) $c_2 < 0$ and $c_3 < 0$. For the cases (i) and (ii), i.e., if $c_2 \geq 0$, it follows from (22) that:

$$I(t) = c_1 e^{-r_1 t} + c_2 e^{-r_2 t} + [I(0) - c_1 - c_2]e^{-r_3 t} = c_1\left(e^{-r_1 t} - e^{-r_3 t}\right) + c_2\left(e^{-r_2 t} - e^{-r_3 t}\right) + I(0)e^{-r_3 t} \geq 0 \tag{27}$$

$\forall t \in \mathbb{R}_{0+}$ where the facts that $I(0) = c_1 + c_2 + c_3 \geq 0$ and, $e^{-r_1 t} - e^{-r_3 t} \geq 0$ and $e^{-r_2 t} - e^{-r_3 t} \geq 0$ $\forall t \in \mathbb{R}_{0+}$ since $r_1 < r_2 < r_3$ have been taken into account. For the case (iii), i.e., if $c_2 < 0$ and $c_3 \geq 0$, it follows that:

$$I(t) = [I(0) - c_2 - c_3]e^{-r_1 t} + c_2 e^{-r_2 t} + c_3 e^{-r_3 t} = [I(0) - c_3]e^{-r_1 t} + c_2\left(e^{-r_2 t} - e^{-r_1 t}\right) + c_3 e^{-r_3 t} \geq 0 \tag{28}$$

$\forall t \in \mathbb{R}_{0+}$ by taking into account that $I(0) = c_1 + c_2 + c_3$, $e^{-r_2 t} - e^{-r_1 t} \leq 0$ $\forall t \in \mathbb{R}_{0+}$ since $r_1 < r_2$ and the fact that:

$$I(0) - c_3 = \dfrac{[(r_3 - r_1)(r_3 - \mu - \gamma) - \sigma\beta_1 S(0)]I(0) + \sigma(\mu + \gamma - r_1)E(0)}{(r_3 - r_1)(r_3 - \mu - \gamma)} \geq 0 \tag{29}$$

where (25), (26), $G(r_1,r_2) = r_1 - \mu - \gamma < 0$, $F(r_1,r_2) = 0$ and the constraints in (a) and (b) have been used. In particular, the coefficient multiplying to $I(0)$ in (29) is non-negative if r_1 and r_3 satisfy the third inequality of the constraints (b) by taking into account $\sigma\beta_1 S(0) = \sigma\beta S(0)/N \le \sigma\beta$ and $S(0) \le N$. This later inequality is directly implied by $N = I(0) + E(0) + S(0) + R(0)$ with $I(0) \ge 0$, $E(0) \ge 0$, $S(0) \ge 0$ and $R(0) \ge 0$. Finally, for the case (iv), i.e., if $c_2 < 0$ and $c_3 < 0$, it follows that:

$$I(t) = [I(0) - c_2 - c_3]e^{-r_1 t} + c_2 e^{-r_2 t} + c_3 e^{-r_3 t} = I(0)e^{-r_1 t} + c_2\left(e^{-r_2 t} - e^{-r_1 t}\right) + c_3\left(e^{-r_3 t} - e^{-r_1 t}\right) \ge 0 \quad (30)$$

$\forall t \in \mathbb{R}_{0+}$, where $e^{-r_2 t} - e^{-r_1 t} \le 0$ and $e^{-r_3 t} - e^{-r_1 t} \le 0$, since $r_1 < r_2 < r_3$, and $I(0) = c_1 + c_2 + c_3 \ge 0$ have been taken into account. In summary, $I(t) \ge 0$ $\forall t \in \mathbb{R}_{0+}$ if all partial populations are initially non-negative and the roots $(-r_j)$, for $j \in \{1, 2, 3\}$, of the closed-loop characteristic polynomial satisfy the constraints in (a) and (b). On the other hand, one obtains from (22) and the reverse coordinate transformation to (8) that:

$$E(t) = \frac{1}{\sigma}\left[\bar{E}(t) + (\mu + \gamma)\bar{I}(t)\right] = \frac{1}{\sigma}\sum_{j=1}^{3} c_j(\mu + \gamma - r_j)e^{-r_j t}$$

$$S(t) = \frac{\bar{S}(t) + (\mu + \sigma)(\mu + \gamma)\bar{I}(t) + (2\mu + \sigma + \gamma)\bar{E}(t)}{\sigma\beta_1\bar{I}(t)} \quad (31)$$

$$= \frac{\sum_{j=1}^{3} c_j\left[r_j^2 - (2\mu + \sigma + \gamma)r_j + (\mu + \sigma)(\mu + \gamma)\right]e^{-r_j t}}{\sigma\beta_1 I(t)}$$

from the facts that $\bar{E}(t) = \dot{I}(t)$ and $\bar{S}(t) = \ddot{I}(t)$. If one fixes $r_2 = \mu + \gamma$ then:

$$E(t) = \frac{1}{\sigma}\left[c_1(\mu + \gamma - r_1)e^{-r_1 t} + c_3(\mu + \gamma - r_3)e^{-r_3 t}\right]$$

$$S(t) = \frac{c_1\left[r_1^2 - (2\mu + \sigma + \gamma)r_1 + (\mu + \sigma)(\mu + \gamma)\right]e^{-r_1 t}}{\sigma\beta_1 I(t)} \quad (32)$$

$$+ \frac{c_3\left[r_3^2 - (2\mu + \sigma + \gamma)r_3 + (\mu + \sigma)(\mu + \gamma)\right]e^{-r_3 t}}{\sigma\beta_1 I(t)}$$

where the function $H : \mathbb{R}_+ \to \mathbb{R}$ defined as:

$$H(v) \triangleq v^2 - (2\mu + \sigma + \gamma)v + (\mu + \sigma)(\mu + \gamma) \quad (33)$$

is zero for $v = r_2 = \mu + \gamma$ has been used. From the first equation in (32), it follows that $c_3(\mu + \gamma - r_3) = \sigma E(0) - c_1(\mu + \gamma - r_1)$ and then:

$$E(t) = \frac{c_1(\mu + \gamma - r_1)\left(e^{-r_1 t} - e^{-r_3 t}\right) + \sigma E(0)e^{-r_3 t}}{\sigma} \ge 0 \quad (34)$$

$\forall t \in \mathbb{R}_{0+}$ by applying such a relation between c_1 and c_3 in (32) and by taking into account that $c_1(\mu + \gamma - r_1) > 0$, $E(0) \ge 0$ and $e^{-r_1 t} - e^{-r_3 t} \ge 0$ $\forall t \in \mathbb{R}_{0+}$ since $r_1 < r_3$.

In this way, the non-negativity of $E(t)$ has been proven. From the second equation in (32), it follows that $c_3 H(r_3) = \sigma \beta_1 I(0) S(0) - c_1 H(r_1)$ and then:

$$S(t) = \frac{c_1 H(r_1)(e^{-r_1 t} - e^{-r_3 t}) + \sigma \beta_1 I(0) S(0) e^{-r_3 t}}{\sigma \beta_1 I(t)} \geq 0 \tag{35}$$

$\forall t \in \mathbb{R}_{0+}$ by applying such a relation between c_1 and c_3 in (32) and by taking into account that $c_1 H(r_1) > 0$ since $r_1 < \mu + \text{Min}\{\sigma, \gamma\}$, $S(0) \geq 0$ and, $I(t) \geq 0$ and $e^{-r_1 t} - e^{-r_3 t} \geq 0$ $\forall t \in \mathbb{R}_{0+}$ since $r_1 < r_3$. In this way, the non-negativity of $S(t)$ has been proven. Note that the function $H(v)$ in (33) is an upper-open parabola zero-valued for $v_1 = \mu + \sigma$ and $v_2 = \mu + \gamma$ so $H(r_1) > 0$ since $r_1 < \mu + \text{Min}\{\sigma, \gamma\}$.

(ii) On one hand, if the control law (20) is used instead of that in (12) then the time evolution of the infectious population is also given by (22) while the control action is active. Thus, $I(t) \to 0$ as $t \to \infty$ in (22) implies directly the existence of a finite time instant t_f at which the control (20) switches off. Obviously, the non-negativity of $I(t)$, $E(t)$ and $S(t)$ $\forall t \in [0, t_f]$ is proved by following the same reasoning used in the part (i) of the current theorem. The non-negativity of $R(t)$ $\forall t \in [0, t_f]$ is proven by using continuity arguments. In this sense, if $R(t)$ reaches negative values for some $t \in [0, t_f]$ starting from an initial condition $R(0) \geq 0$ then $R(t)$ passes through zero, i.e., there exists at least a time instant $t_0 \in [0, t_f)$ such that $R(t_0) = 0$. Then, it follows from (4) that:

$$\dot{R}(t_0) = \gamma I(t_0) + \mu N V(t_0) = \gamma I(t_0) + \frac{\mu \sigma \beta + \lambda_0 - \lambda_1(\mu + \gamma) + \lambda_2(\mu + \gamma)^2 - (\mu + \gamma)^3}{\sigma \beta} N$$
$$+ (\lambda_2 + \omega - 3\mu - \sigma - 2\gamma) S(t_0) + \sigma \frac{E(t_0) S(t_0)}{I(t_0)} - \frac{\beta}{N} I(t_0) S(t_0) \tag{36}$$
$$+ \frac{(\mu + \gamma)^2 + (2\mu + \sigma + \gamma)(\mu + \sigma) + \lambda_1 - \lambda_2(2\mu + \sigma + \gamma)}{\beta} N \frac{E(t_0)}{I(t_0)}$$

by introducing the control law (20), taking into account that $V(t) = u(t)$ and where the fact that $I(t_0) + E(t_0) + S(t_0) = N$, since $R(t_0) = 0$, has been used. Moreover, the non-negativity of $I(t)$, $E(t)$ and $S(t)$ $\forall t \in [0, t_f]$ as it has been previously proven, implies that $I(t_0) \leq N$, $E(t_0) \leq N$ and $S(t_0) \leq N$. Also, $I(t_0) \geq \delta > 0$ since $t_0 < t_f$ and from the definition of t_f in (21). Then, one obtains:

$$\dot{R}(t_0) \geq \gamma I(t_0) + \frac{\mu \sigma \beta + \lambda_0 - \lambda_1(\mu + \gamma) + \lambda_2(\mu + \gamma)^2 - (\mu + \gamma)^3}{\sigma \beta} N$$
$$+ (\lambda_2 + \omega - 3\mu - \sigma - 2\gamma - \beta) S(t_0) + \sigma \frac{E(t_0) S(t_0)}{I(t_0)} \tag{37}$$
$$+ \frac{(\mu + \gamma)^2 + (2\mu + \sigma + \gamma)(\mu + \sigma) + \lambda_1 - \lambda_2(2\mu + \sigma + \gamma)}{\beta} N \frac{E(t_0)}{I(t_0)}$$

from (36). The controller tuning parameter λ_i for $i \in \{0, 1, 2\}$ are related to the roots $(-r_j)$, for $j \in \{1, 2, 3\}$, of the closed-loop characteristic polynomial $P(s)$, see Remark 1, by:

$$\lambda_0 = r_1 r_2 r_3 \quad ; \quad \lambda_1 = r_1 r_2 + r_1 r_3 + r_2 r_3 \quad ; \quad \lambda_2 = r_1 + r_2 + r_3 . \tag{38}$$

Then, the assignment of r_j for $j \in \{1, 2, 3\}$ such that the constraints (a) and (b) are fulfilled implies that:

$$\begin{aligned} &\lambda_2 + \omega - 3\mu - \sigma - 2\gamma - \beta \geq 0 \\ &(\mu + \gamma)^2 + (2\mu + \sigma + \gamma)(\mu + \sigma) + \lambda_1 - \lambda_2(2\mu + \sigma + \gamma) \geq 0 . \\ &\mu\sigma\beta + \lambda_0 - \lambda_1(\mu + \gamma) + \lambda_2(\mu + \gamma)^2 - (\mu + \gamma)^3 = \mu\sigma\beta \geq 0 \end{aligned} \tag{39}$$

Then, $\dot{R}(t_0) \geq 0$ from (39) and (37). The facts that $R(t) \geq 0$ $\forall t \in [0, t_0)$, $R(t_0) = 0$ and $\dot{R}(t_0) \geq 0$ imply that $R(t) \geq 0$ $\forall t \in [0, t_f]$ via complete induction.

On the other hand, from (19) and (20), it follows:

$$\begin{aligned} u(t) = &\frac{\mu\sigma\beta - (\mu + \gamma)^3 + \lambda_0 - \lambda_1(\mu + \gamma) + \lambda_2(\mu + \gamma)^2}{\mu\sigma\beta} + \frac{\omega}{\mu N} R(t) - \frac{(3\mu + \sigma + 2\gamma - \omega - \lambda_2)}{\mu N} S(t) \\ &+ \frac{\sigma}{\mu N} \frac{E(t)S(t)}{I(t)} - \frac{\beta}{\mu N^2} I(t)S(t) + \frac{(\mu + \gamma)^2 + (2\mu + \sigma + \gamma)(\mu + \sigma) + \lambda_1 - \lambda_2(2\mu + \sigma + \gamma)}{\mu\beta} \frac{E(t)}{I(t)} \end{aligned} \tag{40}$$

$\forall t \in [0, t_f]$ by taking into account that $S(t) + E(t) + I(t) + R(t) = N$. Moreover:

$$\begin{aligned} u(t) \geq &\frac{\mu\sigma\beta - (\mu + \gamma)^3 + \lambda_0 - \lambda_1(\mu + \gamma) + \lambda_2(\mu + \gamma)^2}{\mu\sigma\beta} + \frac{\lambda_2 + \omega - 3\mu - \sigma - 2\gamma - \beta}{\mu N} S(t) \\ &+ \frac{(\mu + \gamma)^2 + (2\mu + \sigma + \gamma)(\mu + \sigma) + \lambda_1 - \lambda_2(2\mu + \sigma + \gamma)}{\mu\beta} \frac{E(t)}{I(t)} \qquad \forall t \in [0, t_f] \end{aligned} \tag{41}$$

where the facts that $0 < \delta \leq I(t) \leq N$, $E(t) \geq 0$, $S(t) \geq 0$ and $R(t) \geq 0$ $\forall t \in [0, t_f]$ have been used. If the roots of the polynomial $P(s)$ satisfy the conditions in (a) and (b), it follows from (41) that:

$$\begin{aligned} u(t) \geq &1 + \frac{\lambda_2 + \omega - 3\mu - \sigma - 2\gamma - \beta}{\mu N} S(t) \\ &+ \frac{(\mu + \gamma)^2 + (2\mu + \sigma + \gamma)(\mu + \sigma) + \lambda_1 - \lambda_2(2\mu + \sigma + \gamma)}{\mu\beta} \frac{E(t)}{I(t)} \geq 1 \end{aligned} \tag{42}$$

$\forall t \in [0, t_f]$ by taking into account (39) and the non-negativity of $S(t)$, $E(t)$ and $I(t)$ $\forall t \in [0, t_f]$. Finally, it follows that $u(t) \geq 0$ $\forall t \in \mathbb{R}_0^+$ from (20) and (42).   ***

4   Simulation Results

An example based on an outbreak of influenza in a British boarding school in early 1978 [2] is used to illustrate the presented theoretical results. Such an epidemic can be described by the SEIR mathematical model (1)-(4) with $\mu^{-1} = 70$ years $= 25550$ days, $\beta = 1.66$ per day, $\sigma^{-1} = \gamma^{-1} = 2.2$ days and $\omega^{-1} = 15$ days. A total population of $N = 1000$ boys is considered with the initial condition given by $S(0) = 800$ boys, $E(0) = 100$ boys, $I(0) = 60$ boys and $R(0) = 40$ boys. Two sets of simulation results are presented to compare the evolution of the SEIR model populations in two different situations, namely: when no vaccination control actions are applied and if a control action based on the feedback input-output linearization approach is applied.

4.1   Epidemic Evolution without Vaccination

The time evolution of the respective populations is displayed in Fig. 1. The model tends to its endemic equilibrium point as time tends to infinity. There are susceptible, infected and infectious populations at such an equilibrium point. As a consequence, a vaccination control action has to be applied in order to eradicate the epidemics.

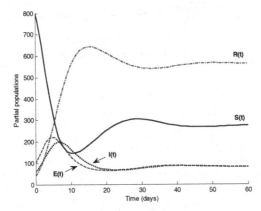

Fig. 1. Time evolution of the individual populations without vaccination

4.2   Epidemic Evolution with a Feedback Control Law

The control law given by (20)-(21) is applied with $\delta = 0.001$ and the free-design controller parameters λ_i, for $i \in \{0, 1, 2\}$, being chosen so that the roots of the characteristic polynomial $P(s)$ associated with the closed-loop dynamics (13) are $-r_1 = -\gamma$, $-r_2 = -(\mu + \gamma)$ and $-r_3 = -(2\mu + \gamma)$. Such values for λ_i are obtained from (38). The time evolution of the respective populations is displayed in Fig. 2 and the vaccination function in Fig. 3.

The vaccination control action achieves the control objectives as it is seen in Fig. 2. The infection is eradicated from the population since both infectious and infected populations converge rapidly to zero. Also, the susceptible population converges to zero while the removed-by-immunity population tracks asymptotically the whole population as time tends to infinity. Such a result is coherent with the result proved in Theorem 3. Moreover, the positivity of the system is maintained for all time as it can be seen from such figures. Such a property is satisfied although all constraints of the assumption (b) of Theorem 3 are not fulfilled by the system parameters and the chosen control parameters. However, such a result is coherent since such constraints are sufficient but not necessary to prove the positivity of the system. The switched off time instant for the vaccination is $t_f \approx 30$ days .

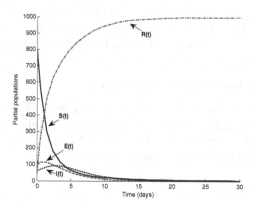

Fig. 2. Time evolution of the individual populations with the vaccination control action

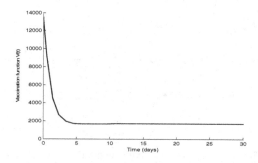

Fig. 3. Time evolution of the vaccination function

The time evolution of the respective partial populations under the application of the developed control strategy is similar to that obtained under the use of other vaccination strategies proposed in [1]. The main novelty of the current research is the use of a systematic method to design a vaccination strategy based on a control technique for input-output linearization of the SEIR epidemic model by exact state feedback.

5   Conclusions

A vaccination control strategy based on feedback input-output linearization techniques has been proposed to fight against the propagation of epidemic diseases. A SEIR model with known parameters is used to describe the propagation of the disease. The stability and the positivity properties of the closed-loop system as well as the eradication of the epidemics have been proved. *Such a strategy has a main drawback, namely, the control law needs the knowledge of the true values of the susceptible, infected and infectious populations at all time instants which may not be available in certain real situations.* However, such a drawback can be overcome by adding an observer to estimate online all the partial populations.

Acknowledgements. The authors thank to the Spanish Ministry of Education by its support through Grant DPI2009-07197 and to the Basque Government by its support through Grants IT378-10, SAIOTEK SPE07UN04 and SAIOTEK SPE09UN12.

References

1. De la Sen, M., Alonso-Quesada, S.: On Vaccination Control Tools for a General SEIR-Epidemic Model. In: 18th Mediterranean Conference on Control & Automation, MED 2010, pp. 1322–1328 (2010), doi:10.1109/MED.2010.5547865
2. Keeling, M.J., Rohani, P.: Modeling Infectious Diseases in Humans and Animals. Princeton University Press, Princeton and Oxford (2008)
3. Li, M.Y., Graef, J.R., Wang, L., Karsai, J.: Global Dynamics of a SEIR Model with Varying Total Population Size. Mathematical Biosciences 160, 191–213 (1999)
4. Makinde, O.D.: Adomian Decomposition Approach to a SIR Epidemic Model with Constant Vaccination Strategy. Applied Mathematics and Computation 184, 842–848 (2007)
5. Mollison, D.: Epidemic Models: Their Structure and Relation to Data. Publications of the Newton Institute, Cambridge University Press (2003)
6. De la Sen, M., Agarwal, R. P., Ibeas, A., Alonso-Quesada, S.: On the Existence of Equilibrium Points, Boundedness, Oscillating Behavior and Positivity of a SVEIRS Epidemic Model Under Constant and Impulsive Vaccination. Advances in Difference Equations 2011, Article ID 748608, 32 pages (2011), doi:10.1155/2011/748608
7. Song, X.Y., Jiang, Y., Wei, H.M.: Analysis of a Saturation Incidence SVEIRS Epidemic Model with Pulse and Two Time Delays. Applied Mathematics and Computation 214, 381–390 (2009)
8. Jumpen, W., Orankitjaroen, S., Boonkrong, P., Wiwatanapataphee, B.: SEIQR-SIS Epidemic Network Model and Its Stability. International Journal of Mathematics and Computers in Simulation 5, 326–333 (2011)
9. Safi, M.A., Gumel, A.B.: Mathematical Analysis of a Disease Transmission Model with Quarantine, Isolation and an Imperfect Vaccine. Computers and Mathematics with Applications 61, 3044–3070 (2011)
10. De la Sen, M., Agarwal, R.P., Ibeas, A., Alonso-Quesada, S.: On a Generalized Time-Varying SEIR Epidemic Model with Mixed Point and Distributed Time-Varying Delays and Combined Regular and Impulsive Vaccination Controls. Advances in Difference Equations 2010, Article ID 281612, 42 pages (2010), doi:10.1155/2010/281612
11. Zhang, T.L., Liu, J.L., Teng, Z.D.: Dynamic Behaviour for a Nonautonomous SIRS Epidemic Model with Distributed Delays. Applied Mathematics and Computation 214, 624–631 (2009)
12. Xu, R., Ma, Z., Wang, Z.: Global Stability of a Delayed SIRS Epidemic Model with Saturation Incidence and Temporary Immunity. Computers and Mathematics with Applications 59, 3211–3221 (2010)
13. Mukhopadhyay, B., Bhattacharyya, R.: Existence of Epidemic Waves in a Disease Transmission Model with Two-Habitat Population. International Journal of Systems Science 38, 699–707 (2007)
14. Isidori, A.: Nonlinear Control Systems. Springer, London (1995)
15. Fulton, W., Harris, J.D.: Representation Theory. Springer, New York (1991)

Medical Imaging as a Bone Quality Determinant and Strength Indicator of the Femoral Neck

Alexander Tsouknidas[1,*], Nikolaos Michailidis[1], and Kleovoulos Anagnostidis[2]

[1] Physical Metallurgy Laboratory, Mechanical Engineering Department,
Aristoteles University of Thessaloniki, Thessaloniki, Greece
alextso@auth.gr, nmichail@eng.auth.gr
[2] 3rd Orthopaedic Department "Papageorgiou" General Hospital,
Aristoteles University of Thessaloniki, Thessaloniki, Greece
kanagn@auth.gr

Abstract. Early diagnosis of osteoporosis is a key factor in preventive medicine of this clinically silent bone pathology. The most severe manifestation of osteoporotic bone loss is encountered in hip fractures and therfore, this study represents an effort in associating bone quality of the femoral neck region to fragility fracture risks through FEA supported imaging techniques. The concepts of Magnetic resonance imaging (MRI), Computer Tomography (CT) and Dual-energy X-ray absorptiometry (DXA) are introduced, along with their limitations in defining bone quality and calculating the apparent bone strength. As DXA dominates surgeons' preference in evaluating bone mineral density in the hip region, in vivo measurements of this method, sustained by ex-vivo uniaxial compression tests and FEA supported calculations are employed to determine a fracture risk indicator of the femoral neck versus bone mineral density (BMD).

Keywords: Medical imaging, Osteoporosis, Femoral neck, FEA, Fracture risk.

1 Introduction

Medical imaging has revolutionalized modern medicine, providing an accurate but foremost non-invasive overview of the patient's condition. This fosters, preoperative recognition of abnormalities or determination of the progression of pathogenesis like osteoporosis.

Osteoporosis is a multifactorial bone disease concerning roughly 4% of the human population [1]. Due to its high morbidity and global nature, osteoporosis is considered a pathology with a significant socioeconomic impact [2]. As an asymptomatic condition, osteoporosis fails to exhibit noticeable symptoms, particularly at early stages and thus is usually underdiagnosed. Untreated however, this clinically silent disease, is likely to increase the risk of fragility fractures [3,4,5].

An osteoporotic patient's BMD is drastically reduced, deteriorating the bones' micostructural characteristics as a result of excessive bone resorption followed by

---

* Corresponding author.

J. Gabriel et al. (Eds.): BIOSTEC 2012, CCIS 357, pp. 210–221, 2013.

insufficient bone formation during remodeling [6,7]. The pathogenesis has been associated to dietary aspects [8], immobilization [9], hyper-parathyroidism [10], vitamin D deficiency [11], alteration of biochemical markers like hormone [12,13] and aging [14]. Regardless etiology, decreased bone mineral density renders the skeletal system susceptibility to fracture, predominantly occurring at the hip [15], the vertebral column [16] and wrist [17].

Early diagnosis allows for tertiary prevention, thus reducing the progression and restricting osteoporosis-related complications. Several methods have been introduced to act as a screening process, aiming to identify individuals who exhibit early signs of loss in BMD and thus demonstrate osteopenia or osteoporosis. The dominating techniques however, due to their nonintrusive nature, are DXA, CT and MRI [18,19]. Techniques like peripheral quantitative computed tomography (pQCT) may also be accurate in measuring BMD at peripheral skeletal sites, exhibit however restrictions that prohibit measurements at the proximal femur [20,21]. According to the World Health Organization, osteopenia and osteoporosis are defined by the patient's bone mass deviation, when compared to that of an average, young and healthy adult [22].

Even though DXA can accurately determine the minerals and lean soft tissue of the examined area, the overall accuracy of the measurement is impaired by the subtraction of the indirectly calculated fat mass [23]. Furthermore, DXA results are represented as mass per area, thus not considering the anisotropy of the bone tissue. This renders DXA as a quantitative and not qualitative index of the bone structure [24] and thus associating DXA measurements to the apparent fracture risk, remains a complex problem requiring heuristic data and FEA supported calculations.

In contrast to these restrictions, CT and MRI measurements can be used to reconstruct an accurate volumetric data set of the examined bone structure, that can be used further on as input to FEA software in order to calculate the strength characteristics of the anatomy. This approach however, requires extensive data processing, thus being rather time consuming.

In the present paper three noninvasive medical imaging techniques (DXA, CT and MRI), with potential applications in correlating BMD in the femoral neck to the fracture risk, will be examined. Their concepts strengths and limitations will be introduced, followed by FEA simulations that can be employed, directly (CT and MRI) or indirectly (DXA), as an indicator of fragility fracture risks.

2   Image Based Reconstruction of Bone Tissue to Determine Its Strength Characteristics

2.1   CT Based Reconstruction

CT is capable of producing 2D images of various body structures, based on their ability to withstand the emitted X-radiation. As bone has a unique spectrum of X-ray permeability within the human body (ranging from 200 to 2000), it shades white in a CT slice, thus allowing its relatively unhindered segmentation of the bone from soft tissue, since no other body part exhibits overlapping CT numbers [25]. This results in a 2D outline of the scanned bone and the 3D data set, is generated by overlaying consecutive measurements. Contemporary CT units facilitate slice spacing in the magnitude of 0.5mm, or even lower [26] while data acquisition follows DICOM (Digital Imaging and Communications in Medicine). DICOM is the standard communication protocol for structuring and encoding medical reports [27] used by commercial image manipulation software, required for converting multiple 2D images into a 3D one.

After the representation of the model's outer surfaces, cortical and cancellous bone properties have to be assigned. This can be directly achieved within the segmentation software through accurate distinction of the two types based on their CT spectrum or by considering both, the inner and outer cortex of the cortical bone.

A reverse engineered cancellous bone model is illustrated in figure 1.

Consecutive measurements   CT reconstructed geometry

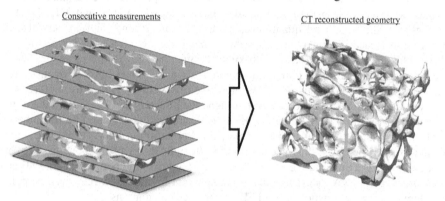

Fig. 1. micro-CT based reconstruction of a cancellous bone model

2.2   MRI Based Reconstruction

MRI scanning is based on atoms magnetization rather than ionizing radiation used by CT. This is particularly useful when trying to encapture tissue with many hydrogen nuclei and little density contrast thus providing high accuracy even for soft tissue images [28]. It is however a disadvantage, when attempting to reconstruct skeletal models, as bone cannot generate a magnetic resonance signal, due to its severely short ^{1}H transverse relaxation times. Therefore the cortical bone is indirectly approximated through the signal generated from the surrounding soft tissue which in most cases leads to accurate results. This is even more complex, when attempting to reconstruct cancellous bone, which must be considered through the reconstruction of the medullar canal, which reproduces a notable contrast within the image [29]. This however, requires extreme caution and experience in cases where interposing soft tissue is present.

Alike to CT, data encoding in MRI, follows DICOM and the basic concept of reconstructing 3D geometries remains similar [30] following two distinctive steps, thresh-holding of the desired spectrums, to determine the bone contour within every scan and overlaying of those to constitute the 3D geometry. The slice spacing however is slightly distanced when compared to CT imaging techniques [31]. This imposes a further restriction as high accuracy models, like the one presented in figure 1, as MRI models facilitate precise representation of cortical bone, whereas cancellous bone must be considered as a homogene material with anisotropic behavior.

2.3   Model Accuracy and Strength Characteristics

The accuracy of both methods, has been quantitatively validated [32] and CT seems to be slightly more accurate in the reconstruction of skeletal characteristics, especially when examining longer bones.

The reverse engineered models can be employed to determine the strength characteristics of the examined area through FEA simulations, indicating apparent fracture risks [33]. The main advantage of these methods is associated to the high degree of geometrical customization, as this approach considers patient specific characteristics.

Figure 2 (left side) indicates the compressive response of a cancellous bone sample [34] under compressive load at a 2.5% strain. The porosity of the sample amounted to 88% and an apparent density to 0.247 gr/cm^3. These calculations ease the determination of critical areas, where strut buckling will gradually deteriorate the strength characteristics of the sample ultimately causing fractures. It is however notable that CT reconstructed models require extensive data manipulation and simulation expertise to acquire the desired result, while each model is highly customized and thus not applicable to anyone but the examined individual.

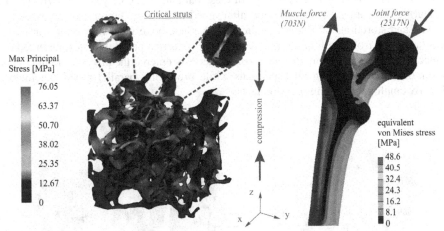

Fig. 2. FEA calculated strength characteristics of a CT reconstructed bone sample and MRI reconstructed femur

The right side of the figure 2, indicates the stress distribution on a MRI reverse engineered human femur. The model was bound at a lower section of the femur and loaded by a 2317N joint force (inclined by 24° to the coronal plane and 6° to the sagittal one) and an abductor muscle force of 703N (inclined by 28° to the coronal plane and 15° to the sagittal one).

Such a model can indicate the apparent fracture risk in the femoral neck area, and even though MRI is widely considered as a state of the art imaging technique, this approach is subject to further restrictions. These limitations are not associated to the geometrical characteristics of the described anatomy, but rather to the mechanical properties of the reverse engineered tissue. As MRI reconstructed models merely represent cortical/cancellous bone allocation and not the micostructural characteristics of trabecular bone, they have to consider bulk material properties for the cancellous bone. As the mechanical properties of bone vary significantly among the human population, a model not considering these patient specific strength characteristics, may prove to be highly inaccurate.

3   DXA Strength Indicator of the Femoral Neck

3.1   Materials and Methods

This part of the study, was conducted on femoral neck samples, harvested from patients undergoing total hip replacement due to osteoarthritis. In order to determine the samples' structural integrity, standard X-rays (anterior- posterior) of the pelvis were taken preoperatively in all cases. Patients with a sort femoral neck, large cysts in neck region or previous surgeries in proximal femur were excluded from the study.

Overall 30 patients (27 female and 3 male) were considered as representative candidates for this study and thus subjected to DXA, to catalogue their proximal femur bone mineral density. The average age of these patients was 63.7 years (57- 76 years).

During the surgical procedure and after a 45° osteotomy, femoral heads were removed and stored at -60°C until evaluation. A plane bone slice with 6mm thickness was harvested from the femoral neck (see figure 3) as two parallel blades, mounted on a mechanical saw at a 6mm distance, simultaneously entered the proximal femur. This ensured similarity among all specimens while producing parallel piped specimens, directly employable in compression tests.

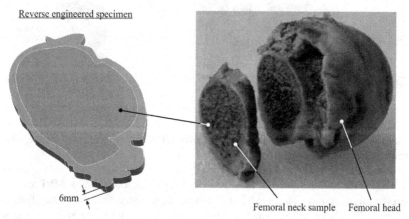

Reverse engineered specimen

6mm

Femoral neck sample   Femoral head

Fig. 3. Considered bone specimen and reverse engineered model

Mechanical testing was performed on an electric INSTRON Testing system. To determine the specimens' strength characteristics, all samples were subjected to uniaxial compression, until failure. A cross-head traveling speed of 0.6mm/s was selected and the maximum travelling distance (upon contact) was set to 5mm in order to avoid contact of the moving cross-head and the fixed base plate. To reduce friction, the sample-actuator contact areas were lubricated. The displacement of the cross-head was measured by means of an inductive sensor, at an accuracy of 1 μm.

The biomechanical parameters were correlated with BMD using the Pearson correlation coefficient (r) and a linear regression model.

30 experiments were conducted to determine both, compressive yield strength and modulus of elasticity and associate these to the DXA determined T-scores. The T-score compares the measured BMD to that of a young adult (at the age of 35) of the same gender with peak bone mass, while considering statistical values.

3.2   Results

A correlation of characteristic and mean values (BMD and T-score) determined by DXA measurements, to the corresponding mechanical properties (yield stress and elastic modulus) of the examined specimens are reflected in figure 4. These values are in good coherence with previously presented data [35, 36]. The offset in the determined values can be attributed to the different sampling sites and techniques of the compared studies.

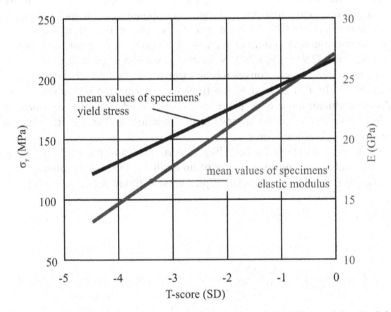

Fig. 4. Equivalent T-score values versus yield stress σy and elasticity modulus (p<0.001)

The introduced experimental investigation revealed a significant correlation of BMD to the mechanical properties of the femoral neck. A strong enslavement of the ultimate material strength to BMD (r=0.838) was found, while the correlation to elastic modulus, calculated based on the linear elastic region of the determined stress-strain curves [37], was weaker (r= 0.689).

A limitation however of the introduced process, is based on the assumption of the material's isotropy and the determination of universal properties of a bone segment comprising of both, cortical and cancellous tissue. This methodology was adopted, as DXA measurements reflect a combined BMA encapturing both bone types by default and thus the assumption of a compound material is beneficiary to the approach.

3.3   FEA Simulation

In order to associate the ultimate compression strength of the samples, to fragility fracture risks of the femoral neck, the geometry of the specimens was reverse engineered and employed in a linear elastic simulation of a gait type loading scenario considering combined multiaxial forces [38].

During the simulation, the specimens were once again considered as a uniform-isotropic material, comprising of cortical and cancellous bone tissue, to directly facilitate the correlation of the DXA measurements to the fracture risk of the femoral neck. The experimentally determined mechanical properties were adopted as bulk properties of the compound material and assigned as such in the simulation. The Poisson ratio was assigned as 0,3 corresponding to a mean value of cortical and cancellous bone [39,40] regardless DXA value.

The acting loads on the femur, comprised of a 2317N joint force [41], evenly distributed over the femoral head (inclined by 24° to the frontal plane and 6° to the sagittal one). This force was remotely applied on the upper surface of the reverse engineered specimens at a distance of 46mm corresponding to the mean distance from the tip of the femoral head at which the specimens were severed from the femur. This, based on the coordination system of the model, resulted in a vector force comprising of F_x= 689N, F_y= 942N and F_z= 2001N for axis x, y and z respectively.

The abductor muscle was considered as inactive, as this muscle force acts during the lift up of the foot, thus loading the trochanter during the relaxation of the joint force. As the abductor muscle force has been documented to amount to approximately 703N, the worst case scenario during normal loading of the femur, relates to the aforementioned 2317N joint force.

The acting force and boundary conditions were chosen to mimic the average loading history encountered during walking of an adult human, corresponding to 10.000 daily cycles [41] and are schematically represented in figure 5.

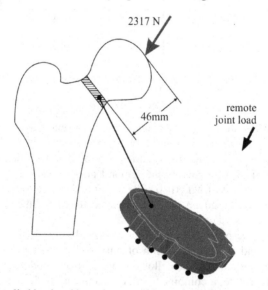

Fig. 5. Applied load and boundary conditions of the developed FEA model

There exists skepticism concerning the ability of compression tests in predicting the hip fracture risk, as fractures in the hip region are the effect of complex dynamic force application, comprising of shear, tension and compression. Based on the forgoing description of the model, it becomes evident that the conducted compression experiments encapture the loading scenario in a realistic manner, as the compressive strength of the femoral neck exerts a dominant impact on the structural integrity of the femur. Furthermore, the compression tests were identically performed in all cases while the only variation between samples was based on the bone mineral density.

A characteristic stress field developing on a femoral neck sample (T-score=-4.47, σ_y=109.448MPa and E=12.6GPa) is demonstrated in figure 6.

equivalent von Mises stress [Pa]

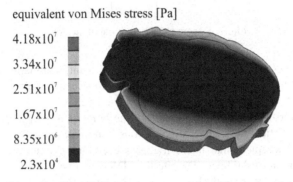

4.18×10^7	
3.34×10^7	
2.51×10^7	
1.67×10^7	
8.35×10^6	
2.3×10^4	

Fig. 6. Calculated stress field on a reverse engineered femoral neck sample

4   Discussion

Musculoskeletal models, accurately describing features of the human body are constantly evolving, improving their ability to mimic structural characteristics, functions or the mobility of the anatomy they are simulating. Medical imaging techniques allow the customized development of such models based on the individual patient's characteristics. Even though, this approach can provide highly accurate results, it should not be considered the method of choice when assessing the fracture risk of the femoral neck due to osteoporosis, mainly due to the reason that simpler methods (i.e. DXA) can provide similar results as demonstrated in the previous paragraphs.

DXA scans in the hip region, are conventionally performed in the trochanter, the Ward's triangle and the femoral neck (in an orthogonal area of 6 by 10mm). The present paper, correlates the BMD obtained from such measurements, to the mechanical strength characteristics of the examined area, as to provide surgeons with a DXA based risk assessment, concerning fragility fractures.

There exists a consensus throughout literature, that bone density can be considered as a strong independent predictor of failure strength [42]. By overlaying fatigue stress value with the experimentally determined fracture strength of the examined specimens, a correlation between T-score and fracture risk can be determined as demonstrated in figure 7.

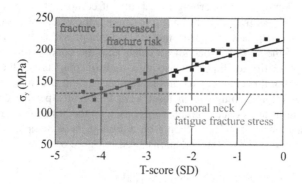

Fig. 7. T-score as a fracture risk indicator

Even though DXA is a cost efficient BMD determinant, dominating the preference of surgeons due to its simplicity, there are some limitations associated to the method that may affect the accuracy of the introduced procedure.

As DXA quantifies the bone mass and not the bone quality of a specific site, micro-fractures in vertical trabeculae of cancellous bone will maintain undetected. It is however widely accepted, that micro-fractures exert an important influence on the mechanical strength of the bone. Despite this, DXA can be treated as a macroscopically indicator of bone strength. Especially in the hip region, where gait like loading ensures constant remodeling and thus the probability of micro fractures is considered as rather low.

Another possible limitation of our study is associated to the patients, the samples were harvested from, as all of them were diagnosed with osteoarthritis. This might have a twofold effect on the BMD-bone properties correlation.

Primary, it has not been established if the most common musculoskeletal disorders of the elderly (osteoarthritisand osteoporosis) may be treated as independent, studies have shown that the presence of one disease may act protective against the other [43,44]. The effect however of this on the presented results, can be neglected as the selected patients exhibited significant differences in terms of BMD.

Secondary, osteoarthritis has been associated to subchondral scleroses in femoral head; the femoral neck and the trochander region however, are rarely affected by the condition [45]. In order to circumvent this aspect, our methodology considered DXA scans in femoral neck, trochanter and Ward's triangle and was determinedas reliable. Additionally, osteoarthritic patients undergoing total hip arthroplasty were the only group of patients from whom, we could receive bone samples from the femoral neck region.

Studies have indicated that the femur carries a 30% of the applied loads in the subcapital region, while the base of the neck is subjected by 96% of the total load [46]. This strengthens the vital role of the femoral neck's capacity to transmit the compressive stress from the joint to the shaft of the femur. Although the etiology of osteoporotic hip fracture is complex and multifactorial [47,48], bone quality is, without a doubt, a major risk factor.

5   Conclusions

Bone mineral density measured by DXA, regardless limitations associated to the technique's ability to encapture bone quality, is a strong predictor of bone strength in the femoral neck region. Supported by an adequate FEA simulation, DXA may be regarded as a valuable tool during the prediction of BMD spectrums which present a significant risk of fragility fractures.

References

1. Melton III, L.J., Chrischilles, E.A., Cooper, C., Lane, A.W., Riggs, B.L.: Perspective. How many women have osteoporosis? Journal of Bone and Mineral Research 7, 1005–1010 (1992)
2. Ray, N.F., Chan, J.K., Thamer, M., Melton, L.J.: Medical expenditures for the treatment of osteoporotic fractures in the United States in 1995: Report from the National Osteoporosis Foundation. Journal of Bone and Mineral Research 12, 24–35 (1997)
3. Ettinger, B.: A personal perspective on fracture risk assessment tools. Menopause. The Journal of The North American Menopause Society 15(5), 1023–1026 (2008)
4. Rockwood, P.R., Horne, J.G., Cryer, C.: Hip Fractures: a future epidemic? Journal of Orthopaedic Trauma 4, 388–396 (1990)
5. Cooper, C., Campion, G., Melton, L.J.: Hip fractures in the elderly: A word-wide projection. Osteoporosis International 2, 285–289 (1992)
6. Frost, H.M., Thomas, C.C.: Bone Remodeling Dynamics. Springfield, IL (1963)
7. Raisz, L.: Pathogenesis of osteoporosis: concepts, conflicts, and prospects. Journal of Clinical Investigation 115(12), 3318–3325 (2005)
8. Hackett, E.S., MacLeay, J.M., Green, M., Enns, R.M., Pechey, C.L., Les, C.M., Turner, A.S.: Femoral Cortical Bone Mineral Density and Biomechanical. Properties in Sheep Consuming an Acidifying Diet. Nutrition and Metabolic Insights. 1, 11–16 (2009)
9. Minaire, P.: Immobilization osteoporosis: a review. Clinical Rheumatology 8(2), 95–103 (1989)
10. Dupree, K., Dobs, A.: Osteopenia and Male Hypogonadism. Reviews in Urology 6(6), S30–S34 (2004)
11. Holick, M.F.: Vitamin D: importance in the prevention of cancers, type 1 diabetes, heart disease, and osteoporosis. American Journal of Clinical Nutrition 79(3), 362–371 (2004)
12. Parfitt, A.M., Villanueva, A.R., Foldes, J., Rao, D.S.: Relations between histologic indices of bone formation: implications for the pathogenesis of spinal osteoporosis. Journal of Bone and Mineral Research 10(3), 466–473 (1995)
13. Black, D.M., Greenspan, S.L., Ensrud, K.E., Palermo, L., McGowan, J.A., Lang, T.F., Garnero, P., Bouxsein, M.L., Bilezikian, J.P., Rosen, C.J.: The effects of parathyroid hormone and alendronate alone or in combination in postmenopausal osteoporosis. The New England Journal of Medicine 349(13), 1207–1215 (2003)
14. Newton-John, H.F., Morgan, D.B.: The loss of bone with age, osteoporosis, and fractures. Clinical Orthopaedics and Related Research 71, 229–252 (1970)
15. Bohr, H., Schaadt, O.: Bone mineral content of the femoral neck and shaft: relation between cortical and trabecular bone. Calcified Tissue International 37, 340–344 (1985)
16. Old, J.L., Calvert, M.: Vertebral compression fractures in the elderly. American Family Physician 69(1), 111–116 (2004)

17. Dempster, D.W.: Osteoporosis and the burden of osteoporosis-related fractures. American Journal of Managed Care 17(6), S164–S169 (2011)
18. Genant, H.K., Engelke, K., Fuerst, T., Gluer, C.C., Gramp, S., Harris, S.T., et al.: Noninvasive assessment of bone mineral and structure: State of the art. Journal of Bone and Mineral Research 11, 707–730 (1996)
19. Braun, M.J., Meta, M.D., Schneider, P., Reiners, C.: Clinical evaluation of a high-resolution new peripheral quantitative computerized tomography (pQCT) scanner for the bone densitometry at the lower limb. Physics in Medicine and Biology 43, 2279–2294 (1998)
20. Augat, P., Reeb, H., Claes, L.E.: Prediction of fracture load at different skeletal sites by geometric properties of the cortical shell. Journal of Bone and Mineral Research 11, 1356–1363 (1996)
21. Augat, P., Fan, B., Lane, N.E., Lang, T.F., LeHir, P., Lu, Y., Uffmann, M., Genant, H.K.: Assesment of bone mineral at appendicular sites in females with fractures of the proximal femur. Bone 22, 395–402 (1998)
22. World Health Organisation. Assessment of fracture risk and its application to screening for postmenopausal osteoporosis. Technical Report Series. WHO, Geneva (1994)
23. St-Onge, M.P., Wang, J., Shen, W., Wang, Z., Allison, D.B., Heshka, S., Pierson, R.N., Heymsfield, S.B.: Dual-Energy X-Ray Absorptiometry-Measured Lean Soft Tissue Mass: Differing Relation to Body Cell Mass Across the Adult Life Span. Journals of Gerontology Series A: Biological Sciences and Medical Sciences 59(8), 796–800 (2004)
24. Lochmuller, E.M., Miller, P., Burklein, D., Wehr, U., Rambeck, W., Eckstein, F.: In situ femoral DXA related to ash weight, bone size and density and its relationship with mechanical failure loads of the proximal femur. Osteoporosis International 11, 361–367 (2000)
25. Winder, J., Bibb, R.: Medical Rapid Prototyping Technologies: State of the Art and Current Limitations for Application in Oral and Maxillofacial Surgery. Journal of Oral and Maxillofacial Surgery 63(7), 1006–1015 (2005)
26. Schmutz, B., Wullschleger, M.E., Schuetz, M.A.: The effect of CT slice spacing on the geometry of 3D models. In: Proceedings 6th Australasian Biomechanics Conference. The University of Auckland, Auckland (2007)
27. Riesmeier, J., Eichelberg, M., Kleber, K., Oosterwijk, H., von Gehlen, S., Grönemeyer, D.H.W., Jensch, P.: DICOM Structured Reporting—a prototype implementation. International Congress Series (Computer Assisted Radiology and Surgery) 1230, 795–800 (2001)
28. Tuncbilek, N., Karakas, H.M., Okten, O.O.: Dynamic contrast enhanced MRI in the differential diagnosis of soft tissue tumors. European Journal of Radiology 53, 500–505 (2005)
29. Rathnayaka, K., Coulthard, A., Momot, K., Volp, A., Sahama, T., Schuetz, M., et al.: Improved image contrast of the bone-muscle interface with 3 T MRI compared to 1.5 T MRI. In: 6th World Congress on Biomechanics, Singapore (2010)
30. Chirani, R.A., Jacq, J.J., Meriot, F., Roux, C.: Temporomandibular joint: a methodology of magnetic resonance imaging 3-D reconstruction. Oral Surgery, Oral Medicine, Oral Pathology, Oral Radiology & Endodontics 97(6), 756–761 (2004)
31. Drapikowski, P.: Surface modeling—uncertainty estimation and visualization. Computerized Medical Imaging and Graphics 32(2), 134–139 (2008)
32. Rathnayaka, K., Momot, K.I., Noser, H., Volp, A., Schuetz, M.A., Sahama, T., Schmutz, B.: Quantification of the accuracy of MRI generated 3D models of long bones compared to CT generated 3D models. Medical Engineering & Physics 34(3), 357–363 (2012)
33. Tsouknidas, A., Maropoulos, S., Savvakis, S., Michailidis, N.: FEM assisted evaluation of PMMA and Ti6Al4V as materials for cranioplasty resulting mechanical behaviour and the neurocranial protection. Bio-Medical Materials and Engineering 21(3), 139–147 (2011)

34. Savvakis, S., Maliaris, G., Tsouknidas, A.: Correlation of trabecular bone structure to its stress-strain dependent biomechanical response. In: ESB 2012: 18th Congress of the European Society of Biomechanics, Lisbon, Portugal, July 1-4 (2012) (in press)

35. Keller, T.S., Mao, Z., Spengler, D.M.: Young's modulus, bending strength and tissue physical properties of human compact bone. Journal of Orthopaedic Research 8, 592–603 (1990)

36. Reilly, D.T., Burstein, A.H.: The elastic and ultimate properties of compact bone tissue. Journal of Biomechanics 8, 393–405 (1997)

37. Turner, C.H., Burr, D.B.: Basic biomechanical measurements of bone: a tutorial. Bone 14, 595–608 (1993)

38. Jacobs, C.R., Simo, J.C., Beaupre, G.S., Carter, D.R.: Adaptive bone remodeling incorporating simountaneous density and anisotropy considerations. Journal of Biomechanics 30(6), 603–613 (1997)

39. Lu, Y.M., Hutton, W.C., Gharpuray, V.M.: Do bending, twisting and diurnal fluid change in the disc affect the propensity to prolapse? A viscoelastic finite element model. Spine 21, 2570–2579 (1996)

40. Smit, T.H., Odgaard, A., Schneider, E.: Structure and function of vertebral trabecular bone. Spine 22, 2823–2833 (1997)

41. Sarikat, M., Yildiz, H.: Determination of bone density distribution in proximal femur by using the 3D orthotropic bone adaption model. Proceedings of the Institute of Mechanical Engineering, Part H: Journal of Engineering in Medicine 225, 365–375 (2011)

42. Stankewich, C.L., Chapman, J., Muthusamy, R.: Relationship of mechanical factors to the strength of proximal femur fractures fixed with cancellous screws. Journal of Orthopaedic Trauma 10, 248–257 (1996)

43. Solomon, L., Schnitzler, C.M., Browett, J.P.: Osteoarthritis of the hip: the patient behind the disease. Annals of the Rheumatic Diseases 41, 118–125 (1982)

44. Cooper, C., Cook, P.L., Osmond, C., Cawley, M.I.D.: Osteoarthritis of the hip and osteoporosis of the proximal femur. Annals of the Rheumatic Diseases 50, 540–542 (1991)

45. Li, B., Aspden, R.M.: Material properties of bone from the femoral neck and calcarfemorale of patients with osteoporosis or osteoarthritis. Osteoporosis International 7, 450–456 (1997)

46. Lotz, J.C., Cheal, E.J., Hayes, W.C.: Stress distributions within the proximal femur during gait and falls: implications for osteoporotic fracture. Osteoporosis International 5, 252–261 (1995)

47. Melton, L.J., Riggs, B.L.: Risk factors for injury after a fall. Symposium on falls in the elderly: biological and behavioral aspects. Clinics in Geriatric Medicine 1, 525–539 (1985)

48. Greenspan, S.L., Myers, E.R., Maitland, L.A., Resnick, N.M., Hayes, W.C.: Fall severity and bone mineral density as risk factor for hip fracture in ambulatory elderly. Journal of the American Medical Association 271, 128–133 (1994)

Parallel GPGPU Evaluation of Small Angle X-Ray Scattering Profiles in a Markov Chain Monte Carlo Framework

Lubomir D. Antonov*, Christian Andreetta*, and Thomas Hamelryck

The Bioinformatics Section, Department of Biology, University of Copenhagen, Denmark
thamelry@binf.ku.dk

Abstract. Inference of protein structure from experimental data is of crucial interest in science, medicine and biotechnology. Low-resolution methods, such as small angle X-ray scattering (SAXS), play a major role in investigating important biological questions regarding the structure of proteins in solution.

To infer protein structure from SAXS data, it is necessary to calculate the expected experimental observations given a protein structure, by making use of a so-called forward model. This calculation needs to be performed many times during a conformational search. Therefore, computational efficiency directly determines the complexity of the systems that can be explored.

We present an efficient implementation of the forward model for SAXS with full hardware utilization of Graphics Processor Units (GPUs). The proposed algorithm is orders of magnitude faster than an efficient CPU implementation, and implements a caching procedure employed in the partial forward model evaluations within a Markov chain Monte Carlo framework.

Keywords: SAXS, GPU, GPGPU, MCMC, Protein Structure Determination, OpenCL.

1 Introduction

Proteins play a crucial role in science, medicine and biotechnology: without them, cellular activities such as catalysis, signaling and regulation would be next to impossible. Protein function is determined by protein structure, which has been proven to be determined by the amino acid sequence [1].

Despite encouraging improvements, determining the ensemble of possible conformations in solution is far from an accomplished goal. High resolution experimental methods, notably X-ray crystallography and Nuclear Magnetic Resonance (NMR), can only partially provide information on such ensembles, and encounter several limitations in fully describing the flexibility of large systems in physiological conditions [2].

Low resolution methods, on the other hand, can more easily provide information on such ensembles. In particular, Small Angle X-ray Scattering (SAXS) provides information on the excess electron density of the sample versus the surrounding environment. Recently, with the advent of automated high-throughput SAXS analysis of

---

* These authors contributed equally to this work.

J. Gabriel et al. (Eds.): BIOSTEC 2012, CCIS 357, pp. 222–235, 2013.

biomolecules [3, 4], high-throughput data acquisition is within reach. Since SAXS data only describes the spherical averaging of the electron density of the average conformation of the ensemble, additional information is needed to assist structural interpretation.

Typical *in silico* protein structure determination methods, such as ones based on Markov chain Monte Carlo (MCMC) simulations, propose plausible structural conformations, and compute their associated simulated data by means of a forward model. Then, the simulated and the experimental data are compared using an error model of the experiment. For this procedure to be successful, an efficient method for both sampling protein structures and calculating the simulated data is required.

One approach for the calculation of a SAXS curve from a given structure makes use of the Debye formula [5], which is calculated from a set of spherical scatterers [6–9]. Another, more recent approach, is based on spherical harmonics expansions [10]. This approach is faster, but becomes problematic for certain structures, such as those with internal cavities [11]. Here, we present an efficient application of the Debye formula, based on a simplified representation of protein structure and the computational power provided by Graphics Processor Units (GPUs).

In recent publications, our group developed probabilistic models for the proposal of protein-like conformations, in full atomic detail, for both backbone and side chains [12, 13]. These models were used for the inference of protein structure from NMR data [14]. We also developed a forward model of the scattering profile evaluation, that includes the experimental error associated with SAXS data [15]. The forward model consists of a coarse-grained computation based on the Debye formula. Our main aim is the study of proteins consisting of multiple domains connected by flexible linkers. Such proteins play a major role in the regulation of gene expression, cell growth, cell cycle, metabolic pathways, signal transduction, protein folding and transport [16, 17]. With this aim, a computationally efficient forward model for the calculation of SAXS curves is paramount.

We ported our original implementation of the Debye formula to General Purpose computing on Graphics Processing Units (GPGPU). GPUs are parallel computing engines that can offer great advantages in terms of cost-efficiency and low power consumption [18]. One of the emerging standards of choice for their programming is the Open Computing Language (OpenCL), an open standard that provides an abstraction layer over multi- and many-core computational hardware [19]. The OpenCL Debye implementation was utilized as a likelihood term in an MCMC simulation, providing the basis for efficient protein structure determination from low-resolution SAXS data.

2 Methods

2.1 Forward SAXS Computation

The observed data in a SAXS experiment is a one-dimensional intensity curve, $I(q)$, measured at discretized scattering momenta $q = 4\pi \sin(\theta)/\lambda$, with λ the wavelength of the incoming radiation, and 2θ the angle between this beam and the scattered rays. The calculation of a theoretical SAXS profile from a given atomic structure is based on the well-known Debye formula [5]:

$$I(q) = \sum_{i=1}^{M} \sum_{j=1}^{M} F_i(q) F_j(q) \frac{\sin(q \cdot r_{ij})}{q \cdot r_{ij}} \tag{1}$$

where F_i and F_j are the scattering form factors of the individual particles i and j, and r_{ij} is the Euclidean distance between them. The summations run over all the M scattering particles.

2.2 Coarse-Grained Protein Models

If some scatterers are fixed relative to each other, they can be approximated by a single large scattering body (*dummy body*). This approximation, more precise at low q, fades with the progression of the scattering angle up to a resolution equal to the scattering diameter of the dummy body. We found that the amino acids constituent to the protein chain can be approximated by one or two large bodies (*dummy atoms*), and that this approximation holds up to scattering angles normally not measured in the current experimental standards [15].

In the two body model, the amino acids are individually represented by two dummy atoms; one representing the backbone, and the other representing the side chain. Glycine and alanine, lacking a side chain with conformational freedom, are represented by a single dummy atom. The dummy atoms are placed at the respective centers of mass (see Fig. 1). A total of 21 form factors need to be estimated for the two body model: one for alanine, one for glycine, one for the generic backbone and 18 for the remaining side chains.

For the one body model, the single dummy atom is placed at the center of mass of the amino acid. Hence, 20 form factors need to be estimated; one for each amino acid type. For a given protein, the one body model allows to represent the molecule with roughly half the number of scattering bodies employed in the two body model. If the experimental data is recorded at low resolutions only, the former is thus clearly preferable for reasons of computational efficiency.

2.3 Form Factor Descriptors

Due to the lack of publicly available high-quality experimental data needed for the estimation of the form factors, artificial data curves were generated for known high-resolution protein structures using the state-of-the-art program CRYSOL [21]. This program computes the theoretical scattering curve from a given full-atom structure using spherical harmonics expansions, therefore limiting its applications at studying compact quasi-globular proteins. We can however use this input to feed a learning protocol, and make use of the Debye formula in eq. 1 to overcome structural assumptions.

Therefore, a large scale Monte Carlo simulation has been conducted to estimate the values of the form factors of the dummy atoms [15]. The resulting profiles for these descriptors are shown in Fig. 2.

In Fig. 3 we show a SAXS curve generated with our method, and the theoretical scattering computed by CRYSOL as a reference.

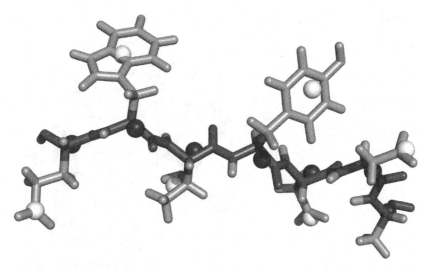

Fig. 1. Coarse-grained models of protein structure. Example of a protein backbone stretch (dark gray) with side chain atoms (light gray). The placement of the dummy bodies for the center of mass of the backbone atoms (dark spheres) and for the side chains (light spheres) are indicated. Figure prepared with PyMOL [20], adapted from [15].

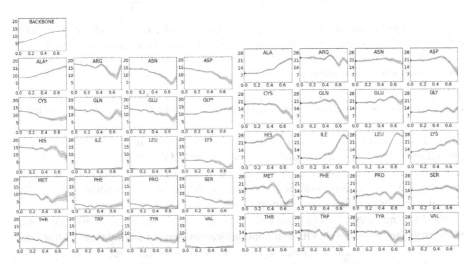

Fig. 2. Form factors. Mean (dark curve) and standard deviations (shaded areas) for the form factors (Y-axis) as a function of q (X-axis). *Left*: backbone and side-chains. An asterisk indicates that this form factor describes both the backbone and side chain atoms of the residue. *Right*: the single body form factors. Figure adapted from [15].

2.4 OpenCL Programming Model

An OpenCL program contains a host program that executes on the CPU, and kernels that execute on the abstracted parallel device. The device consists of one or more compute

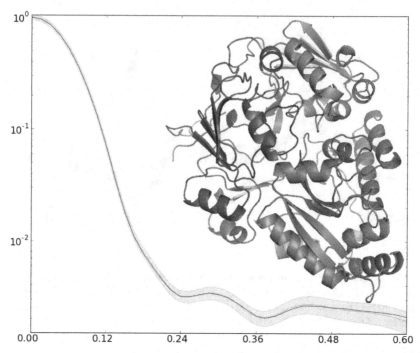

Fig. 3. SAXS profile reconstruction example. Comparison of the reference profiles $I(q)$ computed by CRYSOL from the all atom structure (light gray) and by the two body model (dark line). Error shade indicates the simulated "experimental" error. PDB code 1JET (520 residues). Cartoon made with PyMOL [20].

units, which are composed of one or more processing elements and, in some cases, local memory.

The host program coordinates the execution of the kernels, and can be written in any programming language. Kernels are written in a variant of the latest released C language standard (C99) and are compiled at run time to device-specific instructions. A kernel describes the operations of a single work-item, or thread, and is run simultaneously by a set of work-items called a work-group.

The local memory of a compute unit, if present, is shared by all work-items in a work-group and provides an efficient communication channel among them. It has very low latency and is usually implemented with a full crossbar interface, but is limited in size and does not retain its state between kernel executions.

Kernels execute most efficiently when the size of the work-group matches the size of the compute unit on the OpenCL device and when all work-items in a work-group follow the same execution path.

2.5   Efficient GPGPU Implementation

Parallel Page-Tile SAXS Algorithm. The computation of a SAXS profile is experimentally discretized in a set of q points, and thus naturally provides the first level of

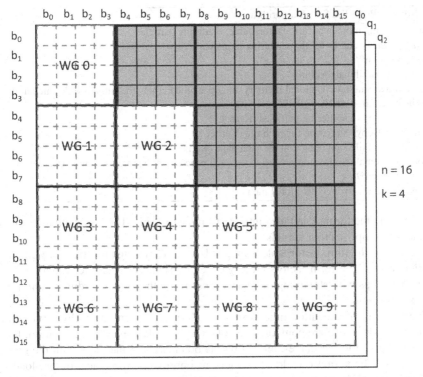

Fig. 4. Domain decomposition for the Page-Tile algorithm. Work-groups operate on square tiles from the matrix. Only tiles in the lower-left part and the diagonal are evaluated.

parallelization, into pages. Each page represents the computation of the intensity curve $I(q)$ for a single value of q. A page can be visualized as a square problem matrix of side equal to the number of scattering particles M, with each cell representing the contribution of a single term of the Debye formula for particular i and j.

For performance considerations and direct mapping to the hardware, pages are partitioned into square tiles of side k, where k is set to the specific compute unit size of the OpenCL device. Since each problem matrix is symmetrical, only the tiles encompassing the lower-left triangle and the diagonal are computed and their value is simply duplicated for the mirror tiles in the upper-right triangle of the matrix. The domain decomposition is illustrated in Fig. 4 for an example of 16 scattering particles and work-group size of 4.

GPUs suffer performance penalties when they have to work with data that is not aligned to their native architecture. The algorithm therefore pads the data and aligns it to the specified work-group size. The resultant dummy particles participate in the Debye calculations, but they are assigned a form factor of 0, so their contribution to the intensity $I(q)$ is null.

Algorithm 1 presents the pseudocode for the Page-Tile SAXS algorithm. The form factors table, supplied as input, is packed and organized by scattering momentum and particle type. The scattering particles, in addition to their position in three dimensions,

---

Algorithm 1. Page-Tile SAXS algorithm

Input: scattering momenta, form factors table, scattering particles
Output: intensity curves for the scattering momenta
/*Host program*/
Initialize the parallel algorithm
Transfer input data to GPU global memory and queue the kernels for initial profile calculation
/*Kernels executed on the GPU*/
Map form factors to scattering particles (Kernel 1)
Compute the Debye sum term for each tile (Kernel 2)
Perform vertical tile sum reduction for each page (Kernel 3)
Perform horizontal margin sum reduction for each page to get the intensity curve (Kernel 4)
/*Host program*/
Retrieve the results from GPU global memory

---

have a type in the form of an index into the form factor table. The initial intensity curve calculation comprises four kernels.

In Kernel 1 the form factor table is mapped into a form suited for hardware-efficient parallel access. The form factors are organized by scattering particle, which enables the work-groups with streaming memory access for both center coordinates and form factors.

The majority of execution time is spent in Kernel 2, where the Debye sum terms for the individual tiles are computed. The Debye formula is used for each term, but i and j are limited to the ranges defined by the boundaries of the tile within the global index space. The kernel uses local memory to improve performance, by pre-loading the particles and their form factors, and by performing an in-place parallel reduction to produce the partial sum for the tile. During the Debye calculation, 4x loop unrolling utilizes local registers to further optimize this stage.

Kernel 3 reduces the tile partial sums, which are stored in a global cache, to bottom margin sums that are further reduced by Kernel 4 to yield the final intensity curve.

Tile Recalculation. Markov chain Monte Carlo simulations explore the conformational space of the protein structures by applying partial modifications to an accepted proposal. The average SAXS computation is therefore a partial re-evaluation of a previously computed profile, where only a subset of the bodies changed their position.

It is therefore possible to identify the subset of tiles that needs to be updated. Since the Page-Tile algorithm caches the partial contribution of each tile to the global summation, we can impose a partial recalculation of only the affected tiles (see Fig. 5). This leads to a substantial reduction in the time necessary to derive an intensity curve after a Monte Carlo transition (move).

Algorithm 2 illustrates the pseudocode for tile recalculation. The form factor table from the initial calculation is reused, so execution starts directly with Kernel 5, which identifies the affected tiles and invokes Kernel 2 for them. Kernel 3 and Kernel 4 perform the reductions as in the initial calculation.

Floating Point Precision. Floating point numbers can be stored and manipulated with single precision (SP) or double precision (DP). Mathematical operations on floating

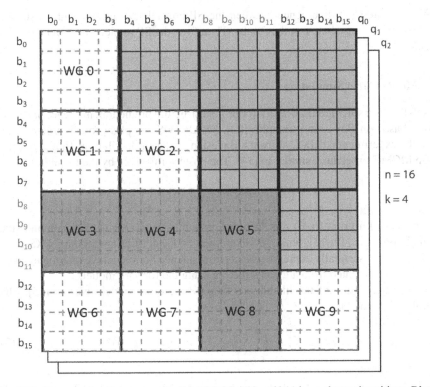

Fig. 5. Problem matrix after a move. Particles b8, b9, b10 and b11 have changed positions. Blocks WG3, WG4, WG5 and WG8 will be recalculated.

Algorithm 2. Tile recalculation

Input: moved scattering particles
Output: updated intensity curve for the scattering momenta
/*Host program*/
Transfer input data to the GPU global memory and queue the kernels involved in the profile recalculation
/*Kernels executed on the GPU*/
Compute the Debye sum term for the changed tiles (Kernel 5)
Perform the vertical tile sum reduction for each page (Kernel 3)
Perform horizontal margin sum reduction for each page to get the intensity curve (Kernel 4)
/*Host program*/
Retrieve the results from the GPU global memory

---

point numbers introduce errors, due to the finite precision available. Those errors tend to accumulate when a large number of operations is performed, as is the case with the double sum of the Debye formula. However, the Page-Tile algorithm significantly reduces this error growth, because its successive partitioning of the problem space results in an execution pattern resembling pairwise summation [22]. The algorithm can be executed with SP or DP, paying a performance penalty of a factor of 2 to 4 with DP.

We measured the divergence between the SP and DP executions, and no significant differences arise between the results. Therefore, the SP implementation is used by default.

2.6 Monte Carlo Simulation

PHAISTOS is a software framework for protein structure prediction, inference and simulation based on Bayesian principles [23]. PHAISTOS samples protein structures X given the experimental data $I_{\exp}$ from a Bayesian posterior distribution $P(X|I_{\exp})$ using an MCMC procedure, similar to [14]. The posterior is given by the formula:

$$P(X|I_{\exp}) \propto P(I_{\exp}|X)P(X) \tag{2}$$

and consists of a prior $P(X)$ that includes probabilistic models of the main and side chains in proteins [12, 13], while the likelihood $P(I_{\exp}|X)$ brings in the SAXS data. The likelihood essentially expresses the correspondence between the experimental data and the data calculated from a given structure using the forward model.

For the calculation of the likelihood, we used the error model given in [15]. The resulting likelihood is:

$$P(I_{\exp}|X) = \prod_q \mathcal{N}(I_{\exp}(q)|I_{\text{calc}}(q), \sigma(q)), \tag{3}$$

where $\mathcal{N}(\cdot)$ is the normal distribution with mean $I_{\text{calc}}(q)$ and standard deviation $\sigma(q)$, controlled by scaling parameters α and β:

$$\sigma(q) = I_{\exp}(q) \cdot (q + \alpha) \cdot \beta \tag{4}$$

The prior $P(X)$ is brought in indirectly by sampling from the proposal distribution for protein conformations [14].

The majority of time dedicated to each simulation step is spent on computing the energy function for the proposed structure. The GPGPU SAXS algorithm directly reduces this time. Furthermore, at each MCMC step, PHAISTOS performs local moves on a portion of the protein, which allows the Page-Tile algorithm to use the fast tile recalculation path.

2.7 Performance Test Configuration

Performance was measured on a system with a Core i7-920 CPU (4 cores / 8 hardware threads), 12GB of DDR3 RAM and a NVIDIA GeForce GTX 560 Ti GPU with 1GB of GDDR5 RAM. The GTX 560 Ti has 8 compute units with 32 processing elements each, comprising 384 processing elements, with 32KB 32-bit registers and 48KB of local memory for each compute unit.

3 Results and Discussion

3.1 Computational Efficiency of the SAXS Modeling

The Debye formula (Equation 1) leads to a computational complexity of $O(M^2)$, with M the number of scatterers in the structure under examination. Our coarse-grained approach reduces M by representing several atoms by one scattering body (a dummy atom), thereby lowering the complexity to $O\left(\left(M/k\right)^2\right)$, with k the number of scatterers (atoms) described by a dummy body.

The precise value of k is dependent on the primary sequence of the protein. On large datasets, the two dummy model leads to an average k of 4.24 (with a performance increase of $k^2 \simeq 18$). The single body model leads to $k \simeq 7.8$, allowing for a $k^2 \simeq 60$ times faster execution.

3.2 GPGPU Implementation

The performance of the Page-Tile algorithm was measured against a test protein of over a thousand amino acids, modeled with 1888 scattering bodies in the dual dummy atom representation, and a discretization of the q space in 51 scattering momenta. Protein moves were modeled by a random mutation of 40% of the particles, to approximate the asymptotic move rate in a Monte Carlo simulation. The execution times for the model test case are presented in Table 1.

Table 1. Execution times for SAXS curve calculation for a protein with 1888 bodies, 51 scattering momenta and 21 form factors per momentum. Execution times from the top are for a single-core CPU implementation, a parallel GPGPU full computation, and GPGPU partial computation, respectively. Partial computations mimic the costs in a Monte Carlo simulation, where at each step around 40% of the proposal structure is updated.

Algorithm	Time (ms)
CPU SP Time	2408
GPGPU full calculation	9
GPGPU recalculation	6.484

The performance of the algorithm was also measured for protein sizes ranging from 64 to 8192 scattering particles. Each protein was moved 1000 times, in order to obtain an average of the recalculation steps. Figure 6 shows the speed increase, relative to the CPU single precision implementation, calculated as t_{cpu}/t_{gpu}.

Figure 6 also illustrates the hardware utilization of the parallel Page-Tile algorithm. The plot shows an asymptotic behavior around problem sizes of 2000 scattering bodies. The GTX 560 Ti GPU employed in the tests is composed of 8 compute units operating on 8 cascading work-groups, allowing for a theoretical peak of 64 active work-groups. The work-group size is 32, therefore the card would reach theoretical peak processing power at 2048 bodies. Our tests show saturation at the same level, thus indicating optimal use of the hardware.

OpenCL is thread-safe and allows access to the same device from multiple processes and threads, so by creating multiple instances of the Page-Tile algorithm, more than one

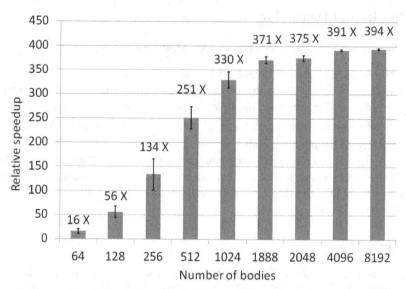

Fig. 6. Algorithm performance for problem sizes ranging from 64 to 8192 scattering bodies, including the model test case of 1888, with error bars showing standard deviation. All SAXS computations involve the summations over 51 scattering momenta. The asymptotic behavior indicates hardware saturation at around 2000 bodies, which is the theoretical maximum for the GPU model used in the tests.

calculation can be run at the same time. This is especially relevant in the case of problem sizes that would not lead to a full GPU saturation, therefore allowing for multi-threaded Monte Carlo simulations.

3.3 Monte Carlo Performance

The effect of the GPGPU SAXS curve calculation on the overall MCMC simulation of the 1888-body test protein was measured in PHAISTOS, by gathering performance data for 1 million MCMC steps with varying number of concurrent CPU threads, with and without GPU acceleration. The first 100,000 steps were discarded as burn-in and the remaining 900,000 were used for analysis.

The best-performing CPU SAXS algorithm was used in the evaluation. It caches the terms of the Debye formula and uses a lookup table for the sine function, resulting in a five-fold speed increase, at the expense of numerical precision. The MCMC simulation was executed with one to six threads, the maximum possible under the test configuration, due to the significant memory footprint per thread. The lowest execution time was achieved when using four threads, and was used as a basis for comparison with the GPGPU version.

The MCMC simulation, using the GPGPU Page-Tile algorithm, was executed with one to eight CPU threads. Execution time was compared to the multi-threaded CPU version (Fig. 7).

The GPU-accelerated MCMC simulation exhibits consistently better performance compared to the best-performing multi-threaded CPU version. The speed increase scales

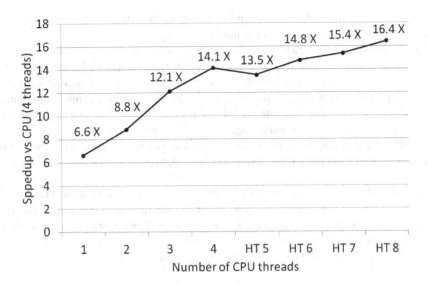

Fig. 7. Relative speedup of PHAISTOS when using the GPGPU SAXS energy term vs. the best-performing CPU configuration (four threads). Threads above four use hyper-threaded CPU cores, so lower performance scaling is expected.

up with the number of CPU threads, due to two factors: 1) incomplete loading of the GPU by each thread; and 2) concurrent MCMC calculations outside of the SAXS energy term.

When invoked from within a CPU thread, the GPU calculates the SAXS intensity profile and is then idle, while the host thread processes the result and queues a new structure for evaluation. Multiple CPU threads can take advantage of these idle GPU cycles by queueing additional SAXS curve calculations, thus leading to the performance scaling observed.

While the GPU accelerates the calculation of the SAXS energy term, the CPU host still has to propose a new structure in each MCMC step and to process the result from the energy function calculation. The work done by the CPU limits the speed increase for the overall step. Conversely, running multiple host threads parallelizes these operations, further contributing to the performance scaling.

The theoretical possible speedup under PHAISTOS is 18.6 ×, and can be calculated from the Page-Tile algorithm speedup (371 × from Fig. 6) adjusted for the number of CPU threads (4) and the speedup from using a cache and a sine lookup table (5 ×). The observed speed increase of 16.4 × approaches the theoretical maximum and is clearly limited by the CPU-bound portions of the MCMC simulation.

4   Conclusions

We have presented an efficient implementation of the forward model for the computation of Small Angle X-ray Scattering profiles, utilizing Graphics Processing Units.

The application is multi-thread safe, and benchmarks show that the algorithm delivers the full theoretical output of which the hardware is capable.

Parallelization is achieved on multiple levels by taking advantage of the structure of the Debye formula. The first level divides the SAXS evaluation in multiple independent computations according to the binning of the scattering momenta. A nested level then makes full use of the work-groups in the hardware by splitting the inner summation of the Debye formula into separate partial sums. The resulting program runs orders of magnitude faster than an optimized single core CPU implementation.

A caching algorithm on the inner contributions allows for the efficient re-evaluation of SAXS profiles from partially updated structures, delivering even greater performance benefits.

The GPGPU algorithm was integrated into an energy term within the PHAISTOS software framework for protein structure determination, inference and simulation. This yielded a 16-fold speed increase of the Markov chain Monte Carlo simulation, compared to the best multi-threaded CPU implementation, enabling its application to important biological targets. An open source implementation is now available.

Acknowledgements. This research was funded by the Danish Council for Independent Research (FTP, 274-09-0184). CA acknowledges partial funding from the Danish Strategic Research Council (contract number 10-092299), for the HIPERFIT research center (http://hiperfit.dk).

Source Code Availability

The source code of our implementation can be found online at: http://www.phaistos.org and http://sourceforge.net/projects/phaistos/files.

References

1. Anfinsen, C.B.: Principles that govern the folding of protein chains. Science 181, 223–230 (1973)
2. Zheng, W., Doniach, S.: Fold recognition aided by constraints from small angle X-ray scattering data. Protein Eng. Des. Sel. 18, 209–219 (2005)
3. Toft, K., Vestergaard, B., Nielsen, S., Snakenborg, D.: High-throughput small angle X-ray scattering from proteins in solution using a microfluidic front-end. Anal. Chem. 80, 3648–3654 (2008)
4. Hura, G.L., Menon, A.L., Hammel, M., Rambo, R.P., Poole, F.L., Tsutakawa, S.E., Jenney, F.E., Classen, S., Frankel, K.A., Hopkins, R.C., Jae Yang, S., Scott, J.W., Dillard, B.D., Adams, M.W.W., Tainer, J.A.: Robust, high-throughput solution structural analyses by small angle X-ray scattering (SAXS). Nat. Methods 6, 606–614 (2009)
5. Debye, P.: Zerstreuung von röntgenstrahlen. Ann. Phys. 351, 809–823 (1915)
6. Chacon, P., Moran, F., Diaz, J., Pantos, E., Andreu, J.: Low-resolution structures of proteins in solution retrieved from X-ray scattering with a genetic algorithm. Biophys. J. 74, 2760–2775 (1998)
7. Svergun, D., Petoukhov, M., Koch, M.: Determination of domain structure of proteins from X-ray solution scattering. Biophys. J. 80, 2946–2953 (2001)

8. Tjioe, E., Heller, W.: ORNL_SAS: software for calculation of small-angle scattering intensities of proteins and protein complexes. J. Appl. Crystallogr. 40, 782–785 (2007)
9. Förster, F., Webb, B., Krukenberg, K., Tsuruta, H.: Integration of small-angle X-ray scattering data into structural modeling of proteins and their assemblies. J. Mol. Biol. 382, 1089–1106 (2008)
10. Svergun, D.I., Stuhrmann, H.B.: New developments in direct shape determination from small-angle scattering. 1. Theory and model calculations. Acta Crystallogr. A 47, 736–744 (1991)
11. Koch, M., Vachette, P., Svergun, D.: Small-angle scattering: a view on the properties, structures and structural changes of biological macromolecules in solution. Q Rev. Biophys. 36, 147–227 (2003)
12. Boomsma, W., Mardia, K., Taylor, C., Ferkinghoff-Borg, J., Krogh, A., Hamelryck, T.: A generative, probabilistic model of local protein structure. Proc. Natl. Acad. Sci. U S A 105, 8932–8937 (2008)
13. Harder, T., Boomsma, W., Paluszewski, M., Frellsen, J., Johansson, K.E., Hamelryck, T.: Beyond rotamers: a generative, probabilistic model of side chains in proteins. BMC Bioinformatics 11, 306 (2010)
14. Olsson, S., Boomsma, W., Frellsen, J., Bottaro, S., Harder, T., Ferkinghoff-Borg, J., Hamelryck, T.: Generative probabilistic models extend the scope of inferential structure determination. J. Magn. Reson. 213, 182–186 (2011)
15. Stovgaard, K., Andreetta, C., Ferkinghoff-Borg, J., Hamelryck, T.: Calculation of accurate small angle X-ray scattering curves from coarse-grained protein models. BMC Bioinformatics 11, 429 (2010)
16. Levitt, M.: Nature of the protein universe. Proc. Natl. Acad. Sci. U S A 106, 11079–11084 (2009)
17. Madl, T., Gabel, F., Sattler, M.: NMR and small-angle scattering-based structural analysis of protein complexes in solution. J. Struct. Biol., 1–11 (2010)
18. Göddeke, D., Strzodka, R.: Cyclic reduction tridiagonal solvers on GPUs applied to mixed precision multigrid. IEEE Transactions on Parallel and Distributed Systems 22, 22–32 (2011), doi:10.1109/TPDS.2010.61
19. Stone, J.E., Gohara, D., Shi, G.: OpenCL: A parallel programming standard for heterogeneous computing systems. Computing in Science and Engineering 12, 66–73 (2010)
20. Schrödinger, L.: The PyMOL molecular graphics system, version 1.3r1 (2010)
21. Svergun, D., Barberato, C., Koch, M.: CRYSOL - a program to evaluate X-ray solution scattering of biological macromolecules from atomic coordinates. J. Appl. Crystallogr. 28, 768–773 (1995)
22. Higham, N.J.: The accuracy of floating point summation. SIAM J. Sci. Comput. 14, 783–799 (1993)
23. Borg, M., Mardia, K., Boomsma, W., Frellsen, J., Harder, T., Stovgaard, K., Ferkinghoff-Borg, J., Røgen, P., Hamelryck, T.: A probabilistic approach to protein structure prediction: PHAISTOS in CASP9. In: Gusnanto, A., Mardia, K., Fallaize, C. (eds.) LASR2009 - Statistical Tools for Challenges in Bioinformatics, pp. 65–70. Leeds University Press, Leeds (2009)

Part III

Bio-inspired Systems
and Signal Processing

Assessment of Gait Symmetry and Gait Normality Using Inertial Sensors: In-Lab and In-Situ Evaluation

Anita Sant'Anna[1,*], Nicholas Wickström[1],
Helene Eklund[2], Roland Zügner[3], and Roy Tranberg[3]

[1] Intelligent Systems Lab, Halmstad University, Sweden
[2] Center for Person-Centered Care, Sahlgrenska Academy, Sweden
[3] Department of Orthopedics, Sahlgrenska Academy, Sweden
anita.santanna@hh.se

Abstract. Quantitative gait analysis is a powerful tool for the assessment of a number of physical and cognitive conditions. Unfortunately, the costs involved in providing in-lab 3D kinematic analysis to all patients is prohibitive. Inertial sensors such as accelerometers and gyroscopes may complement in-lab analysis by providing cheaper gait analysis systems that can be deployed anywhere. The present study investigates the use of inertial sensors to quantify gait symmetry and gait normality. The system was evaluated in-lab, against 3D kinematic measurements; and also in-situ, against clinical assessments of hip-replacement patients. Results show that the system not only correlates well with kinematic measurements but it also corroborates various quantitative and qualitative measures of recovery and health status of hip-replacement patients.

Keywords: Gait analysis, Symmetry, Normality, Accelerometer, Gyroscope, Inertial sensors.

1 Introduction

Quantitative gait analysis (GA) can improve the assessment of a number of physical and cognitive conditions. The importance of GA in the treatment of children with cerebral palsy is well known and documented [1]. The use of GA to monitor and assess Parkinson's Disease [2], stroke [3], and orthopedic [4] patients have also been investigated.

Despite many positive results, GA is still not routinely used in the clinical setting. Several factors contribute to the low adoption of GA as a routine clinical tool. Perhaps the most significant factor is that the accepted gold standard for GA, in-lab 3D motion capture (MOCAP), is simply not available to all patients. The costs involved in equipping a gait lab and training personnel are prohibitive for many clinical institutions, especially in underprivileged areas and developing countries.

The current alternative to MOCAP is observational gait analysis (OGA), which is intrinsically subjective and sensitive to the observer's experience [5]. Recently, however, large efforts have been employed in developing low-cost, inertial sensor systems that can complement OGA with objective and reliable information. The success of such systems will hopefully incur in a wide-spread adoption of quantitative GA as a clinical tool.

---

* Corresponding author.

J. Gabriel et al. (Eds.): BIOSTEC 2012, CCIS 357, pp. 239–254, 2013.
© Springer-Verlag Berlin Heidelberg 2013

The present study concerns the development of a inertial sensor system for GA that can be successfully deployed in the clinical setting. This system, composed of wearable inertial sensor units, can complement OGA by providing reliable quantitative measures of gait symmetry and gait normality. This study evaluates the proposed symmetry and normality measures both in-lab, against measures derived from MOCAP; and in-situ, compared to clinical assessments of hip-replacement patients.

2   Related Work

2.1   Observational Gait Analysis

Observational gait analysis (OGA) is frequently aided by video recordings, which provide certain interesting functions to the observer such as pause and slow motion. In some cases, quantitative measurements, such as joint angles, can be directly calculated from the video image [6]. This type of analysis is often accompanied by a form or questionnaire that facilitates the extraction of relevant information from the video. Two such forms have been more thoroughly investigated and more widely adopted: the Visual Gait Assessment Scale (VGAS) [7] and the Edinburgh Visual Gait Score (EVGS) [8]. Both questionnaires target the assessment of children with cerebral palsy.

OGA can be complemented by other more quantitative measurements, such as average gait speed, average step length and other gait parameters. These are typically measured during walking tests, such as the 10-m walking test [9], or the timed up and go test (TUG) [10]. The TUG is normally employed in studies where balance and risk of fall are of interest, as it requires that the subject stand up and sit down on a chair without help. The 10-m walking test is a simple way of determining, average gait speed, stride length and cadence. Average gait speed, for example, has been identified as an indicator of: activity of daily living function in geriatric patients [11]; high risk of health-related outcomes in well-functioning older people [12]; and leg strength in older people [13]. Stride length is another interesting measure that has been associated with, for example, metabolic cost and impact during walking [14].

2.2   MOCAP Gait Analysis

Gait may be studied in the spatio-temporal, kinetic or kinematic domains. In order to simplify the analysis of this extremely rich source of information, one may focus on measures of symmetry and normality. Symmetry refers to the similarity between the movements of the right and left sides of the body. Normality refers to the similarity between the movements of one individual compared to average movements of a population that is judged healthy or normal.

The symmetry of kinematic gait data is usually evaluated by visual inspection of superimposed curves from right and left sides. Few quantitative symmetry measures have been proposed which take into account complete joint angle curves. A measure of trend symmetry based on the variance around the 1st principal component of a right-side vs. left-side plot has been suggested [15]. This trend symmetry measure is insensitive to scaling, and must be compensated by an additional measure, the range amplitude ratio. In contrast, the present study introduces a quantitative symmetry measure based on kinematic data which can be expressed as one index.

The Gillette Functional Assessment Questionnaire (GFAQ) Walking Scale is a widely accepted gait normality measure based on observation. Considerable efforts have been put into deriving an equivalent measure from kinematic data. Principal component analysis (PCA) on 16 discrete gait variables has been used to create a representation of the data in a different space. The magnitude of the projection of an abnormal data set onto this space is used as a normality index, known as the Gillette Gait Index (GGI) [16]. A very similar PCA approach named the Gait Deviation Index (GDI) has been introduced by [17]. One advantage of PCA approaches is that they transform the possibly dependent gait variables into a new set of independent variables. The disadvantage is that results cannot be traced back to the original gait variables.

A much simpler method, the Gait Profile Score (GPS) and Movement Analysis Profile (MAP), has been suggested [18]. The MAP is created by taking the root mean square error (RMS) between a reference joint angle curve and the corresponding curve from a subject. This creates one normality index for each joint angle curve. A unique index, the GPS, can be derived by concatenating all joint angle curves end to end, and taking the RMS of this aggregated curve. Although the GDI presents some nice properties such as normal distributions across GFAQ levels, the GPS is more easily interpreted because the original variables suffer no transformations and results are given in degrees. It has been shown that the GPS correlates significantly with clinical judgment [19].

2.3 Instrumented Gait Analysis

Inertial sensors, such as accelerometers and gyroscopes, can complement MOCAP systems and OGA by providing quantitative and objective gait measurements outside the gait lab, and for a fraction of the cost.

Most symmetry measures calculated from inertial sensor data take into account only discrete spatio-temporal variables, e.g. [20], [21]. Although discrete symmetry indices have been shown useful, a more informative measure of symmetry may be obtained using the entire continuous sensor data. Few approaches to calculating symmetry using continuous accelerometer data have been introduced. One example is an unbiased autocorrelation method using trunk acceleration data [22]. Although this may provide a good general estimate of gait symmetry, it lacks information about each individual limb.

More recently, gyroscopes data obtained from shanks and thighs was used to calculate symmetry using a normalized cross correlation approach [23]. This method segments and normalizes the data to individual strides. As a result, only the shape of the signal and not its relative temporal characteristics are taken into account. A symbolic method for estimating gait symmetry using accelerometers [24] or gyroscopes [25] has been suggested, which takes into account not only the shape but also the temporal characteristics of the signal. This symmetry measure is used in the present paper.

Based on this symmetry measure, the authors proposed a normality measure based on symbolized inertial sensor data, described in the present paper. No other normality measures based on inertial sensor data were found in the literature.

3   Method

The study was conducted in two phases. The first phase was an in-lab evaluation that compared the proposed mobile GA system with 3D kinematic analysis. The second, was an in-situ evaluation of the proposed system against quantitative and qualitative clinical assessments of hip-replacement patients. This study was approved by the Regional Ethics Board in Gothenburg, Sweden.

3.1   In-Lab Data Collection

The data collection took place at the clinical gait lab at Sahlgrenska University Hospital, Gothenburg, Sweden. A group of 19 healthy volunteers participated in the study. The average hight of the group was 172.1 ± 7.6 cm; and the average weight was 71.8 ± 17.2 Kg. Seven participants were male and twelve female, averaging an age of 34 ± 13 years.

Kinematic and kinetic data were recorded with a 3D motion capture (MOCAP) system, Qualisys MCU 240, sampling at 240Hz. A total of 15 spherical reflective markers, of 19 mm in diameter, were placed on the sacrum, anterior superior iliac spine, lateral knee-joint line, proximal to the superior border of the patella, tibial tubercle, heel, lateral malleolus and between the second and third metatarsals [26].

Subjects were also equipped with 3 Shimmer® sensor nodes, each containing one 3-axis accelerometer and one 3-axis gyroscope, sampling at 128Hz. One node was placed on each outer shank, about 3cm above the lateral malleolus, Figure 1. The remaining node was placed mid-way between the anterior superior iliac spine markers, Figure 2. Sensors were synchronized using a beacon signal from the host computer, and the data was stored on-board each sensor node.

Fig. 1. Shank sensor node approximately 3cm above the lateral malleolus

Fig. 2. Waist sensor node mid-way between the anterior superior iliac spine

The subjects' movements were simultaneously recorded with the sensor nodes and with the Qualisys system. They were instructed to walk in 3 different ways: 1) normally at a comfortable speed; 2) with a limp, as if injured; and 3) slowly, as if very tired. All subjects performed three tests for each type of walk. One test of each type was then randomly chosen for further analysis.

3.2   In-Situ Data Collection

This data collection took place at the orthopedic ward at Sahlgrenska University Hospital, Mölndal, Sweden. Eleven patients were included in the study. All patients had undergone unilateral hip-arthroplasty for the first time and presented no other physical or cognitive conditions. The group was composed of four women and seven men, the mean age was 69±15 years, mean weight was 81±20 Kg, and mean height was 172±9 cm.

All subjects were equipped with sensor nodes similarly to the in-lab data collection. They were then asked to walk by themselves along a 10-meter walkway at a comfortable speed, twice. The walkway was marked with black tape on the floor. The time and number of steps taken to complete the walkway were recorded.

This procedure took place on the day the patient was discharged from the hospital, and a few months later, when the patient came back for a follow-up evaluation. The average number of days spent at the ward after surgery was 4±1 day. The time between baseline and follow-up measurements was 108±15 days. All patients employed a walking aid during baseline measurements, six used two crutches and five used a walker with wheels. During follow-up measurements six patients used one crutch and five patients walked without any aiding device.

Patients filled out an EQ-5D^{TM} health questionnaire (Swedish version) approximately two weeks before surgery, and soon after their follow-up session. The EQ-5D^{TM} is a standardized instrument for measuring health outcome, developed by the EuroQol Group (www.euroqol.org). The English version of the questionnaire, validated for Ireland, is shown in Figure 3. Each answer is given a value from 1 to 3, lower values indicate better health.

3.3 Observational Gait Analysis

The time, Tm, and number of steps, $NumSteps$, taken to complete the 10-meter walk test were used to compute average speed, $Speed = 10/Tm$ (m/s), and average step length, $StepLeng = 10/NumSteps$ (m). In addition, step length was normalized by the patient's height. These variables were used as reference for the improvement of the patient, under the assumption that average speed and step length should increase as the patient recovers.

3.4 MOCAP Gait Analysis

Gait normality and symmetry measures were calculated from the 3D kinematic data. The normality index used for the kinematic data was the GPS and the MAP [18]. However, the mean value was removed from all curves before calculating the score, and foot progression was not used because it was not available in the reference data set. Removing the curves' mean values makes the normalcy measure more robust to offset errors, while preserving the shape and range of the curves.

The reference data set was an ensemble of 34 randomly selected adult subjects presenting no known pathologies, previously acquired at the clinical gait lab at Sahlgrenska University Hospital, Gothenburg, Sweden. Joint angle curves were calculated for each individual and normalized to stride time. The ensemble average of the normalized curves was used as a reference curve.

Each MAP component, Eq. 1, was calculated as the RMS difference between the reference curve, C_{ref}, and the subject's curve, C_{subj}, where N is the number of points

By placing a tick in one box in each group below, please indicate which statements best describe your own health state today.

Mobility

I have no problems in walking about	1
I have some problems in walking about	2
I am confined to bed	3

Self-Care

I have no problems with self-care	1
I have some problems washing or dressing	2
I am unable to wash or dress myself	3

Usual Activities

I have no problems with performing my usual activities	1
I have some problems with performing my usual activities	2
I am unable to perform my usual activities	3

Pain/Discomfort

I have no pain or discomfort	1
I have moderate pain or discomfort	2
I have extreme pain or discomfort	3

Anxiety/Depression

I am not anxious or depressed	1
I am moderately anxious or depressed	2
I am extremely anxious or depressed	3

Fig. 3. EQ-5D^{TM} English version validated for Ireland. ©*1990 EuroQol Group EQ-5D^{TM} is a trademark of the EuroQol Group.*

in the curve. The GPS was calculated similarly by concatenating all joint curves end to end. For each subject, MAP and GPS results were calculated as the average between right and left sides.

$$MAP = \sqrt{\frac{1}{N} \sum_{n=1}^{N} (C_{subj}(n) - C_{ref}(n))^2} \tag{1}$$

Based on the GPS, a measure of symmetry was derived for the kinematic data. In this case, the components of MAP-symmetry were calculated as the RMS error between the curves for the right and left sides, after removing their corresponding mean values. Similarly, GPS-symmetry was calculated by concatenating all joint curves end to end and calculating the RMS difference between left and right sides.

3.5 Instrumented Gait Analysis

The symmetry measure used in this paper was presented in [25]. The sensor signal, accelerometer or gyroscope, was standardized to zero mean and unitary standard deviation, then segmented into N symbols. Symbolization is done by quantization into N levels. The quantization levels are chosen based on the empirical probability distribution of the signal, so as to produce equi-probable symbols.

The period between consecutive occurrences of the same symbol are calculated and stored in a *period histogram* [25]. Similarly, the period between symbol transitions of the same type may be stored in a *transition histogram*. The symmetry index is a measure of the similarity between symbol period (transition) histograms for the right and left sides. Histograms are compared using a relative error measure shown in Eq. 2, where Z is the number of symbols; K is the number of bins in the histograms; n_i is the number of non-empty histogram bins (for either foot) for symbol i; $h_{Ri}(k)$ is the normalized value for bin k in the period histogram i for the *right* foot; and $h_{Li}(k)$ is the normalized value for bin k in the period histogram i for the *left* foot.

$$SI_{symb} = \frac{\sum_{i=1}^{Z} \frac{1}{n_i} \sum_{k=1}^{K} |h_{Ri}(k) - h_{Li}(k)|}{\sum_{i=1}^{Z} \frac{1}{n_i} \sum_{k=1}^{K} |h_{Ri}(k) + h_{Li}(k)|} 100 \qquad (2)$$

The normality measure for the inertial sensor data was derived from the symmetry measure. Instead of comparing the histograms for right and left sides, one subject's histograms are compared to histograms derived from a reference data set. The reference data set was formed by selecting the (in-lab) subjects that presented the smallest GPS based on the normal walk kinematic data.

The normal walk inertial sensor data from these reference subjects was standardized to zero mean and unitary standard deviation, symbolized, and symbol periods were calculated. The symbol periods were normalized to stride time. That is, a period that coincides with stride time is represented as 1 and all other periods are scaled correspondingly. This normalization is common when dealing with kinematic data, and it ensures that the analysis is not affected by gait speed. The symbol periods from all reference subjects were used to create reference histograms.

3.6 Data Analysis

The data acquired from the MOCAP system was processed in Visual 3D (C-Motion Inc., Germantown, MD) to generate kinematic join angle data and spatio-temporal parameters such as stride time. The data was then exported to MATLAB (MathWorks, Natick, MA) where MAP, GPS, MAP-symmetry and GPS-symmetry were calculated for each subject.

The signals from the shank accelerometers and gyroscopes were low-pass filtered with a Butterworth filter of order 6 and cut-off frequency 20Hz. The waist sensor data was filtered at 10Hz. The signals were filtered once, then reversed and filtered again to avoid any phase shift. The three axes of each accelerometer were combined into a resultant signal, $A_{res} = \sqrt{A_x^2 + A_y^2 + A_z^2}$. For each gyroscope, only pitch and roll rotations were considered, $G_{res} = \sqrt{G_{pitch}^2 + G_{roll}^2}$.

Symmetry was calculated using right and left shank signals. Normality was calculated using both shanks and waist signals. Measures were calculated considering period and transition histograms, using from 5 to 25 symbols. The resulting values outside two standard deviations were considered outliers and removed. The remaining values were used to calculate the Spearman's rank correlation coefficient with kinematic measurements, MAP, GPS, MAP-symmetry and GPS-symmetry. The optimal number of symbols was chosen so as to maximize the correlation coefficients. These optimal parameters were used to calculate symmetry and normality for the hip-replacement patients.

The Spearman's rank correlation coefficient was used to evaluate the correlation between two variables. The non-parametric Wilcoxon rank sum test was used to compare two distributions, and a Kruskal-Wallis test was used to compare more than two distributions. All linear model approximations were calculated based on least mean square errors.

The area under the receiver operating characteristic curve (AUC) was used to evaluate the discriminatory power of the normality index. The ROC curve was constructed based on tests performed on the same individuals. Therefore, any statistically significant comparison between different AUC must take into account the correlated nature of the data. A nonparametric approach based on generalized U-statistics was used to estimate the covariance matrix of the different curves [27].

All measurements of the in-situ data collection included two trials, which were used to assess the test-retest reliability of each index using intra-class correlation coefficient (ICC) type A-1 as a measure of absolute agreement [28]. All tests were bi-directional with confidence level, $\alpha = 0.05$. All data analysis was undertaken in MATLAB (MathWorks, Natick, MA).

4   Results

4.1   In-Lab Evaluation

The best correlation of MAP and GPS with the proposed normality index was achieved with the waist accelerometer sensor, 18 symbols and transition histograms. The best correlated signals are shown in Table 1. The best correlation of the inertial sensor symmetry with MAP-symmetry and GPS-symmetry was achieved with the shank gyroscopes, 20 symbols, and symbol period histograms. The best correlation coefficients are shown in Table 2. These configurations were used to calculate normality and symmetry respectively in the in-situ evaluation.

4.2   In-Situ Evaluation

All but one participant answered the EQ-5D[TM] questionnaire on both occasions. The values of the answers given to each category were added to a single score for that category. Results from before the operation and after the follow-up session are shown in Figure 4. Lower scores correspond to more patients in better health. The biggest changes were regarding mobility, usual activities and pain/discomfort.

Symmetry results for baseline and follow-up measurements are shown in Figure 5 for each subject. Measurements were averaged over both trials of each session. The symmetry index ranges from 0 to 100, a low symmetry index indicates good symmetry whereas a high value indicates asymmetry. According to the proposed index, gait symmetry improved at follow-up for approximately half the subjects. The asymmetry at follow-up may be caused by the use of one crutch. The symmetry index according

Table 1. Correlation of MAP and GPS values with the proposed normality measure (waist sensor node)

sensor	accelerometer	
placement	waist	
histogram	transition	
no. symbols	18	
variable	r	p-value
MAP knee flex.	0.77	<0.0001
MAP hip flex.	0.82	<0.0001
MAP pelv. rot.	0.71	< 0.0001
GPS	0.81	< 0.0001

Table 2. Correlation of MAP-symmetry and GPS-symmetry values with the proposed symmetry measure (shank sensor nodes)

sensor	gyroscope	
placement	shank	
histogram	symbol period	
no. symbols	20	
variable	r	p-value
ankle flex.	0.64	< 0.0001
knee flex.	0.81	< 0.0001
hip flex.	0.68	< 0.0001
all	0.84	< 0.0001

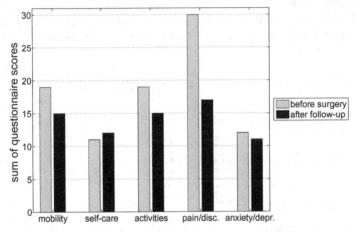

Fig. 4. Questionnaire results from before the surgery and after the follow-up sessions. Lower scores correspond to more patients in better health

walking aid is shown in Figure 6. There is a clear difference between the symmetry of patients using two crutches at baseline and patients walking with no aid at follow-up. However, none of the distributions were significantly different.

Normality results are shown in Figure 7, measurements for each patient were averaged over both trials of each session. Similarly, the normality index ranges between 0 and 100, and a low value indicates good normality. In this case, the follow-up measurements were better than baseline measurements for all patients. A Wilcoxon test indicated that baseline and follow-up groups were statistically significantly different, p<0.0001. Figure 8 illustrates the distribution of the normality index according to walking aid. As expected, the normality index for those patients walking without aid was, on average, better than the others. A Kruskal-Wallis test indicated that the free walking group was statistically different from the walker and crutches groups, and that the one crutch group was statistically different from the walker group.

In order to calculate the correlation between normality and walking aid, each category was represented by a number. In the order shown in Figure 8, walker was

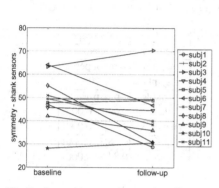

Fig. 5. Symmetry results at baseline and follow-up. The two distributions are not statistically significantly different.

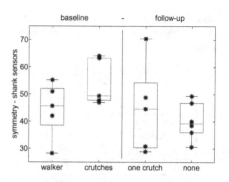

Fig. 6. Symmetry results according to walking aid. Box-plot representations of the distributions. The whiskers represent the smallest and largest observations, the edges of the box correspond to the lower and upper quartiles, the horizontal line indicates the median.

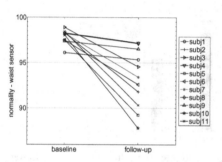

Fig. 7. Normality results at baseline and at follow-up. The two distributions are statistically different according to a Wilcoxon rank sum test, p<0.0001.

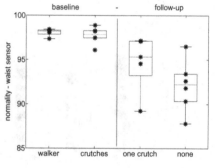

Fig. 8. Normality results according to walking aid. A Kruskal-Wallis test indicates that the distribution of no walking aid is significantly different from distributions of two crutches and walker.

represented by 1 and no-aid was represented by 4. The resulting Spearman's rank correlation coefficient was r=-0.78, p<0.0001.

The normality index also correlates well with both average speed, r=-0.79 p<0.0001, and normalized average step length, r=-0.76 p<0.0001. Normality values for each individual trial are shown against average speed values in Figure 9, and against normalized step length in Figure 10. On both plots the linear model approximation is shown as a solid line, and the 95% confidence interval (CI) for predicted observations is shown as dotted lines. The mean average speed at baseline, 0.46±0.16 m/s, was significantly different from the speed at follow-up, 1.06±0.22 m/s, p<0.0001.

Normality results were also compared to the EQ-5D™ answers that varied the most between before the surgery and after follow-up, namely mobility (Figure 11), usual activities (Figure 12), and pain/discomfort (Figure 11). In all cases, there is a

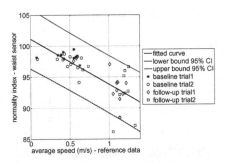 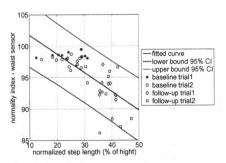

Fig. 9. Normality compared to average speed. Variables are well correlated, Spearman's rank correlation coefficient r=-0.79, p<0.0001. The solid line indicates the linear model approximation a+bx, where a=95.5 with confidence interval (CI) [94.9, 96.3]; and b=-2.6 with CI [-3.3, -1.9]. The dashed and dotted lines indicate the 95% CI of predicted observations.

Fig. 10. Normality compared to average step length. Variables are well correlated, Spearman's rank correlation coefficient r=-0.76, p<0.0001. The solid line indicates the linear model approximation a+bx, where a=95.5 with confidence interval (CI) [94.9, 96.3]; and b=-2.5 with CI [-3.2, -1.8]. The dashed and dotted lines indicate the 95% CI of predicted observations.

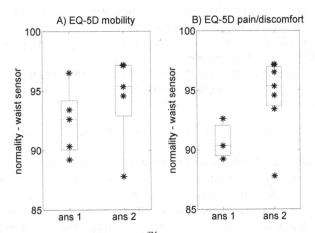

Fig. 11. Normality compared to EQ-5D[TM] answers regarding (A) mobility and (B) pain/discomfort. Mobility answers - ans 1: I have no problems in walking about; ans 2: I have some problems in walking about. Pain/discomfort answers - ans 1: I have no pain or discomfort; ans 2: I have moderate pain or discomfort.

tendency for better health to be accompanied by better normality index. This correlation is particularly strong between normality and usual activities scores, Spearman's r=0.75, p=0.0127. There was no correlation between the health scale in Part B of the questionnaire and normality.

Improvement in normality was calculated as the difference between baseline and follow-up values. Figure 13 shows how improvement in normality correlates with number of days spent at the ward after surgery. Although a Wilcoxon test indicated that there was no statistically significant difference between groups, the Spearman's rank

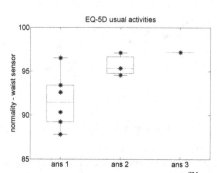

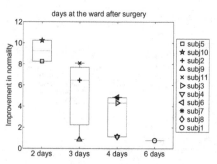

Fig. 12. Normality compared to EQ-5D™ answers regarding usual activities. Ans 1: I have no problems with performing my usual activities; ans 2: I have some problems with performing my usual activities; ans 3: I am unable to perform my usual activities.

Fig. 13. Improvement in normality compared to days spent at ward after surgery. Improvement in normality is the difference between normality values at baseline and at follow-up. Although the distributions are not statistically different, variables are well correlated, Spearman's rank correlation coefficient r=-0.75, p=0.0081.

correlation coefficient was r=-0.75, p=0.0081. There was no correlation between improvement in normality and days between baseline and follow-up sessions.

The normality index can also be evaluated based on its discriminatory values. That is, the ability to differentiate baseline measurements from follow-up measurements. The AUC was 0.94, confidence interval (CI) (0.87, 1.00), p<0.0001. The test-retest reliability was also high, r=0.81, CI (0.60, 0.92), p<0.0001.

5   Discussion

The in-lab evaluation confirmed that the proposed measures of normality and symmetry are well correlated with the kinematic measures, namely MAP, GPS, MAP-symmetry and GPS-symmetry. The found optimal parameters were used for the in-situ evaluation. The remaining of this section discusses the results of the in-situ evaluation.

The average speeds at baseline and follow-up are in agreement with measurements reported in [29], 0.46 m/s less than 16 days after hip replacement surgery and 1.17 m/s more than 20 days after surgery. Average gait speed of approximately 1 m/s three months after surgery were also reported in [30]. According to [31] the greatest improvements in gait speed are observed within the first three months post-op. The follow-up measurement can, therefore, be considered representative of patient's improvement in gait speed. In addition, [32] determined that changes in speed superior to 0.10 m/s are clinically meaningful after hip fracture treatment. The changes in speed observed from baseline to follow-up, 0.60±0.29, are therefore also clinically meaningful.

Measures of gait normality correlate well with both gait speed, Figure 9 and step length, Figure 10. Given that speed and step length are measures related to patient recovery, there is a good chance the normality index is also a good indicator of recovery. Unfortunately, no other quantitative gait parameters were available in the data set to demonstrate that the normality index correlates to recovery when the data is corrected for speed. However, in Part I of this study symmetry and normality measures are shown to correlate to joint-angle curves normalized to stride time, not containing any velocity information. The normality index is also normalized to stride time and as such is independent of walking cadence. It is expected that the normality index would differentiate between normal and abnormal patterns at the same speed. Further investigations are needed to support this assumption.

Another factor supporting the usefulness of the normality index is its correlation with the type of walking aid used during the test, Figure 8. The test-retest reliability and discriminatory power of the index were also satisfactory. Overall, the proposed index can possibly be developed into a reliable and clinically relevant measure of gait normality.

Another interesting result was the correlation between improvement of normality and number of days spent at the ward, Figure 13. Whereas there was no correlation between improvement in normality and number of days between baseline and follow-up. This possibly suggests that the rate of recovery at the ward is indicative of the total rate of recovery, which is little affected by the recovery time at home. This assumption should be further investigated.

Normality results and the answers to the EQ-5D^{TM} questionnaire showed some positive trends. Greater discomfort and difficulties in performing usual activities seem to be accompanied with worse normality, Figure 11. Besides the self-assessment questionnaire, the use of walking aids was also considered an indication of how well the patient's health status was, i.e. patients who did not need any walking aid were, on average, in better condition than those who used one crutch. Another indicator of recovery was the number of days the patient spent at ward, assuming that patients who recovered better or more quickly were discharged sooner. The normality index seems to be in agreement with all the above mentioned qualitative health status assessments.

Symmetry results are difficult to judge due to the variety of walking aids used. The large variety of symmetry at follow-up, Figure 5, was mostly influenced by the patients using one crutch only. This could be explained by the fact that some patients were more dependent on the crutch and consequently leaned more to one side. Whereas some patients barely used the crutch for support.

Due to their recent surgery, patients were very uncomfortable during the baseline measurements. It was important to keep the data collection as simple and quick as possible. No more than five minutes had to be spared by the patient to complete the entire procedure, and they were all willing to participate in the study. Briefness is also important for the staff responsible for the procedure so that the addition of GA is not an extra burden. The placement of the sensors was also quick and easy. However, in the future, the waist sensor should be placed on the lower back so as not to be affected by subjects' different shapes and sizes.

Another issue with the present study is that the number of participants was very small. Any statistical inference on the results is greatly affected by the sample size. However, results are promising and suggest that a larger study will likely produce positive results.

At the ward where the data was collected, gait analysis is not normally used, and most records are based on rough qualitative descriptions. This lack of quantitative measures makes the assessment of patient improvement a difficult and very subjective task. The introduction of a simple 10-meter walk test can already provide quantitative measures of speed and stride length. The addition of a wearable GA system, however, can quickly increase the amount of quantitative data to include more complex measures of symmetry and normality.

6   Conclusions

The present study investigated the use of inertial sensors for quantitative GA, both in-lab and in-situ. The proposed system served as a tool to facilitate the extraction of certain gait characteristics, namely symmetry and normality. The system was evaluated against 3D kinematic measures of symmetry and normality, as well as clinical assessments of hip-replacement patients. Not only was the proposed system in agreement with kinematic variables but it also correlated well with the level of recovery and health status of the patients in a very intuitive way. This study showed that such a system may be deployed in a real clinical environment in order to aid current clinical assessment by incorporating quantitative GA.

Acknowledgements. This study was partially funded by the Promobilia Foundation. The authors would like to thank Lars-Eric Olsson, PhD, for including the present study in his project, funded by the Institute of Health and Care Sciences, Sahlgrenska Academy, University of Gothenburg, Sweden.

References

1. Chang, F.M., Rhodes, J.T., Flynn, K.M., Carollo, J.J.: The role of gait analysis in treating gait abnormalities in cerebral palsy. Orthopedic Clinics of North America 41(4), 489–506 (2010)
2. Salarian, A., Russmann, H., Vingerhoets, F., Dehollain, C., Blanc, Y., Burkhard, P., Aminian, K.: Gait assessment in Parkinson's disease: Toward an ambulatory system for long-term monitoring. IEEE Transactions on Biomedical Engineering 51(8), 1434–1443 (2004)
3. Cruz, T.H., Dhaher, Y.Y.: Evidence of abnormal lower-limb torque coupling after stroke: An isometric study supplemental materials and methods. Stroke 39(1), 139–147 (2008)
4. Benedetti, M., Montanari, E., Catani, F., Vicenzi, G., Leardini, A.: Pre-operative planning and gait analysis of total hip replacement following hip fusion. Computer Methods and Programs in Biomedicine 70(3), 215–221 (2003)
5. Toro, B., Nester, C., Farren, P.: A review of observational gait assessment in clinical practice. Physiotherapy Theory and Practice 19(3), 137–149 (2003)
6. Embrey, D.G., Yates, L., Mott, D.H.: Effects of neuro-developmental treatment and orthoses on knee flexion during gait: A single-subject design. Physical Therapy 70(10), 626–637 (1990)
7. Brown, C., Hillman, S., Richardson, A., Herman, J., Robb, J.: Reliability and validity of the visual gait assessment scale for children with hemiplegic cerebral palsy when used by experienced and inexperienced observers. Gait & Posture 27(4), 648–652 (2008)
8. Ong, A., Hillman, S., Robb, J.: Reliability and validity of the edinburgh visual gait score for cerebral palsy when used by inexperienced observers. Gait & Posture 28(2), 323–326 (2008)

9. Kempen, J., de Groot, V., Knol, D., Polman, C., Lankhorst, G., Beckerman, H.: Community walking can be assessed using a 10 metre timed walk test. Multiple Sclerosis Journal 17(8), 980–990 (2011)

10. Nordin, E., Lindelöf, N., Rosendahl, E., Jensen, J., Lundin-Olsson, L.: Prognostic validity of the timed up-and-go test, a modified get-up-and-go test, staff's global judgement and fall history in evaluating fall risk in residential care facilities. Age and Ageing 37(4), 442–448 (2008)

11. Potter, J.M., Evans, A.L., Duncan, G.: Gait speed and activities of daily living function in geriatric patients. Archives of Physical Medicine and Rehabilitation 76(11), 997–999 (1995)

12. Cesari, M., Kritchevsky, S.B., Penninx, B.W.H.J., Nicklas, B.J., Simonsick, E.M., Newman, A.B., Tylavsky, F.A., Brach, J.S., Satterfield, S., Bauer, D.C., Visser, M., Rubin, S.M., Harris, T.B., Pahor, M.: Prognostic value of usual gait speed in well-functioning older people - results from the health, aging and body composition study. Journal of the American Geriatrics Society 53(10), 1675–1680 (2005)

13. Buchner, D.M., Larson, E.B., Wagner, E.H., Koepsell, T.D., De Lateur, B.J.: Evidence for a non-linear relationship between leg strength and gait speed. Age and Ageing 25(5), 386–391 (1996)

14. Russell, E., Braun, B., Hamill, J.: Does stride length influence metabolic cost and biomechanical risk factors for knee osteoarthritis in obese women? Clinical Biomechanics 25(5), 438–443 (2010)

15. Crenshaw, S.J., Richards, J.G.: A method for analyzing joint symmetry and normalcy, with an application to analyzing gait. Gait & Posture 24(4), 515–521 (2006)

16. Schutte, L.M., Narayanan, U., Stout, J.L., Selber, P., Gage, J.R., Schwartz, M.H.: An index for quantifying deviations from normal gait. Gait & Posture 11(1), 25–31 (2000)

17. Schwartz, M.H., Rozumalski, A.: The gait deviation index: A new comprehensive index of gait pathology. Gait & Posture 28(3), 351–357 (2008)

18. Baker, R., McGinley, J.L., Schwartz, M.H., Beynon, S., Rozumalski, A., Graham, H.K., Tirosh, O.: The gait profile score and movement analysis profile. Gait & Posture 30(3), 265–269 (2009)

19. Beynon, S., McGinley, J.L., Dobson, F., Baker, R.: Correlations of the gait profile score and the movement analysis profile relative to clinical judgments. Gait & Posture 32(1), 129–132 (2010)

20. Sadeghi, H., Allard, P., Prince, F., Labelle, H.: Symmetry and limb dominance in able-bodied gait: A review. Gait & Posture 12(1), 34–45 (2000)

21. Zifchock, R., Davis, I., Higginson, J., Royer, T.: The symmetry angle: A novel, robust method of quantifying asymmetry. Gait & Posture 27(4), 622–627 (2008)

22. Moe-Nilssen, R., Helbostad, J.L.: Estimation of gait cycle characteristics by trunk accelerometry. Journal of Biomechanics 37(1), 121–126 (2004)

23. Gouwanda, D., Senanayake, A.S.M.N.: Identifying gait asymmetry using gyroscopes-a cross-correlation and normalized symmetry index approach. Journal of Biomechanics 44(5), 972–978 (2011)

24. Sant'Anna, A., Wickström, N.: A symbol-based approach to gait analysis from acceleration signals: Identification and detection of gait events and a new measure of gait symmetry. IEEE Transactions on Information Technology in Biomedicine 14(5), 1180–1187 (2010)

25. Sant'Anna, A., Salarian, A., Wickström, N.: A new measure of movement symmetry in early parkinson's disease patients using symbolic processing of inertial sensor data. IEEE Transaction on biomedical Engineering 58(7) (2011)

26. Tranberg, R., Saari, T., Zügner, R., Kärrholm, J.: Simultaneous measurements of knee motion using an optical tracking system and radiostereometric analysis (RSA). Acta Orthopaedica 82(2), 171–176 (2011)

27. DeLong, E.R., DeLong, D.M., Clarke-Pearson, D.L.: Comparing the areas under two or more correlated receiver operating characteristic curves: A nonparametric approach. Biometrics 44(3), 837–845 (1988)
28. McGraw, K.O., Wong, S.P.: Forming inferences about some intraclass correlation coefficients. Psychological Methods 1, 30–46 (1996)
29. Kennedy, D., Stratford, P., Wessel, J., Gollish, J., Penney, D.: Assessing stability and change of four performance measures: a longitudinal study evaluating outcome following total hip and knee arthroplasty. BMC Musculoskeletal Disorders 6(1), 3 (2005)
30. Aminian, K., Rezakhanlou, K., De Andres, E., Fritsch, C., Leyvraz, P.F., Robert, P.: Temporal feature estimation during walking using miniature accelerometers: an analysis of gait improvement after hip arthroplasty. Medical and Biological Engineering and Computing 37, 686–691 (1999)
31. Macnicol, M., McHardy, R., Chalmers, J.: Exercise testing before and after hip arthroplasty. Journal of Bone and Joint Surgery - British 62-B(3), 326–331 (1980)
32. Palombaro, K.M., Craik, R.L., Mangione, K.K., Tomlinson, J.D.: Determining meaningful changes in gait speed after hip fracture. Physical Therapy 86(6), 809–816 (2006)

MRI TV-Rician Denoising

Adrian Martin[1,2,3], Juan-Francisco Garamendi[4], and Emanuele Schiavi[2,3]

[1] Fundación CIEN-Fundación Reina Sofía, Madrid, Spain
[2] Neuroimaging lab., Univ. Politécnica de Madrid-Univ. Rey Juan Carlos,
Center for Biomedical Technology, Madrid, Spain
[3] Departamento de Matemática Aplicada, Universidad Rey Juan Carlos, Madrid, Spain
[4] INRIA/INSERM U746/CNRS, UMR6074/University of Rennes I,
VisAGeS Research Team, Rennes, France
{adrian.martin,emanuele.schiavi}@urjc.es,
juan-francisco.garamendi_bragado@inria.fr

Abstract. Recent research on magnitude Magnetic Resonance Images (MRI) reconstruction from the Fourier inverse transform of complex (gaussian contaminated) data sets focuses on the proper modeling of the resulting Rician noise contaminated data. In this paper we consider a variational Rician denoising model for MRI data sets that we solve by a semi-implicit numerical scheme, which leads to the resolution of a sequence of Rudin, Osher and temi (ROF) models. The (iterated) resolution of these well posed numerical problems is then proposed for Total Variation (TV) Rician denoising. For numerical comparison we also consider a direct semi-implicit approach for the primal problem which amounts to consider some (regularizing) approximating problems. Synthetic and real MR brain images are then denoised and the results show the effectiveness of the new method in both, the accuracy and the speeding up of the algorithm.

Keywords: MRI Rician denoising, Total variation, Numerical resolution, ROF model.

1 Introduction

Modelling MRI denoising, a fundamental step in medical image processing, leads naturally to the assumption that MR magnitude images are corrupted by Rician noise which is a signal dependent noise (see [1], [2] and [3]). In fact this noise is originated in the computation of the magnitude image from the real and imaginary images, that are obtained from the inverse Fourier Transform applied to the original raw data. This process involves a non-linear operation which maps the original Gaussian distribution of the noise to a Rician distribution.

Nevertheless it is usually argued that this bias do not affect seriously the processing and subsequent analysis of MR images and a gaussian noise, identically distributed and not signal dependent, is modeled. To go beyond the unlikely assumption of gaussian noise, we consider, in a variational framework, a denoising model for MR Rician noise contaminated images recently considered (independently) in [4] and in [5], which combines the Total Variation semi-norm with a data fitting term (see also [6] for an application to DT-MRI data denoising where low SNR Diffusion Weighted Images (DWI)

J. Gabriel et al. (Eds.): BIOSTEC 2012, CCIS 357, pp. 255–268, 2013.

are acquired). When the resulting functional is considered for minimization, the variational approach leads to the resolution of a nonlinear degenerate PDE elliptic equation as the Euler Lagrange equation for optimization. This has a number of theoretical problems when the Total Variation operator is considered as a smoother, because the energy functional is not differentiable at the origin (i.e. $\nabla u = \bar{0}$) and regular, approximating problems must be solved. In turn this approach cause a over-smoothing effect in the numerical solutions of the model and accuracy in fine scale details is lost because the edges diffuse. A direct gradient descent method has been used in [4] in order to validate the model assumption of Rician noise but the method is found to be inherently slow because a stabilization at the steady state is needed. Also, that scheme is finally explicit and very small time steps have to be used to avoid numerical oscillations.

Our aim is to present a new framework to solve numerically and efficiently the gradient descent scheme (gradient flow) associated to the Rician energy minimization problem introducing a new semi-implicit formulation. Using a simple Euler discretization of the time derivative, stationary problems of the Rudin, Osher and Fatemi (ROF) type [7] are deduced. This allows to use the well known dual formulation of the ROF model proposed in [8] for a speed up of the computations. As a by-product of this approach the exact Total Variation operator can be computed and this provides accuracy to the solution in so far truly (discontinuous) bounded variation solutions are numerically approximated.

This paper is organized as follows: in section 2 and 3 we present the model equation and the numerical scheme recently proposed in [4]. In section 4 we propose a new framework which leads to a more efficient and accurate numerical scheme. The proposed method is tested in section 5, where we consider synthetic MR brain images to compare it with the method of [4] and some preliminary results of applying this algorithm to real Diffusion Weighted Magnetic Resonance Images (DW-MRI) are shown in subsection 5. Finally in section 6 we present our conclusions.

2 Model Equations

Let Ω be a bounded open subset of $\mathbb{R}^d$, $d = 2, 3$ defining the image domain and let $f : \Omega \to \mathbb{R}$ be a given noisy image representing the data, with $f \in L^\infty(\Omega) \cap [0, 1]$ (otherwise we normalize). Let $BV(\Omega)$ be the space of functions with bounded variation in Ω equipped with the seminorm $|u|_{BV}$ defined as

$$|u|_{BV} = \sup \left\{ \int_\Omega u(x) \operatorname{div} \bar{\xi}(x) \, dx \ : \ \bar{\xi} \in C_c^1(\Omega, \mathbb{R}^d), \ |\bar{\xi}(x)|_\infty \leq 1, \ x \in \Omega \right\} \quad (1)$$

where $|\cdot|_\infty$ denotes the l_∞ norm in $\mathbb{R}^d$, $|\bar{\xi}|_\infty = \max_{1 \leq i \leq d} |\xi_i|$ (details on this space and the related geometric measure theory can be found in [9]). Following a Bayesian modelling approach we consider the minimization problem

$$\min_{u \in BV(\Omega)} J(u) + \lambda H(u, f) \quad (2)$$

where $J(u)$ is the convex nonnegative total variation regularization functional

$$J(u) = |u|_{BV} = \int_\Omega |Du| \tag{3}$$

being $\int_\Omega |Du|$ the Total variation of u with Du its generalized gradient (a vector bounded Radon measure). When $u \in W^{1,1}(\Omega)$ we have $\int_\Omega |Du| = \int_\Omega |\nabla u| dx$. The λ parameter in (2) is a scale parameter tuning the model.

The data term $H(u, f)$ is a fitting functional which is nonnegative with respect to u for fixed f. To model Rician noise the form of $H(u, f)$ has been deduced in [6] in the context of weighed diffusion tensor MR images. The Rician likelihood term is of the form:

$$H(u, f) = \int_\Omega \left(\left(\frac{u^2 + f^2}{2\sigma^2} \right) - \log I_0 \left(\frac{uf}{\sigma^2} \right) - \log \left(\frac{f}{\sigma^2} \right) \right) dx \tag{4}$$

where σ is the standard deviation of the Rician noise of the data and I_0 is the modified zeroth-order Bessel function of the first kind. Notice that the constant terms $(1/2\sigma^2)\|f\|_2^2$ and $\int_\Omega \log (f/\sigma^2)$ appearing in (4) do not affect the minimization problem. Dropping these terms (which do not allows to define the energy $H(u, 0)$ corresponding to a black image $f \equiv 0$) we have:

$$H(u, f) = \frac{1}{2\sigma^2} \int_\Omega u^2 dx - \int_\Omega \log I_0 \left(\frac{uf}{\sigma^2} \right) dx \tag{5}$$

with $H(u, 0) = (1/2\sigma^2)\|u\|_2^2$ and $H(0, f) = 0$ for any given $f \geq 0$. Using (2), (3) and (5) the minimization problem is formulated as:

Fixed λ and σ and given a noisy image $f \in L^\infty(\Omega) \cap [0, 1]$ recover $u \in BV(\Omega) \cap L^\infty(\Omega) \cap [0, 1]$ minimizing the energy:

$$J(u) + \lambda H(u, f) = |Du|(\Omega) + \frac{\lambda}{2\sigma^2} \int_\Omega u^2 dx - \lambda \int_\Omega \log I_0 \left(\frac{uf}{\sigma^2} \right) dx \tag{6}$$

Due to the fact that the functional in (3) (hence in (6)) is not differentiable at the origin we introduce the subdifferential of $J(u)$ at a point u by

$$\partial J(u) = \{p \in BV(\Omega)^* | J(v) \geq J(u) + <p, v - u>\}$$

for all $v \in BV(\Omega)$, to give a (weak and multivalued) meaning to the Euler-Lagrange equation associated to the minimization problem (6). In fact the first order optimality condition reads

$$\partial J(u) + \lambda \partial_u H(u, f) \ni 0 \tag{7}$$

with (Gâteaux) differential

$$\partial_u H(u, f) = \left(\frac{u}{\sigma^2} \right) - \left[I_1 \left(\frac{uf}{\sigma^2} \right) / I_0 \left(\frac{uf}{\sigma^2} \right) \right] \left(\frac{f}{\sigma^2} \right)$$

where I_1 is the modified first-order Bessel function of the first kind and verifies ([10]) $0 \leq I_1(\xi)/I_0(\xi) < 1, \forall \xi > 0$. This model, first proposed in [4], differs from [6] because of the geometric prior (the TV-based regularization term) which substitutes their Gibb's prior model based on the Perona and Malik energy functional [11]. Also it differs from the classical gaussian noise model because of the nonlinear dependence of the solution of the ratio I_1/I_0.

3   The Primal Descent Gradient Numerical Scheme

A number of mathematical difficulties is associated with the multivalued formulation (7) and a regularization of the diffusion term $\text{div}\,(\nabla u/|\nabla u|)$ in form $\text{div}\,(\nabla u/|\nabla u|_\epsilon)$, with $|\nabla u|_\epsilon = \sqrt{|\nabla u|^2 + \epsilon^2}$ and $0 < \epsilon \ll 1$ is implemented to avoid degeneration of the equation where $\nabla u = \bar{0}$. Using this approximation it is possible to give a (weak) meaning to the following formulation:

Fixed λ, σ and (small) ϵ and given $f \in L^\infty(\Omega) \cap [0,1]$ find $u_\epsilon \in W^{1,1}(\Omega) \cap [0,1]$ solving

$$-\text{div}\left(\frac{\nabla u_\epsilon}{|\nabla u_\epsilon|_\epsilon}\right) + \frac{\lambda}{\sigma^2}\left[u_\epsilon - r_\epsilon(u_\epsilon, f)f\right] = 0 \tag{8}$$

complemented with Neumann homogeneous boundary conditions $\partial u_\epsilon/\partial n = 0$ and where, for notational simplicity, we introduced the nonlinear function $r_\epsilon(u_\epsilon, f) = I_1(u_\epsilon f/\sigma^2)/I_0(u_\epsilon f/\sigma^2)$.

This is a nonlinear (in fact quasilinear) elliptic problem that we solve with a gradient descent scheme until stabilization (when $t \to +\infty$) of the evolutionary solution to steady state, i.e. a solution of the elliptic problem (8) which is a minimum of the energy

$$J_\epsilon(u_\epsilon) + \lambda H(u_\epsilon, f) =$$

$$= \int_\Omega \sqrt{|\nabla u_\epsilon|^2 + \epsilon^2}dx + \frac{\lambda}{2\sigma^2}\int_\Omega u^2 dx - \lambda \int_\Omega \log I_0\left(\frac{uf}{\sigma^2}\right)dx \tag{9}$$

When $\epsilon \to 0$ we have $u_\epsilon \to u$, $J_\epsilon(u_\epsilon) \to J(u)$ and the energies in (6) and (9) coincide.

This approach amounts to solve the associated nonlinear parabolic problem:

$$\frac{\partial u_\epsilon}{\partial t} = \text{div}\left(\frac{\nabla u_\epsilon}{|\nabla u_\epsilon|_\epsilon}\right) - \frac{\lambda}{\sigma^2}\left[u_\epsilon - r_\epsilon(u_\epsilon, f)f\right] \tag{10}$$

complemented with Neumann homogeneous boundary conditions $\partial u_\epsilon/\partial n = 0$ and initial condition $u_\epsilon(0, x) = u_0^\epsilon(x)$ whose (weak) solution stabilizes (when $t \to +\infty$) to the steady state of (8), i.e. a minimum of (9) which approximates, for ϵ sufficiently small, a minimum of the energy functional (6). Following [4] and using forward finite difference for the temporal derivative it is straightforward to define a semi-implicit iterative scheme which simplifies to the explicit one:

$$\left(1 + \tau\frac{\lambda}{\sigma^2}\right)u_\epsilon^{n+1} = u_\epsilon^n + \tau\left(\text{div}\left(\frac{\nabla u_\epsilon^n}{|\nabla u_\epsilon^n|_\epsilon}\right) + \frac{\lambda}{\sigma^2}r(u_\epsilon^n, f)f\right) \tag{11}$$

where τ is the time step and spatial discretization for the approximated TV-term is performed as in [12].

4   A Semi-implicit Formulation

In the previous section we considered the approximated Euler-Lagrange equation (8) associated to the minimization of the energy (6). This is a modelling approximation and we can get rid of it. In fact, considering the original Euler-Lagrange equation associated to the energy (6) we have (with abuse of notation for the diffusive term)

$$-\text{div}\left(\frac{\nabla u}{|\nabla u|}\right) + \frac{\lambda}{\sigma^2}\left[u - r(u, f)f\right] = 0 \tag{12}$$

with $r(u, f) = I_1(uf/\sigma^2)/I_0(uf/\sigma^2)$. A rigorous treatment of equation (12) should follow the multivalued formalism of (7).

Using again a gradient descent scheme we have to solve the parabolic problem:

$$\frac{\partial u}{\partial t} = \text{div}\left(\frac{\nabla u}{|\nabla u|}\right) - \frac{\lambda}{\sigma^2}\left[u - r(u, f)f\right] \tag{13}$$

together with Neumann homogeneous boundary conditions $\partial u/\partial n = 0$ and initial condition $u(0, x) = u_0(x)$. For comparison purposes we used $u_0(x) = u_0^\epsilon(x)$ in all numerical tests.

Using forward finite difference for the temporal derivative in (13) and a semi-implicit scheme where only the term depending on the ratio of the Bessel's functions is delayed, results in the numerical scheme:

$$\left(1 + \tau\frac{\lambda}{\sigma^2}\right)u^{n+1} = u^n + \tau\left(\text{div}\left(\frac{\nabla u^{n+1}}{|\nabla u^{n+1}|}\right) + \frac{\lambda}{\sigma^2}r(u^n, f)f\right) \tag{14}$$

where the diffusive term is (formally) exact and implicitly considered (compare with (11)). Defining $\beta = (\tau\lambda)/\sigma^2$, $\gamma = (1 + \beta)/\tau$ and

$$g^n = \left(\frac{1}{1 + \beta}\right)u^n + \left(\frac{\beta}{1 + \beta}\right)r(u^n, f)f \tag{15}$$

we can write:

$$-\text{div}\left(\frac{\nabla u^{n+1}}{|\nabla u^{n+1}|}\right) + \left(\frac{1}{\gamma}\right)(u^{n+1} - g^n) = 0 \tag{16}$$

which is the Euler-Lagrange equation of the ROF energy functional ([7]):

$$E_n(u) = \int_\Omega |Du| + \left(\frac{1}{2\gamma}\right)\int_\Omega (u - g^n)^2 dx \tag{17}$$

for any positive integer $n > 0$, with (artificial) time $t_n = n\tau$. Hence, at each gradient descent step τ, we can solve a ROF problem associated to the minimization of the energy (17) in the space $BV(\Omega) \cap [0, 1]$. This problem is mathematically well-posed and it can be numerically solved by very efficient methods, when formulated using well known duality arguments as can be seen in [8], in order to apply this algorithm we must introduce the discrete setting for the Total Variation Rician Denoising problem.

Numerical Resolution for Rician Denoising

We consider a regular Cartesian grid of size $N \times M$: $(i\,h, j\,h)_{1 \leq i \leq N, 1 \leq j \leq M}$ where h denotes the size of the spacing. The matrix $(u^h_{i,j})$ represents a discrete image where each pixel $u_{i,j}$ is located in the correspondent node $(i\,h, j\,h)$. In what follows, we shall choose $h = 1$ because it only causes a rescaling of the energy. Henceforth we shall drop the dependence of the mesh size and $u^h = u$. Let $X = \mathbb{R}^{N \times M}$ be the vectorial solutions space. We introduce the discrete gradient $\nabla^h = \nabla$ defined as:

$$(\nabla u)_{i,j} = \begin{pmatrix} (\partial^+_x u)_{i,j} \\ (\partial^+_y u)_{i,j} \end{pmatrix} = \begin{pmatrix} u_{i+1,j} - u_{i,j} \\ u_{i,j+1} - u_{i,j} \end{pmatrix} \tag{18}$$

except at the boundaries $i = N$ where $(\partial^+_x u)_{N,j} = 0$, and $j = M$ with $(\partial^+_y u)_{i,j} = 0$. Hence the (discrete) ∇ operator is a linear map from X to $Y = X \times X$. The discrete version of the isotropic Total Variation semi-norm is:

$$\|\nabla u\|_{\ell^1} = \sum_{i,j} |(\nabla u)_{i,j}|,$$

with

$$|(\nabla u)_{i,j}| = \sqrt{((\nabla u)^1_{i,j})^2 + ((\nabla u)^2_{i,j})^2}$$

The discrete energy for Rician denoising (6) reads as:

$$\sum_{i,j} |(\nabla u)_{i,j}| + \lambda \sum_{i,j} \left[\frac{u^2_{i,j}}{2\sigma^2} - \log I_0 \left(\frac{u_{i,j} f_{i,j}}{\sigma^2} \right) \right] \tag{19}$$

where the matrix $(f_{i,j})$ represents the discrete noisy image, with each pixel $f_{i,j}$ located at the node (i, j). In the same way we can define the discrete version of the ROF problem deduced in (17) which is (at a generic time step t^n)

$$\sum_{i,j} |(\nabla u)_{i,j}| + \left(\frac{1}{2\beta} \right) \sum_{i,j} (u_{i,j} - g^n_{i,j})^2 \tag{20}$$

The algorithm presented in [8] is based in the dual formulation of the ROF problem then if we endow the spaces X and Y with the standard Euclidean scalar product, the adjoint operator of the discrete gradient ∇^h (see 18) denoted by $-\mathrm{div}^h = \mathrm{div}$ is defined as

$$< \nabla u, p >_Y = - < u, \mathrm{div}\, p >_X \tag{21}$$

for any $u \in X$ and (with $p^h = p$) say $p = (p^1_{i,j}, p^2_{i,j}) \in Y$, and it is given by the following formulas:

$$(\mathrm{div}\, p)_{i,j} = (p^1_{i,j} - p^1_{i-1,j}) + (p^2_{i,j} - p^2_{i,j-1})$$

for $2 \leq i, j \leq N - 1$. The term $(p^1_{i,j} - p^1_{i-1,j})$ is replaced with $p^1_{i,j}$ if $i = 1$ and with $-p^1_{i-1,j}$ if $i = N$, while the term $(p^2_{i,j} - p^2_{i,j-1})$ is replaced with $p^2_{i,j}$ if $j = 1$ and with $-p^2_{i,j-1}$ if $j = N$. The final algorithm for Rician Denoising is as follows:

TV-Rician Denoising Algorithm

- Initialization: given λ, σ, τ_1, τ_2, ϵ_1, ϵ_2 and f. Set:
  - $u^0 \not\equiv 0$
  - $\beta = (\tau_1 \lambda)/\sigma^2$
  - $\gamma = (1 + \beta)/\tau_1$

- While $\|u^k - u^{k-1}\|_\infty > \epsilon_1$:

  1. $g_{i,j}^k = \left(\dfrac{1}{1+\beta}\right) u_{i,j}^k + \left(\dfrac{\beta}{1+\beta}\right) r(u_{i,j}^k, f_{i,j}) f_{i,j}$

  2. Set $p^0 \equiv 0$. While $\|p^n - p^{n-1}\|_\infty > \epsilon_2$:
     - $p_{i,j}^{n+1} = \dfrac{p_{i,j}^n + \tau_2(\nabla(\operatorname{div} p^n - g^k/\gamma))_{i,j}}{1 + \tau_2|(\nabla(\operatorname{div} p^n - g^k/\gamma))_{i,j}|}$

  3. $u_{i,j}^{k+1} = g_{i,j}^k - \gamma\,(\operatorname{div} p)_{i,j}$

5   Results and Discussion

The theoretical result presented in the previous section have to be numerically confirmed in order to asses the well behaviour of the method and also the advantages it presents when it is compared to the original regularized method which computes the approximating u_ϵ solution. In order to assess the performance of our algorithm we tested it with synthetic and real brain images. The obtained results are presented and discussed below.

Synthetic Brain Images
The synthetic brain images we used for our study were obtained from the BrainWeb Simulated Brain Database[1] at the Montreal Neurological Institute [13]. The original phantoms were contaminated artificially with Rician noise considering the data as a complex image with zero imaginary part and adding random gaussian perturbations to both the real and imaginary part, before computing the magnitude image. This process allows to control the amount and distribution of the Rician noise so providing a gold standard for our study. For this we used different values of the σ parameter which represents the variance of the noise ($\sigma = 0.025$, $\sigma = 0.05$ and $\sigma = 0.1$) and different values of the λ parameter ($\lambda = 0.05$, $\lambda = 0.1$ and $\lambda = 0.125$). Notice that, fixed σ (which can be estimated for real images) the λ parameter is the only one we have to choose for regularization (as in the gaussian case).

We can observe in Figure 1 the performance of the denoising method based on the semi-implicit formulation for $\lambda = 0.05$, $\lambda = 0.1$ and $\lambda = 0.125$. This implicit method solves exactly the total variation operator in (6) due to its dual formulation and not its approximate form as the explicit method which solves the primal formulation, so the solution obtained should be close to the ideal minimum of (6). This behaviour can be

---

[1] Available at http://www.bic.mni.mcgill.ca/brainweb

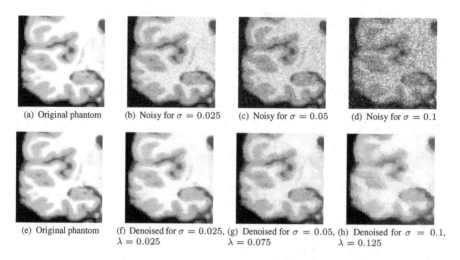

(a) Original phantom   (b) Noisy for $\sigma = 0.025$ (c) Noisy for $\sigma = 0.05$ (d) Noisy for $\sigma = 0.1$

(e) Original phantom   (f) Denoised for $\sigma = 0.025$,   (g) Denoised for $\sigma = 0.05$,   (h) Denoised for $\sigma = 0.1$,
$\lambda = 0.025$ $\lambda = 0.075$ $\lambda = 0.125$

Fig. 1. The original free noise phantom is shown in images a) and e). In b), c) and d) the contaminated phantoms for $\sigma = 0.025$, 0.05 and 0.1 respectively. Below, their respective denoised images e), f) and g) for $\lambda = 0.025$, 0.075 and 0.125.

in fact observed in Figure 2, where using the same values for the algorithms ($\tau = 10^{-3}$ and $\lambda = 0.1$) the proposed method reach a solution whose energy is smaller than the obtained by the solution of the first method. This difference caused by the fact that now we are using the true Total Variation can be also observed in the images of the absolute difference between the original (free of noise) image and the solutions found by the two methods. We can see how the image difference corresponding to the solution of the approximated method (Figure 3 a) presents more structural details than the image corresponding to the implicit method (Figure 3 b), which confirms that this last method recovers more structural details, that are eventually lost by the explicit method because of the ϵ approximation.

The other important characteristic of this new formulation is that the diffusion term is implicitly considered and this provides numerical stability which in turn allows to increase the value of τ compared to those used in the explicit method, so less iterations of the algorithm are necessary for time stabilization. In fact if we increase the value of τ to the value $\tau = 10^{-1}$ the explicit method becomes unstable and it begins to oscillate without reaching the minimum of the energy we obtained with $\tau = 10^{-3}$. Also the implicit method takes less iterations to reach the same minimum. The performance of the two algorithms for $\tau = 10^{-1}$ can be observed in Figure 4 where the energy computed along the iterates of the implicit method is clearly less than the same energy calculated along the approximated iterates.

This behaviour is crucial for the selection of the algorithm in so far even if the new method has more computational cost per iteration (because we solve a ROF problem at each iteration), we can increase the value of τ in order to reach the solution in less iterations than the first method, finding a best solution for our problem (in the sense of figure 4) and spending less time of computation. Finally, in the last figure(figure 5) we show that this framework is robust in the sense that the same solution is obtained

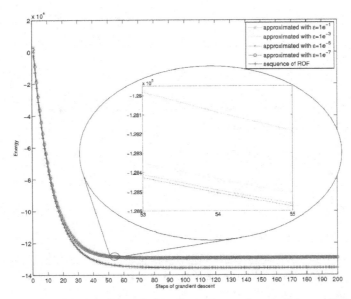

Fig. 2. Energy in functional 6 of the solution obtained at each step of the gradient descent by the approximated method for $\lambda = 0.1$, $\sigma = 0.05$, $\tau = 0.001$ and different values of $\epsilon = 10^{-1}, 10^{-3}, 10^{-5}, 10^{-7}$, and by the new method for $\lambda = 0.1$, $\sigma = 0.05$, $\tau = 0.001$

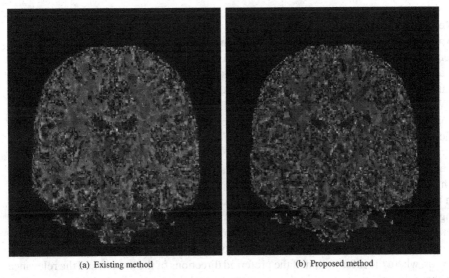

(a) Existing method   (b) Proposed method

Fig. 3. Absolute difference between the original image and the solution of the existing method and the proposed method for the values $\lambda = 0.1$, $\sigma = 0.05$, $\tau = 0.001$ in both methods and $\epsilon = 10^-7$ for the approximated method

when completely different initial condition are used for initialization in the gradient flow schemes we considered. This is suggestive of uniqueness for the non trivial solution of the corresponding elliptic problems.

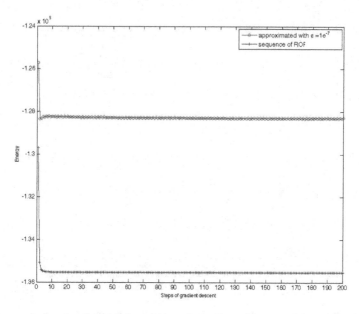

Fig. 4. Energy in functional (6) of the solution obtained at each step of the gradient descent by the approximated method for $\epsilon = 10^{-7}$, $\lambda = 0.1$, $\sigma = 0.05$, $\tau = 0.1$ and by the new method for $\lambda = 0.1$, $\sigma = 0.05$, $\tau = 0.1$

Real Brain Images

Apart from the modelling exercise and the implementation details of the algorithm presented above, our main interest relies in the application of the proposed algorithm to real brain images. In the following we present some preliminary results we are obtaining for Diffusion Weighted Magnetic Resonance Images (DW-MRI) denoising. The DW-MR images are acquired and used for Diffusion Tensor Image (DTI) reconstruction, and the importance of the denoising step is crucial in DW-MRI analysis because their characteristic very low SNR [6]. Diffusion Tensor Imaging is becoming one of the most popular methods for the analysis of the white matter (WM) structure of the brain, where some alterations can be found from early stages in some degenerative diseases. This technique measures the Brownian motion (random motion) of the water molecules in the brain, which is assumed to be isotropic when it is not restricted by the surrounding structures, since the WM regions contains densely packed fibre bundles they cause an anisotropic diffusion of the water molecules along the perpendicular directions to them. At each voxel of a DTI the water diffusion is represented by a symmetric 3×3 tensor, where the information of the preferred directions of the motion and the relevance of these directions is found in the eigenvectors and the eigenvalues of the tensor. This tensorial data can be represented as different scalar measurements, one of them is the Fractional Anistropy (FA) of the tissue, which is defined as

$$FA = \sqrt{\frac{3\left((\hat{\lambda} - \lambda_1)^2 + (\hat{\lambda} - \lambda_2)^2 + (\hat{\lambda} - \lambda_3)^2\right)}{2\left(\lambda_1^2 + \lambda_2^2 + \lambda_3^2\right)}}$$

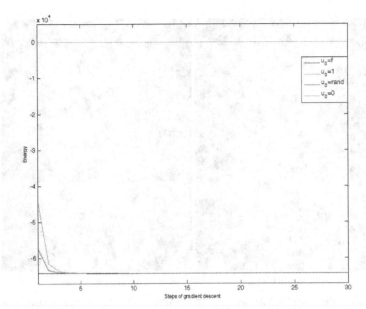

Fig. 5. Energy in functional (6) of the solution obtained at each step of the gradient descent by the new method for $\lambda = 0.05$, $\sigma = 0.05$, $\tau = 0.1$ and different initial data u_0: black image ($u_0 \equiv \bar{0}$), white image ($u_0 \equiv \bar{1}$), the noisy image ($u_0 \equiv f$) and a random image ($u_0 \equiv rand$)

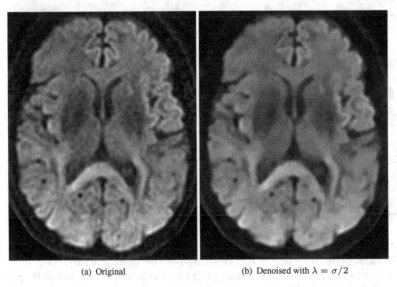

(a) Original   (b) Denoised with $\lambda = \sigma/2$

Fig. 6. A slice of the original Diffusion Weighted Image corresponding to the $(1, 0, 0)$ gradient direction and the corresponding denoised image

where the λ_i are the eigenvalues of the tensor and $\hat{\lambda} = (\lambda_1 + \lambda_2 + \lambda_3)/3$. The FA values vary from 0, (when the motion in the voxel is completely isotropic) to 1 (totally anisotropic). For the reconstruction of the DTI a set of DWI has to be acquired, scanning the tissue in different directions of the space. At least six DWI volumes are

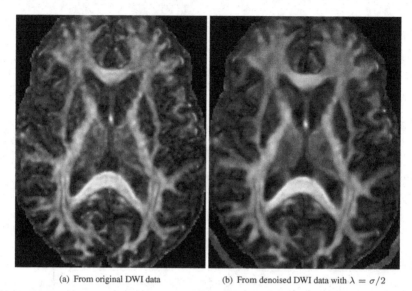

(a) From original DWI data   (b) From denoised DWI data with $\lambda = \sigma/2$

Fig. 7. A slice of the Fractional Anisotropy estimated from the Tensor Image. Dark colour corresponds to values near zero (isotropic regions) and bright color corresponds to values near one (anisotropic regions).

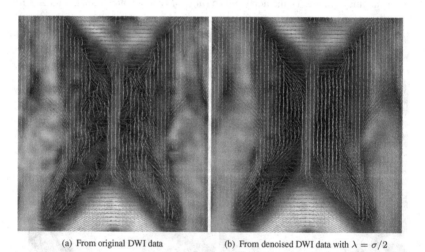

(a) From original DWI data   (b) From denoised DWI data with $\lambda = \sigma/2$

Fig. 8. A detail of the first eigenvectors of the DTI over the FA image. The color is based on the main orientation of the tensorial data. Red means right-left direction, green anterior-posterior and blue inferior-superior. Fibres with an oblique angle have a color that is a mixture of the principal colors and dark color is used for the isotropic regions.

needed in order to be able to calculate the DTI, which is a positive defined matrix. The noise present into the DWI scalar images can generate small, negative eigenvalues. Increasing the number of directions along which the brain is scanned improves the image quality but at the expenses of a longer acquisition time. The importance of pre-processing the DW Images previously to the DTI reconstruction is then two-fold: to improve the DT image quality through accurate Rician denosing so allowing shorter scanning time. The data we used consist of a DW-MR brain volume provided by Fundación CIEN-Fundación Reina Sofía which was acquired with a 3 Tesla General Electric scanner equipped with an 8-channel coil. The DW images have been obtained with a single-shot spin-eco EPI sequence (FOV=24cm, TR=9100, TE=88.9, slice thickness=3mm, spacing=0.3, matrix size=128x128, NEX=2). The DW-MRI data consists on a volume obtained with b=0/mm^2 and 15 volumes with b=1000s/mm^2 corresponding with the gradient directions specified in [14]. These DW-MR images, which represent diffusion measurements along multiples directions, are denoised with the proposed method previously to the Diffusion Tensorial Image reconstruction, which was done with the 3d Slicer tools[2]. In Figure 6(a) we show a slice of the original DWI data corresponding to the $(1, 0, 0)$ gradient direction where the affecting noise is clearly visible. The complete DW-MRI data volume is denoised using the proposed method. The Rician noise standard deviation (σ) has been estimated for each slice of each gradient direction following [3], while we used a value of $\lambda = \sigma/2$ for the denoising. The slice resulting from the denoising process is shown in Figure 6(b). It can be observed how in the denoised images the noise has been removed but the details and the edges have been fully preserved, as we should expect when the exact TV model is solved. The effect of this denoising process over the reconstructed tensor and their derived scalar measurements (obtained with the 3d Slicer tools) is presented in Figures 7 and 8. Figure 7 shows a Fractional Anisotropy image where the structures and details are clearly enhanced if the DW-MRI volume is denoised previously. When finer details are considered the denoising step is yet more crucial. For instance in Figure 8 the main eigenvector of the tensor is represented, where the noise on the original DWI data cause inhomogeneities (see Figure 8(a)) in the eigenvectors field which are product of the noise (Figure 8(b)).

6   Conclusions

In this notes we address the problem of the numerical computation of the solution of the variational formulation of the Rician denoising model proposed in [4]. We deduce a semi-implicit formulation for the gradient flow which leads to the resolution of ROF like-problems at each step of the time discretization. This is accomplished efficiently using a gradient descent for the dual variable associated to the primal ROF model. While our study is preliminary it indicates how to obtain fast numerical solutions for Rician denoising. This is specially interesting when Diffusion Weighted Images (DWI) are considered for Diffusion Tensor Images reconstruction whereas they have poor resolution and low SNR which makes Rician denoising necessary.

Challenging mathematical issues arise about the existence, uniqueness and convergence, when $\epsilon \to 0$, of weak bounded variation solutions of the quasilinear elliptic equations considered in this paper (i.e. (8) and (12)) and the gradient flow analysis of their parabolic counterpart ((10) and (13)) when $t \to +\infty$. A rigorous justification of the above arguments is desired. Nevertheless this approach is the mostly used regularization technique to approximate and compute the minimizer of the total variation energy and its variants (see [15]).

---

[2] Free available in http://www.slicer.org/

The semi-implicit method we propose is well founded mathematically when the time-discretized problems are dealt with and it represents a feasible alternative to gaussian denoising for low SNR MR images. Further study is undoubtedly necessary in order to make automatic the choice of the parameters in real medical images. Other possibilities, such as Inverse scaling, which makes the parameter estimation less crucial and provide contrast enhanced images shall also be explored.

Acknowledgements. This work was supported by project TEC2009-14587-C03-03 of the Spanish Ministry of Science. Also we thank to Mrs. Eva Alfayate, MR-scanner technician of the Fundación Reina Sofía, for her professional and kindly collaboration.

References

1. Henkelman, R.M.: Measurement of signal intensities in the presence of noise in MR images. Med. Phys. 12(2), 232–233 (1985)
2. Gudbjartsson, H., Patz, S.: The Rician distribution of noisy MRI data. J. of Magn. Reson. Med. 34(6), 910–914 (1995)
3. Sijbers, J., Den Dekker, A.J., Van Audekerke, J., Verhoye, M., Van Dyck, D.: Estimation of the noise in magnitude MR images. Magn. Reson. Imaging 16(1), 87–98 (1998)
4. Martín, A., Garamendi, J.F., Schiavi, E.: Iterated Rician Denoising. In: Proceedings of the International Conference on Image Processing, Computer Vision and Pattern Recognition, IPCV 2011, pp. 959–963. CSREA Press, Las Vegas (2011)
5. Getreuer, P., Tong, M., Vese, L.A.: A Variational Model for the Restoration of MR Images Corrupted by Blur and Rician Noise. In: Bebis, G., Boyle, R., Parvin, B., Koracin, D., Wang, S., Kyungnam, K., Benes, B., Moreland, K., Borst, C., DiVerdi, S., Yi-Jen, C., Ming, J. (eds.) ISVC 2011, Part I. LNCS, vol. 6938, pp. 686–698. Springer, Heidelberg (2011)
6. Basu, S., Fletcher, T., Whitaker, R.T.: Rician Noise Removal in Diffusion Tensor MRI. In: Larsen, R., Nielsen, M., Sporring, J. (eds.) MICCAI 2006, Part I. LNCS, vol. 4190, pp. 117–125. Springer, Heidelberg (2006)
7. Rudin, L.I., Osher, S., Fatemi, E.: Nonlinear total variation based noise removal algorithms. Physica D 60, 259–268 (1992)
8. Chambolle, A.: An algorithm for Total variation Minimization and Applications. J. of Mathematical Imaging and Vision 20, 89–97 (2004)
9. Ambrosio, L., Fusco, N., Pallara, D.: Functions of Bounded Variation and free discontinuity problems. Oxford Mathematical Monographs. The Clarendon Press, Oxford University (2000)
10. Lassey, K.R.: On the Computation of Certain Integrals Containing the Modified Bessel Function $I_0(\xi)$. J. of Mathematics of Computation 39(160), 625–637 (1982)
11. Perona, P., Malik, J.: Scale-space and edge detection using anisotropic diffusion. IEEE Transactions on PAMI 12(7), 629–639 (1990)
12. Nikolova, M., Esedoglu, S., Chan, T.F.: Algorithms for Finding Global Minimizers of Image Segmentation and Denoising Models. SIAM Journal of Applied Mathematics 66(5), 1632–1648 (2006)
13. Aubert-Broche, B., Griffin, M., Pike, G., Evans, A., Collins, D.: Twenty New Digital Brain Phantoms for Creation of Validation Image Data Bases. IEEE Transactions on Medical Imaging 24(11), 1410–1416 (2006)
14. Jones, D.K., Horsfield, M.A., Simmons, A.: Optimal strategies for measuring diffusion in anisotropic systems by magnetic resonance imaging. J. of Magn. Reson. Med. 42(3), 515–525 (1999)
15. Casas, E., Kunisch, K., Pola, C.: Some applications of BV functions in optimal control and calculus of variations. ESAIM: Proceedings. Control and Partial Differential Equations 4, 83–96 (1998)

Applying ICA in EEG: Choice of the Window Length and of the Decorrelation Method

Gundars Korats[1,2], Steven Le Cam[1], Radu Ranta[1], and Mohamed Hamid[1]

[1] Université de Lorraine, CRAN UMR 7039, Vandoeuvre-les-Nancy, 54516, France
CNRS, CRAN UMR 7039, Vandoeuvre-les-Nancy, 54516, France
[2] Ventspils University College, 101 Inzenieru iela, LV-3601, Ventspils, Latvia
{steven.le-cam,radu.ranta}@univ-lorraine.fr,
gundars.korats@gmail.com

Abstract. Blind Source Separation (BSS) approaches for multi-channel EEG processing are popular, and in particular Independent Component Analysis (ICA) algorithms have proven their ability for artefacts removal and source extraction for this very specific class of signals. However, the blind aspect of these techniques implies well-known drawbacks. As these methods are based on estimated statistics from the data and rely on an hypothesis of signal stationarity, the length of the window is crucial and has to be chosen carefully: large enough to get reliable estimation and short enough to respect the rather non-stationary nature of the EEG signals. In addition, another issue concerns the plausibility of the resulting separated sources. Indeed, some authors suggested that ICA algorithms give more physiologically plausible results than others. In this paper, we address both issues by comparing four popular ICA algorithms (namely FastICA, Extended InfoMax, JADER and AMICA). First of all, we propose a new criterion aiming to evaluate the quality of the decorrelation step of the ICA algorithms. This criterion leads to a heuristic rule of minimal sample size that guarantees statistically robust results. Next, we show that for this minimal sample size ensuring constant decorrelation quality we obtain quasi-constant ICA performances for some but not all tested algorithms. Extensive tests have been performed on simulated data (i.i.d. sub and super Gaussian sources mixed by random mixing matrices) and plausible data (macroscopic neural population models placed inside a three layers spherical head model). The results globally confirm the proposed rule for minimal data length and show that the use of sphering as decorrelation step might significantly change the global performances for some algorithms.

Keywords: EEG, BSS, ICA, Whitening, Sphering.

1 Introduction

The analysis of electro-physiological signals generated by brain sources leads to a better understanding of brain structures interaction and is useful in many clinical applications or for brain-computer interfaces (BCI) [1]. One of the most commonly used method to collect these signals is the scalp electroencephalogram (EEG). The EEG consists in several signals recorded simultaneously using electrodes placed on the scalp (see fig.1). The electrical activity of the brain sources is propagated through the anatomical

J. Gabriel et al. (Eds.): BIOSTEC 2012, CCIS 357, pp. 269–286, 2013.

structures and the resulting EEG is a linear mixture (with unknown or difficult to model parameters) of brain sources and other electro-physiological disturbances, often with a low signal to noise ratio (SNR) [2]. It is widely assumed that electrical brain potentials recorded by the electrodes mainly arise from synchronous activity of neurons within localized cortical *patches*. The far-field projection of such locally generated activity can be suitably model by the projection of a single equivalent current dipole placed at the center of the patch, resulting in a linear mixing of mostly dipolar sources on the EEG [3].

The blind source separation (BSS) is a nowadays well established method to retrieve original sources from the EEG mixing, as it can estimate both the mixing model and original sources [4]. In particular, approaches based on High Order Statistics (HOS) such as Independent Component Analysis (ICA) are common methods in this context and have been very useful for denoising purpose or brain sources identification. Generally, ICA algorithms include a preliminary decorrelation step based on second order statistics (estimated on a user chosen window length), which serves as an initialization for the next optimization step (independence maximization). Still, there is an infinite number of possible decorrelation matrices (as they are determined up to an arbitrary rotation). The two most popular decorrelation techniques are whitening and sphering, and it seems that they might influence the final separation results, especially in EEG applications [5]. In this paper, two issues will then be evaluated: 1) the accuracy of the decorrelation matrix estimation given the considered data length and 2) the sensitivity to the initialization step using whitening or sphering in the specific context of dipolar sources mixing. Four ICA algorithms based on HOS have been chosen: FastICA [6], Extended InfoMax [7], AMICA [8] and JADER [9].

1) The use of BSS on EEG signals implicitly assumes that the estimated second order statistics are meaningful. In order to ensure the reliability of these statistics, different authors propose optimal sample sizes (i.e. EEG signal time points), generally equal to $k \times n^2$ where n is number of channels and k is some empirical constant varying from 5 to 32 [10–12]. If these assumptions are correct, large amount of channels requires huge sample sizes, processing and time resources. On the other hand, EEG signals are at most short term stationary, so it would be interesting to find a sufficient inferior bound for the number of necessary samples. The first question is then how to define a minimum sample size that provides reliable estimation of sources and mixing model.

2) The second issue addressed in this paper concerns the sensitivity of the BSS/ICA performance given the initial decorrelation step in the dipolar mixing context. In the literature [13], some authors observed that using different initializations (different decorrelation methods like classical whitening or sphering), the results are more or less biologically plausible, meaning that more or less dipolar sources are retrieved from the data. A recent extensive study from the same authors [5] proposed an evaluation of the ability of 18 source separation methods to result in maximally independent component processes with nearly dipolar scalp projection. The results show that AMICA and Extended InfoMax give better performances compared to FastICA and JADER. Both AMICA and Extended InfoMax begin by sphering the data, while FastICA and JADER begin with a classical whitening step. We would like to evaluate the impact of

whitening and sphering on these four ICA algorithm performances. Unlike the previous studies that are directly using real EEG data, this evaluation will be performed on simulated data, giving the possibility of a controlled quantification of the algorithm performances. The evaluation is here proposed in the context of randomly generated data using i.i.d. sub and super Gaussian sources mixed by random mixing matrices, and in the context of plausible EEG data generated by macroscopic neural population models placed inside a three layers spherical head model.

The paper is organized as follow: section 2 exposes the EEG forward problem, explains the basics of the BSS methodology and gives some details on the four evaluated ICA methods. Section 3 proposes a normalized Riemannian likelihood as an evaluation criterion for the accuracy of the covariance matrix estimation and recalls the separability performance index used to evaluate the ICA algorithms. In section 4 both random and biologically plausible data set are described, while estimation and separation performances of the algorithms facing these data set are provided in section 5. In section 6, these results are discussed and future works are considered.

2   Problem Statement

2.1   EEG Mixing Model

Classical EEG generation and acquisition model is presented in Figure 1. It is widely accepted that the signals collected by the sensors are linear mixtures of the sources [2].

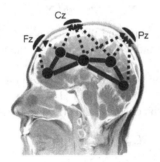

Fig. 1. EEG linear model

Subsequently, the EEG mixture can be written as

$$\mathbf{X} = \mathbf{AS}, \tag{1}$$

where $\mathbf{X}$ are the observations (electrodes), $\mathbf{A}$ is the mixing system (anatomical structure) and $\mathbf{S}$ are the original sources.

2.2   EEG Separation Model

We restrain in this paper to classical well determined mixtures, where the number of channels is equal to the number of underlying sources. In this case, BSS gives the linear transformation (separating) matrix $\mathbf{H}$ and the output signal vector $\mathbf{Y} = \mathbf{HX}$, containing source estimates. Ideally, the global system matrix $\mathbf{G} = \mathbf{HA}$ between the original sources $\mathbf{S}$ and their estimates $\mathbf{Y}$ will be a permuted scaled identity matrix, as it can be proven that the order and the original amplitude of the sources cannot be recovered [4].

In almost all BSS methods, the matrix $\mathbf{H}$ is obtained as a product of two statistically based linear transforms: $\mathbf{H} = \mathbf{JW}$ with

- $\mathbf{W}$ performing data orthogonalization: whitening/sphering,
- $\mathbf{J}$ performing data rotation : independence maximization via higher-order statistics (HOS) or joint decorrelation of several time (frequency) intervals

The first step (data decorrelation) can be seen as an initialization for the second step. In theory any orthogonalization technique can be used to initialize the second step but in this paper we will focus on two popular decorrelation techniques: whitening (classical solution) and sphering (assumed to be more biologically plausible [13]).

BSS Initialization: Whitening/Sphering

Whitening In general EEG signals $\mathbf{X}$ are correlated so their covariance Σ will not be a diagonal matrix and their variances will not be normalized. Data whitening means projection in the eigenspace and normalisation of variances. The whitening transform can be computed from the eigen-decomposition of the data covariance matrix $\Sigma = \Phi\Lambda\Phi^T$:

$$\mathbf{X_w} = \Lambda^{-\frac{1}{2}}\Phi^T\mathbf{X}, \tag{2}$$

where Λ and Φ are the eigenvalues and the eigenvectors matrices respectively. After (2), the signals are orthogonal and with unit variances (Figure 2(c)).

Sphering. completes whitening by rotating data back to the coordinate system defined by principal components of the original data [14]. In other words, sphered data are turned as close as possible to the observed data (Figure 2(d)):

$$\mathbf{X_{sph}} = \Phi\Lambda^{-\frac{1}{2}}\Phi^T\mathbf{X}. \tag{3}$$

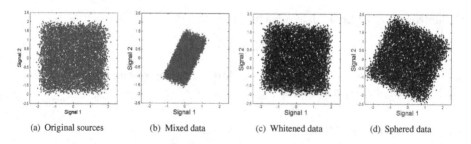

| (a) Original sources | (b) Mixed data | (c) Whitened data | (d) Sphered data |

Fig. 2. Example of different decorrelation approaches for two signals

2.3 Optimization: Data Rotation

Second step would be finding a rotation matrix $\mathbf{J}$ to be applied to the decorrelated data (whitened or sphered) in order to maximize their independence. Rotation can be done using second order statistics (SOS) using joint decorrelations and/or using HOS cost functions. We restrain here to the second (HOS) approach[1]. Several cost functions and optimization techniques were described in the literature (see for example [4, 12]). Among the most well known and used in EEG applications, we can cite FastICA (negentropy maximization [6]), Extended InfoMax (mutual information minimization [7]) and JADER (joint diagonalization of fourth order cumulant matrices [9]). Another recent algorithm has been proposed by Palmer *et al.* and is called AMICA [8]. Based on the modeling of each source component as a sum of extended Gaussians, this method has shown very promising results in the context of EEG data [5].

Specifically, in this paper we test the performances and the robustness of these four ICA algorithms with respect to the sample size and the initialization step in both contexts of random unstructured data mixing and biologically plausible data mixing.

3 Performance Evaluation Criteria

3.1 Reliable Estimate of the Covariance: Riemannian Likelihood

As noted before, BSS model consists of decorrelation and rotation. Both steps are based on statistical estimates. The first step is common for all algorithms and relies on the estimation of the covariance matrix. Therefore it is necessary to have reliable estimates of this matrix. In other words, given a known covariance matrix Σ, we want to evaluate the minimum sample size N necessary to obtain a covariance matrix estimation $\hat{\Sigma}_N$ close enough to the original one with respect to a distance that we have to define.

We propose here an original distance measure between the true and the estimated covariance matrices, inspired from digital image processing and computer vision techniques [15]. In the context of object tracking and texture description, a distance measure is used to estimate whether an observed object or region corresponds to a given covariance descriptor. To estimate similarity between matrices respectively corresponding to the target model and the candidate, and knowing that covariance matrices are symmetric positive definite, the following general[2] distance measure can be used:

$$d^2(\hat{\Sigma}_N, \Sigma) = \mathrm{tr}\left(\log^2\left(\hat{\Sigma}_N^{-\frac{1}{2}}\Sigma\hat{\Sigma}_N^{-\frac{1}{2}}\right)\right) \tag{4}$$

In the ideal case of a perfect estimation, the matrix $\mathbf{C} = \hat{\Sigma}_N^{-\frac{1}{2}}\Sigma\hat{\Sigma}_N^{-\frac{1}{2}}$ equals the identity matrix $\mathbf{I}_n$ and d becomes 0 (n being the number of measured signals, equal to the source number in our case). In real cases though, assuming that the covariance

---

[1] As described in the next section, in our simulations we used random non-Gaussian stationary data, without any time-frequency structure. Therefore algorithms based on SOS as SOBI, SOBI-RO and AMUSE were not used.

[2] On Riemannian manifolds.

estimate is not very far from the real covariance matrix, $\mathbf{C} = \mathbf{I}_n + \epsilon$, ϵ being a symmetric error matrix. In this case, using the eigenvalues decomposition and the properties of the trace of a symmetric matrix, equation (4) can be rewritten as:

$$d^2(\hat{\boldsymbol{\Sigma}}_N, \boldsymbol{\Sigma}) = \text{tr}\left(\log^2(\mathbf{I}_n + \epsilon)\right) \tag{5}$$

$$= \text{tr}\left(\mathbf{U}\log^2(\mathbf{D}_{\mathbf{I}_n+\epsilon})\mathbf{U}^T\right) \tag{6}$$

$$= \text{tr}\left(\log^2(\mathbf{D}_{\mathbf{I}_n+\epsilon})\right) \tag{7}$$

$$\approx \Sigma_{i=1}^n \log^2(1 + \epsilon_{ii}) \tag{8}$$

Now, using the fact that for small ϵ, $\log(1 + \epsilon) \approx \epsilon$ and assuming that the errors are equally distributed over the diagonal $\mathbf{D}_{\mathbf{I}_n+\epsilon}$, (5) becomes:

$$d^2(\hat{\boldsymbol{\Sigma}}_N, \boldsymbol{\Sigma}) \approx n\epsilon^2 \tag{9}$$

proportional with the matrix dimension n and, in our case, with the number of EEG channels. In order to avoid this channel number effect, we propose to modify the distance by multiplying it by k/n, with k being a user chosen constant ensuring the desired level of the estimation quality:

$$d_n^2(\hat{\boldsymbol{\Sigma}}_N, \boldsymbol{\Sigma}) = \frac{k}{n}\text{tr}\left(\log^2\left(\hat{\boldsymbol{\Sigma}}_N^{-\frac{1}{2}}\boldsymbol{\Sigma}\hat{\boldsymbol{\Sigma}}_N^{-\frac{1}{2}}\right)\right) \tag{10}$$

As in [15], we adopt an exponential function of the modified distance d_n as the local likelihood

$$p(\boldsymbol{\Sigma}_N) \propto exp\{-\boldsymbol{\lambda} \cdot d_n^2(\boldsymbol{\Sigma}, \hat{\boldsymbol{\Sigma}}_N)\}. \tag{11}$$

with the parameter λ fixed to the constant value $\lambda = 0.5$ [15]. This $p(\boldsymbol{\Sigma}_N)$ value varies between 0 and 1, 1 meaning perfect estimation ($\boldsymbol{\Sigma} = \hat{\boldsymbol{\Sigma}}_N$). A $p(\boldsymbol{\Sigma}_N)$ value of 0.95 is considered as a well chosen threshold above which the covariance matrices are considered to be approximately equal.

3.2   Separability Performance Index

In order to measure the global performance of BSS algorithms (orthogonalization plus rotation), we use the performance index (PI) [4] defined by

$$PI = \frac{1}{2n(n-1)}\sum_{i=1}^n\left(\sum_{j=1}^n\frac{|g_{ij}|}{\max_k|g_{ik}|} - 1\right) + \frac{1}{2n(n-1)}\sum_{j=1}^n\left(\sum_{i=1}^n\frac{|g_{ij}|}{\max_k|g_{kj}|} - 1\right) \tag{12}$$

where $\mathbf{g}_{ij}$ is the (i, j)-element of the global $n \times n$ system matrix $\mathbf{G} = \mathbf{HA}$, $\max_k|g_{ik}|$ is the maximum value among the absolute values of the elements in the ith row of $\mathbf{G}$ and $\max_k|g_{kj}|$ is the maximum value among the absolute values of the elements in the jth column of $\mathbf{G}$. Perfect separation yields a null performance index. In practice a PI under 10^{-1} means that the separation result is reliable.

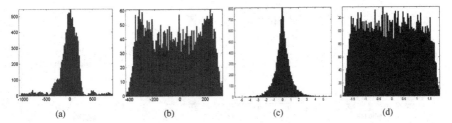

Fig. 3. Histograms of real SEEG samples ((a) background and (b) ictal activities) and histograms of generalized Gaussian simulated data ((c) super-Gaussian and (d) sub-Gaussian data)

4   Simulated Data Set

The algorithms performances are assessed on two types of data. Following our BIOSIG-NALS paper [16], the first data set consists in simulated generalized Gaussian sources mixed by randomly simulated matrices. The second one is obtained by mixing sources given by macroscopic neural populations models [17] with mixing matrices computed from a realistic three layers lead field model.

4.1   Random Data Set

We have chosen to simulate stationary white source signals, as the retrieval of time structures is not the purpose of this work (in fact, in all the tested algorithms, as in most of the HOS type methods, the time structure is ignored). In order to simulate sources with realistic probability distributions, we analysed depth intra-cerebral measures (SEEG). According to our observations (see also [11, 10]), the probability distribution of the electrical brain activity signals can be suitably modelled by Generalized zero-mean Gaussians, as shown in fig. 3(a) and fig. 3(b)). For this reason we used randomly generated both supergaussian (Laplace - Figure 3(c)) and subgaussian (close to uniform (Figure 3(d))) distributions.

Several simulations were made, using 8, 16, 24, 32 and 48 source signals. Half of the sources were generated as supergaussian and half as subgaussian. The sources were afterwards mixed using a randomly generated mixing matrix **A** (uniform distribution in $[-1, 1]$). We then consider here the performance of each of the four ICA algorithms facing simulated stationary non-artefacted data. Such evaluation is likely to give us a rule of a minimum amount of data needed for a reliable source separation in favourable conditions (after an artefact elimination step for example).

4.2   Plausible Data Set

More realistic contexts is to be simulated in order to evaluate the behaviour of the algorithm confronted to the real EEG BSS problem. We then propose a second data set where the sources are simulated by a macroscopic model [17] able to reproduce normal background activity as well as pre-ictal and ictal (epileptic) electromagnetic activity. The mixing matrices are designed using a fast accurate three layers head model

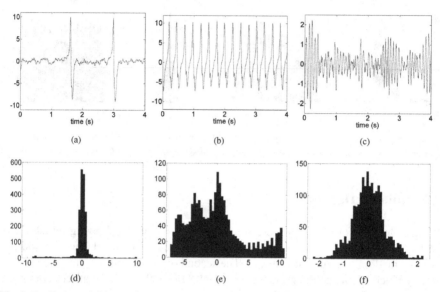

Fig. 4. Realistic activities and corresponding histograms. first pre-ictal: (a) & (d), second pre-ictal: (b) & (e), second ictal: (c) & (f).

[18]. Such mixing matrices guarantee the dipolarity of the underlying sources to be retrieved, providing a framework where the influence issue of the initialization (whitening or sphering) in such context can be addressed.

Realistic Sources. Macroscopic models describe the neuronal activity at the scale of neuronal populations by modelling the interconnection of pyramidal cells with inhibitory or excitatory inter-neurons. They have been particularly used to successfully generate realistic electrophysiological recordings [19, 20]. In this work we have chosen the Wendling's model [17], described by a set of ten differential equations. It has been shown that this model is able to reproduce normal background activity (inter-ictal), first and second pre-ictal activities as well as first and second ictal activities. Parameter values to be chosen in order to get these distinct epileptic activities are detailed in [21]. In this paper, these simulated activities have been introduced as sources (see fig. 4), excepted the normal background and the first ictal activities that have Gaussian-like distribution and are then inadequate for the selected ICA algorithms.

In a realistic situation, the number of ictal sources to be retrieved is limited. In our simulation, the number of realistic sources (from pre-ictal to ictal) has been chosen to be an eighth of the number of channels n (from 1 for 8 channels to 6 for 48 channels). The remaining background activities are simulated randomly as sub and super gaussian like in the previous random data set.

Dipolar Mixing Matrix. The next step for obtaining realistic EEGs, after plausible source generation, is the construction of a realistic mixing matrix. We obtained it using the classical three spheres model of Rush & Driscoll and the Berg surface potential

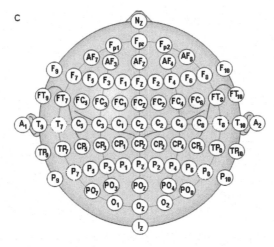

Fig. 5. Electrode placement for the 10-10 montage

estimation method [18]: the realistic sources generated as described in the previous subsection were assumed to be time courses of brain dipoles. The positions and the orientations of these dipoles (8 to 48) were randomly generated inside the inner sphere of a the head model, representing the brain. The six parameters of the dipoles being thus completely defined (Cartesian coordinates, angles and magnitude), one can generate the corresponding scalp map (i.e. the electrical potentials on the outer sphere) using the so-called Lead Field matrix corresponding to the three-spheres forward model (see for example [22]). We used here the Berg technique [18], which accurately and rapidly approximates the scalp potentials generated by a dipole for a three spheres model with the weighted sum of the potentials generated by three dipoles in a single sphere model.

The Lead Field matrix allows the computation of the scalp map (potentials) for every point on the surface of the outer sphere. Still, in the EEG simulation context, we are only interested by the potentials recorder by the scalp electrodes. In our simulation, we used as a basis montage the classical 10-10 EEG montage (figure 5). We have next chosen five subsets consisting of 8, 16, 24, 32 and 48 electrodes corresponding to real EEG applications (sleep studies, brain computer interfaces and clinical setups for epilepsy diagnostics).For example, for the 8-electrodes montage, the chosen electrodes were $F_{pz}, T_7, C_3, C_z, C_4, T_8, O_1$ and O_2; for the 16-electrodes montage, $F_3, F_z, F_4, T_7, C_3, C_z, C_4, T_8, P_7, P_3, P_z, P_4, P_8, PO_7, PO_8, O_z$; and son on.

5   Results and Discussion

The four algorithms are first evaluated on a random data set in order to define a minimum data length rule ensuring accurate separation performances and to analyse the impact of initialization step in the general toy case of unstructured randomly mixed data. An equivalent study is then applied on a more physiologically plausible data set, on which our minimum data length rule is validated. On the same plausible data, the results are evaluated in order to confirm or infirm the superiority of sphering initialization

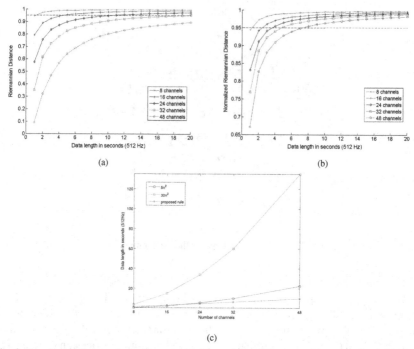

Fig. 6. Results on the random data set: (a) Riemannian Distance, (b) *normalized* Riemannian Distance and (c) Minimum length rules from literature vs proposed (linear) rule

over whitening initialization in the context of dipolar sources separation. For each data set, the four algorithms are evaluated on 8, 16, 24, 32 and 48 source sizes with sample size varying from 1s to 20s with a 1s step (at a 512 Hz sampling rate). The number of iteration for each source size/sample size has been set to 50, a new set of data (random data sources, plausible data sources, mixing matrices) being simulated at each iteration.

5.1 Random Data Set

A Minimum Length Rule. This section presents the results of the covariance estimation accuracy vs the length of the data. In a previous work [16], the distance between covariance matrices was computed using (4) from different sample sizes starting from 100 till 5000 by 100 points step, and number of channels taking values in the set $\{8, 10, 12, 14, 16, 18\}$. The likelihood was further evaluated using (11). A constant threshold was empirically fixed to $p = 0.95$ (Figure 6(a)): likelihood values above this threshold was assumed to guarantee good estimation of covariances as stated in section 3.1. However, this previous study already outlined that this rule might be too strong, leading in a ever decreasing PI with the number of channels. This observation has been confirmed when studying this evolution of PI on larger number of channels. Therefore,we proposed the *normalized* Riemannian distance defined by (10) with the objective to refine our minimum length rule. Again, likelihood values above a threshold of 0.95 were assumed to guarantee good estimation of second order statistics

Table 1. Perfomance index (PI) values for random mixtures: mean and standard deviation for the four algorithms, with two initializations (whitening and sphering) and using the couple n/N (nb. of channels / data length) given by the heuristic rule derived from figure 6(b).

	$n = 8$ $N = 2 \times 512$		$n = 16$ $N = 3 \times 512$		$n = 24$ $N = 4 \times 512$		$n = 32$ $N = 5 \times 512$		$n = 48$ $N = 7 \times 512$	
	W	S	W	S	W	S	W	S	W	S
FastICA	0.048	0.049	0.043	0.043	0.041	0.040	0.040	0.040	0.038	0.038
	(0.014)	(0.012)	(0.006)	(0.007)	(0.006)	(0.006)	(0.005)	(0.005)	(0.004)	(0.006)
AMICA	0.024	0.029	0.018	0.019	0.021	0.029	0.036	0.061	0.113	0.170
	(0.011)	(0.031)	(0.004)	(0.005)	(0.021)	(0.051)	(0.048)	(0.056)	(0.050)	(0.060)
Extended	0.167	0.199	0.274	0.326	0.288	0.334	0.252	0.300	0.274	0.310
Infomax	(0.091)	(0.124)	(0.059)	(0.055)	(0.022)	(0.013)	(0.019)	(0.014)	(0.008)	(0.007)
JADER	0.044	0.044	0.038	0.038	0.035	0.035	0.035	0.035	0.130	0.132
	(0.010)	(0.010)	(0.005)	(0.005)	(0.004)	(0.004)	(0.003)	(0.003)	(0.029)	(0.032)

(Figure 6(a)). The normalization factor k has been experimentally set to 8, taking the 8 channels mean error ϵ as a reference on which higher number of channels configurations are scaled.

This rule is reported on the figure 6(c). The proposed rule is rather linear, thus being in contradiction with current literature suggestions, rather proportional to n^2. Our rule is then between the bounds given in the literature for low number of channels, but is increasing much slower and gives lower bounds for number of channels above 24. A possible way to interpret the figure 6(c) is to use it as a decision rule: for a given number of channels, one can estimate the minimum number of data points necessary to have a reliable estimate of the covariance matrix and thus a reliable whitening. This decision rule leads to data lengths between approximately 2s (1024 data points) for 8 channels to 7s (3584 data points) for 48 channels. This range of time length is more compatible with the stationarity hypothesis than the values obtained using the $30n^2$ rule [12, 11]. Indeed, with this rule, we get from 1920 (3.75s) to 9720 (2min15s) data points respectively for 8 and 48 channels, which is rather contradictory (at least in a realistic EEG setup) with the assumption of stationarity on which most of BSS/ICA algorithms are based[3].

In order to experimentally validate this length rule, we computed the performance index PI (12) for the resulting data length. Mean (over 50 realizations) of the PI as well as its standard deviation for each algorithm (with whitening and sphering initializations) are reported in the Table 1. As it can be seen, FastICA (with either whitening or sphering) gives rather stable PI under 0.05 for this given rule (from 0.048 for 8 channels to 0.038 for 48 channels). It has to be noticed that PI values indicate better performances when the number of channels is increasing with respect to our empirical rule. This could suggest that our proposed criterion could be relaxed and the number of points could be reduced further for FastICA. On the other hand, one must take into account that these tests are performed on simulated stationary random data: if outliers are present, HOS estimates are more affected than the SOS estimations used to define our threshold, thus a higher amount of points might be needed for HOS reliable estimation.

---

[3] This observation is important especially for high resolution EEGs, having a high number of channels.

This observation holds when it comes to AMICA and JADER for number of channels from 8 to 32, the rule being not verified for 48 channels in this particular case of random data set. A quick look at the figure 7(e) (PI vs sample size for 48 channels) shows a rupture of the JADER curves around 6s to 8s, 9s being required to get a PI under 0.1. Such rule seems thus to be inadequate for JADER algorithm for number of channels over 48.

Extended Infomax is showing bad performances for these data length, requiring much more sample size to converge, confirming the results presented in [23]. This phenomenon appears because of our choice of the simulated data. Indeed, because of the use of subgaussian sources, the algorithm (even in the extended version) needs more data points in order to give reliable results. Empirically, one can say that the $30n^2$ rule seems to be adapted for the Extended InfoMax algorithm, but apparently too strong for the three others.

In the case of AMICA, the initialization parameter has to be considered in order to fully understand its misadequation with our minimum data length rule in the random data set case for 48 channels (see below the analysis for the plausible data set).

Impact of Initialization. Our second objective is to analyse the sensitivity of the ICA algorithms to the initialization step with whitening or sphering. The curves of figure 7 are showing evolution of PI with the data sample size for the five channel number configurations considered. Whitening and sphering curves are difficult to distinguish for FastICA and JADER, allowing to conclude that these methods are not sensitive to the decorrelation step. This has to be explained by the optimization strategy of these methods, based respectively on a fixed point and a Jacobi technique, both techniques reputed to have fastest convergence and being more reliable than the gradient technique [9]. AMICA is doing better than FastICA and JADER in most configurations when the amount of data is enough. This algorithm is based on the fitting of extended Gaussian (mixtures of scaled Gaussians) for each source time course, thus needing more data and execution time for accurate estimation and convergence (see the note below on time convergence). Besides, results appear to be quite deceiving for AMICA when it comes to the 48 channels configuration, with a PI around 0.06 for lengths greater than 10s for whitening (with a large standard deviation around 0.05), but a PI well above 0.1 for sphering. In this case initialization shows to have a noticeable impact on AMICA. This observation can be done also for Extended InfoMax for all five channel size cases. For this specific data set of randomly mixed sources, whitening initialization (solid curves) is resulting globally in better PI than sphering initialization (dashed curves).

As pointed out in [5], Extended InfoMax and AMICA are based on a natural gradient descent optimization scheme, initialization is then a major issue for these algorithms: the farthest from the solution the initialization is, the longest will take the optimization procedure. In the case of random mixing matrices, the solutions are distributed widely over the optimization space, making it difficult to define an adequate initialization point. In this context, whitening seems to be on average more appropriate than sphering. It has to be noticed that in our simulation no iteration or convergence criteria parameter has been changed in the Extended InfoMax algorithm, while maximum number of iterations has been set to 300 for the AMICA procedure (some numerical issues have been experienced with the default 100 value for short length data (<=2s)).

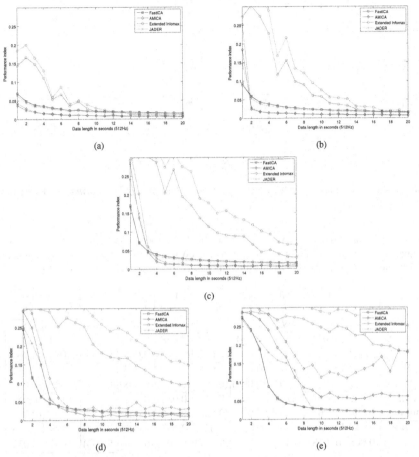

Fig. 7. Results on the random data set: performance index (PI) curves vs data length for (a) 8 channels, (b) 16 channels, (c) 24 channels, (d) 32 channels and (e) 48 channels. Initialization with whitening (solid lines) and sphering (dashed lines)

Note on Time Convergence: Due to the large amount of parameters to be estimated by AMICA, it might be important to notice that this method is extremely time consuming compared to JADER and FastICA, and also much slower than Extended InfoMax. To give an idea: while FastICA is taking less than 1s on 32 channels and $20s$ data length (mean time observed for 50 iterations on random data), JADER requires no more than 3s, Extended InfoMax needs up to 3 minutes, and AMICA requires almost 4 minutes. On the other hand, AMICA is *per se* much more flexible than Extended InfoMax which only try to fit a single generalized Gaussian distribution on each source, explaining the better performances of AMICA when compared to Extended InfoMax.

5.2 Plausible Data Set

Minimum Length Rule Validation. As it can be observed by comparing the computed (normalized) Riemannian distance, covariance estimations on this data set (figure 8)

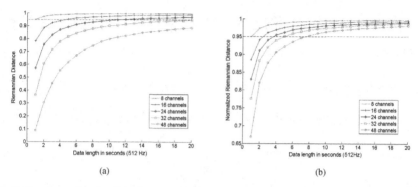

Fig. 8. Results on the plausible data set: (a) Riemannian Distance and (b) *normalized* Riemannian Distance

Table 2. Perfomance index (PI) values for plausible EEG: mean and standard deviation for the four algorithms, with two initializations (whitening (W) and sphering (S)) and using the couple n/N (nb. of channels / data length) given by the heuristic rule derived from figure 8(b).

	$n = 8$ $N = 2 \times 512$		$n = 16$ $N = 3 \times 512$		$n = 24$ $N = 4 \times 512$		$n = 32$ $N = 5 \times 512$		$n = 48$ $N = 7 \times 512$	
	W	S	W	S	W	S	W	S	W	S
FastICA	0.048	0.047	0.044	0.044	0.038	0.038	0.036	0.037	0.034	0.033
	(0.014)	(0.012)	(0.007)	(0.008)	(0.006)	(0.006)	(0.004)	(0.004)	(0.004)	(0.004)
AMICA	0.024	0.023	0.019	0.018	0.015	0.015	0.015	0.013	0.020	0.011
	(0.012)	(0.009)	(0.004)	(0.003)	(0.001)	(0.002)	(0.008)	(0.001)	(0.021)	(0.001)
Extended	0.159	0.089	0.207	0.097	0.218	0.085	0.162	0.055	0.173	0.058
Infomax	(0.099)	(0.046)	(0.061)	(0.038)	(0.038)	(0.026)	(0.024)	(0.015)	(0.017)	(0.011)
JADER	0.040	0.040	0.039	0.039	0.036	0.036	0.037	0.037	0.123	0.123
	(0.008)	(0.008)	(0.006)	(0.006)	(0.004)	(0.004)	(0.003)	(0.003)	(0.032)	(0.032)

show to be very similar to the results obtained on the previous random data set. Thus, plausible source time courses and mixture do not show to have high influence on these second order statistics estimation, allowing to keep the same minimum data length rule derived from figure 6(c). Table 2 gives mean PI related to this proposed decision rule on the plausible data set, where it can be seen that these minimum data length bounds appear to be adequate for all the algorithms in most channel size configurations, even for Extended InfoMax when a sphering initialization is considered. Some lower performances are observed for JADER for 48 channels, confirming the observation made in the previous section that this decision rule might be inadequate for this method for high number of channels. Relative better performances observed on Extended InfoMax, and over all impressive results given by AMICA with sphering for channel size over 24 (mean $PI < 0.015$ with minimal standard deviation of 0.001) have to be explained in the light of the initialization parameter.

Sphering Is Better Than Whitening for Dipolar Sources Separation. Figure 9 displays the evolution of PI with the data sample size for the five considered configurations (channel number). A quick look at these curves let us conclude that FastICA and

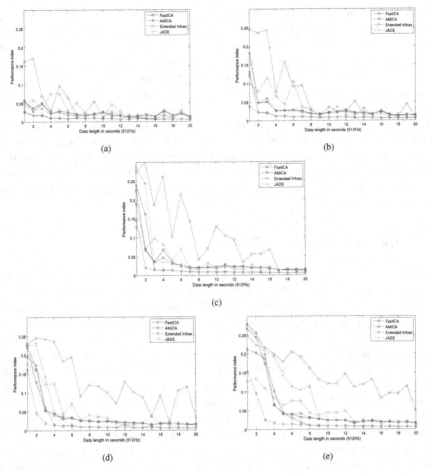

Fig. 9. Results on the plausible data set: performance index (PI) curves vs data length for (a) 8 channels, (b) 16 channels, (c) 24 channels, (d) 32 channels and (e) 48 channels. Initialization with whitening (solid lines) and sphering (dashed lines).

JADER are unsensitive to their initialization as expected and explained in the previous section. No major differences on the performances can be noticed between the random data set and the plausible data set, confirming the reputation of stability and reliability of these techniques in various situations. Concerning natural gradient descent based algorithms (Extended InfoMax and AMICA), the behaviour changes radically from the first data set to the second one. Results improve for both methods, especially when a sphered initialization is used. Extended InfoMax show convergence with PI under 0.1 in the five configurations when data is used. AMICA is found out to show very good performance facing mixtures of dipolar sources, with a high robustness to low sample sizes for high number of channels, especially when initialized with sphering: 3s appears to be enough to get a PI under 0.03 for 48 channels, with a standard deviation of 0.01.

In this particular case of dipolar mixing, the reasons of the superiority of sphering over whitening can be found in [5]: *The objective of Principal Component Analysis (PCA)*[4] *is to lump together as much variance as possible into each successive principal component, whose scalp maps must then be orthogonal to all the others and therefore are not free to model a scalp source projection resembling a single dipole.* In other words, whitening initialize the algorithm far from the solution, leading to a more difficult convergence for methods based on natural gradient descent like Extended InfoMax and AMICA. On the other hand, Delorme et al. [5] emphasize that: *Sphering components, in particular, most often have stereotyped scalp maps consisting of a focal projection peaking at each respective data channel and thus resembling the projection of a radial equivalent dipole.*. Consequently, sphering leads to initialization point much more closer to the solution than whitening in the dipolar case, as it is confirmed and quantified by our results.

6 Conclusions and Future Work

The first goal of this paper was to define a low bound of data length for robust separation results. Four ICA algorithms often used to analyse EEG signals were tested on different data lengths (1 to 20 seconds at 512 Hz sampling rate) and number of signals (8, 16, 24, 32 and 48 sources/channels). A rule of minimum sample size is derived from separation results on a random data set consisting of subgaussian and supergaussian source signals mixed by random mixing matrices, and is validated on a plausible data set in which sources were simulated by a macroscopic model of neuronal population and mixed by dipolar mixing matrices obtained from a three layers head model. This low bound is based on an original, normalized distance measure inspired by the computer vision community and leads to a reasonable minimum time length. According to our results (Tables 1 and 2), the proposed minimal data length rule guarantees a good source separation performance with performance indexes (PI) under 0.05 in most configurations for FastICA, JADER and AMICA (at least in the plausible case). Extended InfoMax has to be considered separately, as this algorithm requires much more data points. Our decision rule gives minimum data length much smaller than those recommended in literature (over $5n^2$) for high number of channels n, being thus more in adequation with the short time stationarity hypothesis accepted for EEG signals and needed for most ICA algorithms.

A second objective was to evaluate the impact of initialization on the separation performances using whitening or sphering in the first step of these algorithms. Due to the optimization strategy on which they are based, FastICA and JADER show no sensitivity to initialization (decorrelation method). Conversely, natural gradient descent based algorithms AMICA and Extended-Infomax show high sensitivity to initialization. Due to their optimization strategy, these algorithms are much more time consuming and less robust facing outliers, thus requesting an adequate initialization for a reliable convergence with acceptable number of iterations. In the particular case of EEG, modelled as a mixture of dipolar sources, it is possible to initialize the algorithm "near" the solution by sphering. Consequently, the performances of these algorithms improve and they can

---
[4] Equivalent to whitening.

be reliably applied. In particular, for dipolar mixtures and using sphering as initialization, AMICA showed impressive performances with very low data length even for high number of channels: 3s of data length (512 Hz sampling rate) are sufficient to get an excellent PI below 0.03 for the separation of 48 sources. The main drawback of AMICA is its time consumption: it requests more than 60 seconds for this particular example, while FastICA converges in less than 1s to get similar PI, although needing 7s of data.

An immediate perspective to this work would be to use more realistic time-structured data, obtained using only modelled neural sources and realistic mixtures (head models). Besides confirming our conclusions for the studied algorithms, this type of simulation setup would allow the evaluation of second order statistics BSS algorithms (SOBI and similar, also widely used for EEG analysis). It might be also useful if algorithms could be tested on more data channels, in order to asses their performances in the context of high-resolution EEG (more than 64 channels).

An interesting perspective, for the specific case of EEG source separation, is to consider new contrasts for BSS algorithms, balancing between dipolarity and independence. Indeed, source independence is known to be often unrealistic for EEGs, as strong synchrony is very likely to appear between distant areas in the brain. A relaxation of the independence constraint might then enhance the EEG source separation performance.

References

1. Schomer, D., Lopes da Silva, F. (eds.): Niedermeyers's Electroenephalography: Basic Principles, Clinical Applications and Related Fields. Wolters Kluwer, Lippincott Willimas & Wilkins (2011)
2. Sanei, S., Chambers, J.: EEG Signal Processing. John Wiley & Sons (2007)
3. Scherg, M., Berg, P.: Use of prior knowledge in brain electromagnetic source analysis. Brain Topography 4, 143–150 (1991)
4. Cichocki, A., Amari, S.: Adaptive Blind Signal and Image Processing Learning Algorithms and Applications. John Wiley & Sons, New York (2002)
5. Delorme, A., Palmer, J., Onton, J., Oostenveld, R., Makeig, S.: Independent eeg sources are dipolar. PLoS ONE 7(2), e30135 (2012)
6. Hyvärinen, A.: Fast and robust fixed-point algorithms for independent component analysis. IEEE Transactions on Neural Networks 10(3), 626–634 (1999)
7. Bell, A.J., Sejnowski, T.J.: An information-maximization approach to blind separation and blind deconvolution. Neural Computation 7, 1129–1159 (1995)
8. Palmer, J., Makeig, S., Delgado, K., Rao, B.: Newton method for the ICA mixture model. In: IEEE International Conference on Acoustics, Speech and Signal Processing, ICASSP 2008, March 31-April 4, pp. 1805–1808 (2008)
9. Cardoso, J.: High-order contrasts for independent component analysis. Neural Computation 11(1), 157–192 (1999)
10. Särelä, J., Vigário, R.: Overlearning in marginal distribution-based ICA: analysis and solutions. J. Mach. Learn. Res. 4, 1447–1469 (2003)
11. Onton, J., Makeig, S.: Information-based modeling of event-related brain dynamics. Progress in Brain Research 159, 99–120 (2006)
12. Delorme, A., Makeig, S.: EEGLab: an open source toolbox for analysis of single-trial EEG dynamics including independent component analysis. Journal of Neuroscience Methods 134(1), 9–21 (2004)

13. Palmer, J., Makeig, S., Delorme, A., Onton, J., Acar, Z.A., Kreutz-Delgado, K., Rao, B.D.: Independent Component Analysis of High-density Scalp EEG Recordings. In: 10th EEGLAB Workshop, Jyväskylä, Finland, June 14-17 (2010)
14. Vaseghi, S., Jetelova, H.: Principal and independent component aanalysis in image processing (2008)
15. Wu, Y., Wu, B., Liu, J., Lu, H.: Probabilistic tracking on riemannian manifolds. In: 19th International Conference on Pattern Recognition, ICPR 2008, pp. 1–4 (December 2008)
16. Korats, G., Le-Cam, S., Ranta, R.: Impact of window length and decorrelation step on ICA algorithms for EEG blind source separation. In: Biosignals/Biostec INSTICC Annual Conference (2012)
17. Wendling, F., Bartolomei, F., Bellanger, J.J., Chauvel, P.: Epileptic fast activity can be explained by a model of impaired GABAergic dendritic inhibition. European Journal of Neuroscience 15(9), 1499–1508 (2002)
18. Berg, P., Scherg, M.: A fast method for forward computation of multiple-shell spherical head models. Electroencephalography and Clinical Neurophysiology 90(1), 58–64 (1994)
19. Lopes da Silva, F.H., Hoeks, A., Smits, H., Zetterberg, L.H.: Model of brain rhythmic activity. Biological Cybernetics 15, 27–37 (1974)
20. Jansen, B., Rit, V.: Electroencephalogram and visual evoked potential generation in a mathematical model of coupled cortical columns. Biological Cybernetics 73, 357–366 (1995)
21. Wendling, F., Hernandez, A., Bellanger, J.J., Chauvel, P., Bartolomei, F.: Interictal to ictal transition in human temporal lobe epilepsy: Insights from a computational model of intracerebral EEG. Clinical Neurophysiology 22(2), 343–356 (2005)
22. Hallez, H., Vanrumste, B., Grech, R., Muscat, J., De Clercq, W., Vergult, A., D'Asseler, Y., Camilleri, K., Fabri, S., Van Huffel, S., Lemahieu, I.: Review on solving the forward problem in EEG source analysis. J. Neuroeng. Rehabil. 4, 46 (2007)
23. Ma, J., Gao, D., Ge, F., Amari, S.-I.: A One-Bit-Matching Learning Algorithm for Independent Component Analysis. In: Rosca, J.P., Erdogmus, D., Príncipe, J.C., Haykin, S. (eds.) ICA 2006. LNCS, vol. 3889, pp. 173–180. Springer, Heidelberg (2006)

Electrical Impedance Properties of Deep Brain Stimulation Electrodes during Long-Term In-Vivo Stimulation in the Parkinson Model of the Rat

Kathrin Badstübner[1,*], Thomas Kröger[2,*], Eilhard Mix[1], Ulrike Gimsa[3], Reiner Benecke[1], and Jan Gimsa[2]

[1] Department of Neurology, University of Rostock,
Gehlsheimer Str. 20, 18147 Rostock, Germany
{kathrin.badstuebner,eilhard.mix,
reiner.benecke}@med.uni-rostock.de
[2] Chair of Biophysics, Institute of Biology, University of Rostock,
Gertrudenstr. 11A, 18157 Rostock, Germany
{thomas.kroeger,jan.gimsa}@uni-rostock.de
[3] Research Unit Behavioral Physiology, Leibniz -Institute for Farm Animal Biology,
Wilhelm-Stahl-Allee 2, 18196 Dummerstorf, Germany
gimsa@fbn-dummerstorf.de

Abstract. Deep brain stimulation (DBS) is an invasive therapeutic option for patients with Parkinson's disease (PD) but the mechanisms behind it are not yet fully understood. Animal models are essential for basic DBS research, because cell based *in-vitro* techniques are not complex enough. However, the geometry difference between rodents and humans implicates transfer problems of the stimulation conditions. For rodents, the development of miniaturized mobile stimulators and adapted electrodes are desirable. We implanted uni- and bipolar platinum/iridium electrodes in rats and were able to establish chronical instrumentation of freely moving rats (3 weeks). We measured the impedance of unipolar electrodes *in-vivo* to characterize the influence of electrochemical processes at the electrode-tissue interface. During the encapsulation process, the real part of the electrode impedance at 10 kHz doubled after 12 days and increased almost 10 times after 22 days. An outlook is given on the quantification of the DBS effect by sensorimotor behavioral tests.

Keywords: EIS, Intracerebral electrodes, Basal ganglia, Subthalamic nucleus, Rat brain, Chronic instrumentation, 6-OHDA, Parkinson's disease.

1 Introduction

Parkinson's disease (PD) is a widespread degenerative disorder of the central nervous system that affects motor function, speech, cognition and vegetative functions. The cardinal symptoms such as tremor, rigidity, bradykinesia and postural instability result mainly from the death of dopaminergic cells in the substantia nigra pars compacta

---

[*] Corresponding authors.

J. Gabriel et al. (Eds.): BIOSTEC 2012, CCIS 357, pp. 287–297, 2013.
© Springer-Verlag Berlin Heidelberg 2013

and the subsequent lack of dopaminergic inputs into the striatum. This causes an alteration of the activity pattern in the basal ganglia [2]. Deep brain stimulation (DBS) is a novel therapeutic option for PD as well as an increasing number of neuropsychiatric disorders. Before DBS became a therapeutic intervention, electric stimulation of basal ganglia had been used to guide neurosurgeons to the precise position for a surgical lesion, the ultimate therapy of a late-stage PD. The main advantage of DBS over surgical lesions is the reversibility and possibility to modulate stimulation parameters [3]. The small volume of the target region for DBS in the human brain requires a highly specific adaption of the electrodes which need to be thoroughly tested in animal models, including different materials and geometries. So far, DBS-data of animal models of PD are scarce. During *in-vivo* stimulation, the properties of the DBS electrodes are changing as a function of time caused by electrochemical processes at the surface of the implant and the subsequent tissue response [5]. The tissue response is a foreign substance reaction. Its intensity depends on the material [Grill and Mortimer, 1994] and is correlated with the thickness of the adventitia finally encapsulating the implant [18]. Adventitia formation causes a steady change in the impedance of the electrodes leading to changes in the attenuation of the stimulating signal. As a result, the efficiency of the surrounding tissue stimulation is changing [12]; [13]; [7]. One opportunity to minimize this problem is to choose an appropriate electrode material. Previous investigations of our group [6]; [8] have shown that the use of stainless steel electrodes is not appropriate because of the corrosion and erosion processes intensified by electrolytic electrode processes. Electrochemically induced alterations are negligible for inert platinum electrodes, even though electrode processes may still influence the surrounding tissue [5]. For an optimal adjustment of the DBS signal, the kinetics of the electrode-impedance alterations caused by the adventitia formation must be taken into account [12]; [13].

2   Materials and Methods

2.1   Animal Treatment

Forty, adult, male Wistar Han rats (240-260 g) were obtained from Charles River Laboratory, Sulzfeld, Germany) and housed under temperature-controlled conditions in a 12 h light-dark cycle with conventional rodent chow and water provided ad libitum. The rats were subject to the following treatments:

- anesthesia (40 rats)
- 6-OHDA-lesioning (40 rats, 2 rats died while surgery)
- electrode implantation (38 rats (2 rats died while surgery): 15 unipolar electrodes, 21 bipolar electrodes)
- chronical instrumentation (26 rats: 21 rats with bipolar electrodes, 5 rats with unipolar electrodes)
- impedance measurement without chronical instrumentation (10 rats with unipolar electrodes)

The study was carried out in accordance with European Community Council directive 86/609/EEC for the care of laboratory animals and was approved by Rostock's Animal Care Committee (LALLF M-V/TSD/7221.3-1.2-043/06).

Anesthesia. The rats were anesthetized by ketamine-hydrochloride (10 mg per 100 g body weight, i.p., Ketanest S®, Pfizer, Karlsruhe, Germany) and xylazine (0,5 mg per 100 g body weight, i.p., Rompun®, Pfizer). Before surgery, the eyes of the rats were medicated with Vidisic (Bausch and Lomb, Berlin Germany). After surgery, the wound was sutured and the rats received 0.1 ml novaminsulfone (Ratiopharm, Ulm, Germany) and 4 ml saline subcutaneously. Rats were exposed to red light (Petra, Burgau, Germany) until normalization of vital functions.

6-Hydroxydopamine (6-OHDA) Lesioning. The 6-OHDA (Sigma, Deisenhofen, Germany) lesioning was performed by stereotactic surgery in adaption to Strauss et al. [17]. The lesions of the right medial forebrain bundle of rats were induced by injection of 26 µg 6-OHDA in 4 µl saline with 1 g/l ascorbic acid delivered over 4 min via a 5 µl hamilton microsyringe (Postnova Analytics, Landsberg/Lech, Germany). The coordinates relative to bregma were: anterior-posterior (AP) = -2.3 mm, medial-lateral (ML) = 1.5 mm and dorsal-ventral (DV) = -8.5 mm [16].

Electrode Implantation. Electrodes were implanted into the subthalamic nucleus (STN), which is the most common target region for treatment of PD patients. The surgical procedure was performed using a stereotactic frame (Stoelting, Wood Dale, IL, USA) modified according to Harnack et al. [9]. To support the surgical procedure, a cold light source (KL 1500 LCD, Schott, Mainz, Germany) was used in combination with a stereo-microscope (Leica, Wetzlar, Germany). The skull was opened by a dental rose-head bur (Kaniedenta, Germany). The coordinates relative to bregma were: anterior-posterior (AP) = -3.5 mm, medial-lateral (ML) = 2.4 mm and dorsal-ventral (DV) = -7.6 mm [16]. A dental drill was used to bore an additional hole in the skull for an anchor screw. The electrode was fixed to the skull by an adhesive-glue bridge (Technovit 5071, Heraeus, Germany) to the anchor screw. After all, a subcutaneous wire was implanted. The suture exit hole was located in the middle of the back of the rat.

The counter-electrodes (dental wires and suture clips in combination with the unipolar electrodes) were implanted into the neck of the rats.

Chronical Instrumentation. Commercial rat jackets (Lomir Biomedical, Quebec, Canada) with a backpack were used to fix all electronic components of the miniaturized custom-made stimulator system (Rückmann und Arndt, Berlin, Germany) to the rat. The DBS stimulator and the battery were located in the backpack of the rat jacket. The DBS electrode and the battery were connected to the DBS stimulator via plug connectors (RS Components GmbH, Mörfelden-Walldorf, Germany). Both were soldered with a lead-free solder tin (RS Components GmbH) and insulated with a biocompatible shrink tubing (RS Components GmbH).

Impedance Measurement. Ten rats were used to measure the kinetics of the electrode impedance alterations caused by the adventitia formation at the surface of unipolar electrodes. During the measurement, the rats were anesthetized. The measuring procedure was performed every day over a measuring period of 12 days (in experiment 1) and over 22 days (in experiment 2). In the first experiment we used a dental wire as counter-electrode and in the second step an array of suture clips (Allgaier Instrumente GmbH, Frittlingen/Tuttlingen, Germany) to evaluate the influence of the counter-electrode (shape and position) on the measurement results of the impedance.

2.2   Stimulation- and Counter-Electrodes

For impedance measurements, we designed unipolar platinum-iridium (Pt/Ir) micro-electrodes which were covered with polyesterimide insulation and custom-made by Polyfil (Zug, Switzerland; Fig. 1). In a first step, dental wires made of biocompatible, nickel-free steel alloy (18% Cr, 18% Mn, 2% Mo, 1% N, remander iron) of 1 mm diameter and 10 mm length were used as counter-electrodes (see: Fig. 4a) in combination with the unipolar DBS electrodes. In a second step, an array of counter-electrodes was pierced into the necks of a number of rats, to evaluate the influence of the counter-electrode on the impedance measurement. For this, suture clips (see: Fig. 4b) were used. Electrochemical electrode effects were negligible at the counter-electrode due to the low current density at its large surface.

We also designed and implanted bipolar Pt/Ir electrodes (Fig. 2) with two stacked tips to test the effects of nonaxial symmetric field distributions. Further, this electrode type does not require the implantation of counter-electrodes.

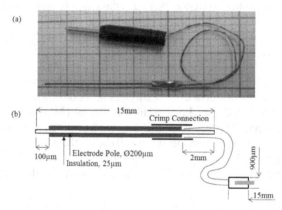

Fig. 1. Photograph (a) and scheme (b) of the unipolar Pt/Ir electrode (Polyfil, Zug, Switzerland). The electrode pole was a round wire made from Pt90Ir10 with a diameter of 200 μm. The length of the non-insulated tip of the electrode pole was 100 μm. The insulation consists of polyesterimide 180 with a thickness of 25 μm [15].

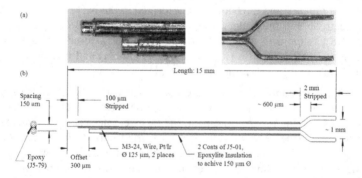

Fig. 2. Photograph (a) and scheme of the bipolar Pt/Ir electrode (FHC, Bowdoin, ME, USA). The two electrode poles were round wires made from Pt90Ir10 with a diameter of 125 μm. The lengths of the non-insulated tips were 100 μm. The thickness of the epoxylite insulation was 25 μm.

2.3   Electric Impedance Spectroscopy (EIS)

Electric impedance spectroscopy (EIS) is a common measuring technique for determining the electrical properties of tissues [4]. It is used in a wide range of applications, such as breast cancer detection [10], the monitoring of the lung volume [1] and in material sciences. EIS is nondestructive and therefore suitable for the characterization of the DBS electrodes during the encapsulation process.

Equipment. The EIS measurements were conducted with an impedance spectrometer Sciospec ISX3 (Sciospec Scientific Instruments, Pausitz, Germany) and a test fixture HP16047D (Hewlett-Packard, Japan) connected to a personal computer with Sciospec-measuring software. The two connectors of the test fixture were connected to the DBS and the counter-electrodes.

Measurement. To characterize the electrode properties during encapsulation, the impedance was recorded in the frequency range from 100 Hz to 10 MHz over a period of two weeks after implantation. Frequency range, amplitude, number of points and the averaging of the impedance spectrometer were programmed by the measuring-software (Sciospec). 401 frequency points were logged which were distributed equidistantly over a logarithmic frequency scale. The measuring voltage (peak to peak) was 12.5 mV$_{PP}$. The measuring-software logged the measuring data of the impedance spectrometer (real and imaginary parts of the impedance vs. frequency) by saving them as a data file.

Before each measurement, the impedance spectrometer was calibrated by open, short and load measurements. Each measurement was repeated three times to improve the statistical significance. The measurements were repeated every day for one week and every second day during the second week.

The stimulation pulse usually applied in DBS has a frequency of 130 Hz and a pulse width of 60 µs. Because of the steep slopes of the needle-shaped pulse, the signal is rich in high harmonic frequencies [5]. For this reason, we measured the impedance within the wide frequency range from 100 Hz to 10 MHz, which is beyond the range of up to 10 kHz reported by Lempka et al. [12].

Impedance Theory. The electrical impedance describes the magnitude ratio between the applied AC voltage and the resulting current flowing with a certain phase shift. Mathematically speaking, the impedance Z* is a complex number with the unit [Ω], which is composed of a real (Z´) and an orthogonal imaginary part (Z´´) marked by the complex unit $j = \sqrt{-1}$:

$$Z^* = \mathrm{Re}(Z^*) + j\cdot\mathrm{Im}(Z^*) = Z' + j\,Z'' \tag{1}$$

For interpretation of the measuring data, an equivalent circuit model is required to be fitted to the measuring data. The aim was to model electrochemical processes and adherent cell growths by combinations of resistors, capacitors and constant phase elements [12].

Data Analysis. The logged data were transferred to Matlab (The MathWorks™, Version 7.9.0.529) to calculate means and standard deviations. For their graphic representation, they were finally copied to Sigma Plot 11.0 (Systat Software, 11.0, Build 11.2.0.5).

2.4   Electron Microscopy Study of Electrode Encapsulation by Tissue

Concentric bipolar microelectrodes with an inner pole diameter of 75 µm and an outer pole diameter of 250 µm (CB CSG75; FHC, Bowdoinham, ME) were placed into the STN of two anesthetized rats. The rats were stimulated for 3 h with biphasic constant-current pulses with a repetition frequency of 130 Hz and pulse duration of 60 µs at 250 µA with a stimulus generator (Multichannel Systems, Reutlingen, Germany). The electrodes were removed and one electrode was incubated in trypsin solution (Trypsin-EDTA (1x) in HBSS W/O CA&MG W/EDTA.4NA, Gibco, UK) at 37°C for 1 h. The other electrode was postfixed overnight in 4% glutaraldehyde (Merck, Germany) in PBS.

Electrodes were washed, fixed in 4% glutaraldehyde in PBS, washed again, postfixed in 1% osmium tetroxide, dehydrated in acetone, and subjected to critical-point drying (EMITECH K850, Ashford, Kent, UK). The samples were sputtered with colloidal gold using a Sputter Coater (BAL-TEC SCD 004, Schalksmühle, Germany) before examination with a scanning electron microscope (DSM 960 A, Zeiss, Oberkochen, Germany).

3   Results

3.1   Electrode Implantation

We implanted 36 Pt/Ir electrodes (15 unipolar electrodes, 21 bipolar electrodes) in Parkinsonian rats and combined the bipolar electrodes with chronical instrumentation. This allowed for stimulation under spontaneous movement conditions without fixing the rats to an apparatus. For an optimal adjustment of the DBS signal in future experiments; we measured in pilot tests (without chronical instrumentation) the kinetics of the electrode impedance alterations caused by adherent cell growth at the surface of the implanted unipolar electrodes.

Verification by Ink Injection. As a first test, the localization of the electrode tip in the target region was verified by ink injection via a 5 µl Hamilton micro syringe of approximately the same size as the implanted stimulation electrode (Fig. 3).

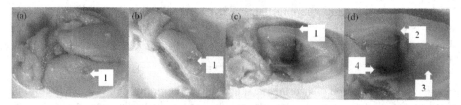

Fig. 3. Rat brain fixed in formalin, (a): top view of a rat brain with a puncture resembling an electrode canal (1), (b): sagittal section of one hemisphere with the puncture (1), (c): hemisphere where tissue was removed to display the injection canal, (d): enlarged photo of (c) with the ink injection canal (2), striatum (3), and other parts of the basal ganglia (4)

Counter-electrode Implantation. For the EIS measurements, counter-electrodes were pierced through the intact scalp close behind the cut for the implantation of the stimulation electrode. In a first step, a dental wire of biocompatible steel alloy used as a counter-electrode was implanted into the neck of two rats, shown in Fig. 4a. In a second step, an array of suture clips was implanted into the neck of 8 rats, shown in the scheme in Fig. 4b.

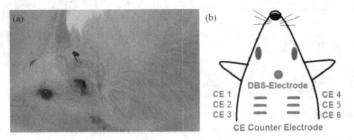

Fig. 4. Photograph of an implanted dental wire of biocompatible steel alloy (a) and scheme of an array of suture clips (b) used as counter-electrodes for the EIS measurement

3.2   Chronic Instrumentation

Chronic instrumentation (in combination with the bipolar DBS electrodes) was established and applied to freely moving Parkinsonian rats (Fig. 5a). For handling, the rats had to be trained every day over a period of 5 weeks. Following electrode implantation, a biocompatible subcutaneous wire was implanted (Fig. 5b). After each stereotactic surgery (6-OHDA lesioning and electrode implantation), the rats were left alone for one week to allow for wound healing. Then, the rat jackets were put on. All electronic components (Fig. 6c) were located in the backpack of the rat jacket. This allowed for a stimulation of up to 3 weeks when the batteries were changed every second day and the jacket once a week. The DBS electrode and the battery were connected to the DBS stimulator via plug connectors. The weight of all components of the chronic instrumentation (including battery) was 5 g.

Previous investigations of our own group [6]; [8] allowed for a stimulation of only 3 h when the rats were fixed to an apparatus.

The custom-made miniaturized DBS stimulator system was manufactured in cooperation with the company Rückmann und Arndt, Berlin, Germany.

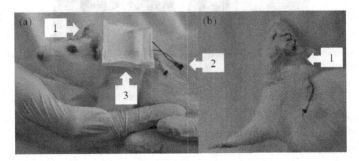

Fig. 5. Photograph of a freely moving rat one week after implantation with an implanted bipolar DBS electrode and chronic instrumentation consisting of (1) bipolar Pt/Ir electrode with subcutaneous cable and plug connector, (2) plug-connector of the DBS stimulator and (3) DBS stimulator and battery in the carrying bag of the rat jacket

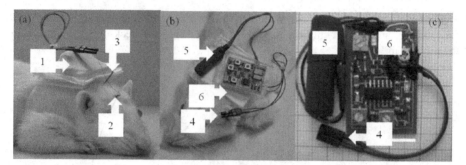

Fig. 6. Photograph of a rat with a unipolar electrode that was anesthetized for implantation. The chronic instrumentation consisted of: (1) rat jacket with carrying bag, (2) unipolar Pt/Ir electrode, (3) counter-electrode made from dental wire, (4) plug connector electrodes, (5) battery, (6) DBS stimulator.

3.3   Impedance Measurement

Our data showed the general tendency of an impedance increase during the encapsulation process. The measuring results are shown in Fig. 8 and 9. The value of Z′ has almost doubled after 12 days (Fig. 8a) and increased almost 10 times (Fig. 9a) after 22 days compared to its initial value at 10 kHz. In parallel, also the absolute value of Z″ (Fig. 8b and 9b) decreased. The results match with findings of Lempka et al. [12] for the monkey brain who described the formation of the adventitia as a foreign substance reaction as the main reason for the impedance increase. Our own electron-microcopy study showed that within 3 h post implantation, a considerable number of cells firmly adhered to the stimulation electrodes (Fig. 7).

The findings we obtained with the implantation of an array of suture clips indicated that the influence of the position, shape and material of the counter-electrode on the impedance measurement results was negligible.

(a) trypsin   (b) glutaraldehyde

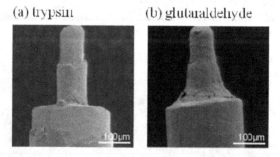

Fig. 7. Scanning electron-microscopical images of electrodes that had been implanted and used for 3 h of stimulation. (a): Electrode cleaned by trypsination prior to electron microscopy. (b): Electrode with fixed adhering tissue.

Our data showed an impedance decrease at the second day after implantation followed by a significant increase from the third day on. Interestingly, the same impedance behavior has already been reported by Lempka et al. [12, p. 6, Fig. 4]. However, no explanation has been given for this phenomenon.

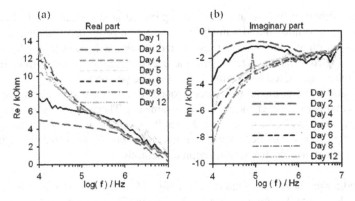

Fig. 8. In *vivo* impedance change of a unipolar Pt/Ir electrode measured against a dental-wire counter-electrode over 12 days ((a) Re = Z′; (b) Im = Z″)

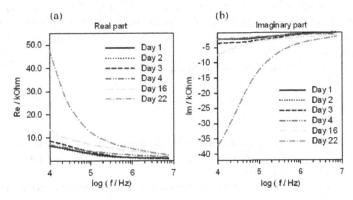

Fig. 9. In *vivo* impedance change of a unipolar Pt/Ir electrode measured against a suture-clip counter-electrode over 22 days ((a) Re = Z′; (b) Im = Z″)

4   Conclusions and Perspectives

Pilot experiments in the rat model have shown that the impedance of unipolar DBS electrodes is significantly increasing with time after implantation. The main reason is the formation of adhering tissue that encapsulates the implant. This tissue response is a foreign body reaction. Its intensity depends on the material. The stimulation efficiency can be improved with an appropriate electrode material. Previous investigations of our own group have shown that stainless steel electrodes are not appropriate. The use of platinum/iridium could improve the stimulation efficiency.

The characterization of the encapsulation process enabled us to readjust our miniaturized prototype stimulator system to adapt the stimulation parameters according to the stimulation duration. To the best of our knowledge, we established for the first time chronic instrumentation of completely freely moving rats for long-term experiments (of up to 3 weeks). Also for the first time, iron-free electrodes have been applied to Parkinsonian rats for performing DBS experiments over the same observation time.

In future experiments, we aim for a comparative study of uni- and bipolar electrodes with a conventional and a modified stimulation program. The quantification of the DBS effect on locomotion, exploration and anxiety will be analyzed by drug- and non-drug induced behavioral tests. For these experiments, the operating life of the battery of the miniaturized DBS stimulator has to be prolonged.

An equivalent circuit model will be developed for a better understanding of the measuring data in order to extract encapsulation parameters. These investigations aim at clarifying the phenomenon of the impedance drop at the second day after implantation.

Potential effects at the electrode-tissue interface will be analyzed by histological, immunochemical and electron-microscopical methods.

Acknowledgements. K.B. is grateful for a stipend of the German Research Foundation (DFG, Research Training Group 1505/1 "welisa"). T.K. acknowledges financing by a project of the Federal Ministery of Economics and Technology (BMWi, V230-630-08-TVMV-S-031). Part of the work was conducted within a project financed by the Federal Ministry of Education and Research (BMBF, FKZ 01EZ0911). The authors are grateful to Dr. R. Arndt for a fruitful cooperation on the stimulator development and to Dr. J. Henning for help with the electron microscopy study. We would like to thank the staff of the electron microscopy center at the University of Rostock's Medical Faculty for outstanding technical support.

References

1. Adler, A., Amyot, R., Guardo, R., Bates, J.H., Berthiaume, Y.: Monitoring changes in lung air and liquid volumes with electrical impedance tomography. Journal of Applied Physiology 83(5), 1762–1767 (1997)
2. Braak, H., Braak, E.: Pathoanatomy of Parkinson's disease. Journal of Neurology 247(2), II3–II10 (2000)
3. Benabid, A.L., Pollak, P., Louveau, A., Henry, S., de Rougemont, J.: Combined (thalamotomy and stimulation) stereotactic surgery of the VIM thalamic nucleus for bilateral Parkinson disease. Applied Neurophysiology 50(1-6), 344–346 (1987)
4. Foster, K.R., Schwan, H.P.: Dielectric properties of tissues and biological materials: a critical review. Critical Reviews in Biomedical Engineering 17(1), 25–104 (1989)
5. Gimsa, J., Habel, B., Schreiber, U., van Rienen, U., Strauss, U., Gimsa, U.: Choosing electrodes for deep brain stimulation experiments – electrochemical considerations. Journal of Neuroscience Methods 142(2), 251–265 (2005)
6. Gimsa, U., Schreiber, U., Habel, B., Flehr, J., van Rienen, U., Gimsa, J.: Matching geometry and simulation parameters of electrodes for deep brain stimulation experiments – numerical considerations. Journal of Neuroscience Methods 150(2), 212–227 (2006)
7. Grill, W.M., Mortimer, J.T.: Electrical properties of implant encapsulation tissue. Annals of Biomedical Engineering 22(1), 23–33 (1994)
8. Henning, J.: Wirkungen der tiefen Hirnstimulation – Analyse der Gen- und Proteinexpression in einem optimierten Rattenmodell. Dissertation. University of Rostock (2007)
9. Harnack, D., Winter, C., Meissner, W., Reum, T., Kupsch, A., Morgenstern, R.: The effects of electrode material, charge density and stimulation duration on the safety of high-frequency stimulation of the subthalamic nucleus in rats. Journal of Neuroscience Methods 138(1-2), 207–216 (2004)

10. Kerner, T.E., Paulson, K.D., Hartov, A., Soho, S.K., Poplack, S.P.: Electrical impedance spectroscopy of the breast: clinical imaging results in 26 subjects. IEEE Transactions on Medical Imaging 21(6), 638–645 (2002)
11. Journal of Neurophysiology 99(6), 2902–2915 (2008)
12. Lempka, S.F., Miocinovic, S., Johnson, M.D., Vitek, J.L., McIntyre, C.C.: In vivo impedance spectroscopy of deep brain stimulation electrodes. Journal of Neural Engineering 6, 046001, 11 (2009)
13. Lempka, S.F., Johnson, M.D., Moffitt, M.A., Otto, K.J., Kipke, D.R., McIntyre, C.C.: Theoretical analysis of intracortical microelectrode recordings. Journal of Neural Engineering 8(4), 045006 (2011)
14. Macdonald, J.R.: Impedance spectroscopy. Annals of Biomedical Engineering 20(3), 289–305 (1992)
15. Nowak, K.A., Mix, E., Gimsa, J., Strauss, U., Sriperumbudur, K.K., Benecke, R., Gimsa, U.: Optimizing a rodent model of Parkinson's disease for exploring the effects and mechanisms of deep brain stimulation. Parkinson's Disease, 414682 (2011)
16. Paxinos, G., Watson, C.: The rat brain in stereotaxic coordinates. Academic Press, San Diego (1998)
17. Strauss, U., Zhou, F.W., Henning, J., Battefeld, A., Wree, A., Köhling, R., Haas, S.J., Benecke, R., Rolfs, A., Gimsa, U.: Increasing extracellular potassium results in subthalamic neuron activity resembling that seen in a 6-hydroxydopamine lesion
18. Wintermantel, E., Ha, S.W.: Medizintechnik mit biokompatiblen Werkstoffen und Verfahren, 3rd edn., pp. 134–135. Springer, Berlin (2002)

Comparison between Thermal and Visible Facial Features on a Verification Approach

Carlos M. Travieso, Marcos del Pozo-Baños, and Jesús B. Alonso

Signals and Communications Department (DSC),
Institute for Technological Development and Innovation in Communications (IDETIC),
University of Las Palmas de Gran Canaria (ULPGC),
Campus Universitario de Tafira, s/n, 35017, Las Palmas de Gran Canaria, Spain
{ctravieso,jalonso}@dsc.ulpgc.es

Abstract. A comprehensible performance analysis of a thermal and visible face verification system based on the Scale-Invariant Feature Transform algorithm (SIFT) with a vocabulary tree is presented in this work, providing a verification scheme that scales efficiently to a large number of features. The image database is formed from front-view thermal images, which contain facial temperature distributions of different individuals in 2-dimensional format and the visible image per subject, containing 1,476 thermal images and 1,476 visible images equally split into two sets of modalities: face and head, respectively. The SIFT features are not only invariant to image scale and rotation but also essential for providing a robust matching across changes in illumination or addition of noise. Descriptors extracted from local regions are hierarchically set in a vocabulary tree using the k-means algorithm as clustering method. That provides a larger and more discriminatory vocabulary, which leads to a performance improvement. The verification quality is evaluated through a series of independent experiments with various results, showing the power of the system, which satisfactorily verifies the identity of the database subjects and overcoming limitations such as dependency on illumination conditions and facial expressions. A comparison between head and face verification is made for both ranges. This approach has reached accuracy rates of 97.60% in thermal head images in relation to 88.20% in thermal face verification. For visible range, 99.05% with visible head images in relation to 97.65% in visible face verification. In this proposal and after experiments, visible range gives better accuracy than thermal range, and with independency of range, head images give the most discriminate information.

Keywords: Thermal face verification, Visible face verification, Face detection, Biometrics, SIFT parameters, Vocabulary tree, k-Means, Image processing, Pattern recognition.

1 Introduction

Human recognition through distinctive facial features supported by an image database is still an appropriate subject of study. We may not forget that this problem presents

J. Gabriel et al. (Eds.): BIOSTEC 2012, CCIS 357, pp. 298–308, 2013.
© Springer-Verlag Berlin Heidelberg 2013

various difficulties. What will occur if the individual's haircut is changed? Is make-up a determining factor in the process of verification? Would it distort significantly facial features?

The use of thermal cameras originally conceived for military purpose has expanded to other fields of application such as control process in production lines, detection/monitoring of fire or applications of security and Anti-terrorism. Therefore, we consider its use in human identification tasks in scenarios where the lack of light restricts the operation of conventional cameras. Different looks of the main role from the film The Saint are shown in figure 1.

Fig. 1. Facial changes of the character played by Val Kilmer in the film The Saint

Val Kilmer modifies his look in this film spectacularly in order to not to be recognized by the enemy.

A correct matching between the test face and that stored in the image database is expected, although it may seem a hard problem to solve even if natural distortion effects such as illumination changes or interference are not considered. The recognition problem should be split in some stages, that is, acquisition of facial images for testing, features extraction from specific facial regions and finally, verification of the individual's identity [1].

Currently, computational face analysis is a very lively research field, in which we observe that new interesting possibilities are being studied. For example, we can quote an approach for improving system performance when working with low resolution images (LR) and decreasing computational load.

In [2], it is presented a facial recognition system, which works with LR images using nonlinear mappings to infer coherent features that favor higher recognition of the nearest neighbour (NN) classifiers for recognition of single LR face image. It is also interesting to cite the approach of [3], in which a multi-resolution feature extraction algorithm for face recognition based on two-dimensional discrete wavelet transform (2D-DWT) is proposed. It exploits local spatial variations in a face image effectively obtaining outstanding results with 2 different databases.

The images of subjects are often taken in different poses or with different modalities, such as thermographic images, presenting different stages of difficulty in their identification.

In [4], results on the use of thermal infrared and visible imagery for face recognition in operational scenarios are presented. These results show that thermal face recognition performance is stable over multiple sessions in outdoor scenarios, and that fusion of modalities increases performance.

In the same year 2004, L. Jiang proposed in [5] an automated thermal imaging system that is able to discriminate frontal from non-frontal face views with the assumption that at any one time, there is only 1 person in the field of view of the camera and no other heat-emitting objects are present. In this approach, the distance from centroid (DFC) shows its suitability for comparing the degree of symmetry of the lower face outline.

The use of correlation filters in [6] has shown its adequacy for face recognition tasks using thermal infrared (IR) face images due to the invariance of this type of images to visible illumination variations. The results with Minimum Average Correlation Energy (MACE) filters and Optimum Trade-off Synthetic Discriminant Function (OTSDF) in low resolution images (20x20 pixels) prove their efficiency in Human Identification at a Distance (HID).

Scale Invariant Feature Transform (SIFT) algorithm [7] are widely used in object recognition. In [8], SIFT has appeared as a suitable method to enhance the recognition of facial expressions under varying poses over 2D images It has been demonstrated how affine transformation consistency between two faces can be used to discard SIFT mismatches.

Gender recognition is another lively research field working with SIFT algorithm. In [9], faces are represented in terms of dense-Scale Invariant Feature Transform (d-SIFT) and shape. Instead of extracting descriptors around interest points only, local feature descriptors are extracted at regular image grid points, allowing dense descriptions of face images.

However, systems generate large number of SIFT features from an image. This huge computational effort associated with feature matching limits its application to face recognition. An approach to this problem has been developed in [10], using a discriminating method. Computational complexity is reduced more than 4 times and accuracy is increased in 1.00% on average by checking irrelevant features.

Constructing methods that scale well with the size of a database and allow finding one element of a large number of objects in acceptable time is an avoidable challenge. This work is inspired by Nister and Stewenius [11], where object recognition by a k-means vocabulary tree is presented. Efficiency is proved by a live demonstration that recognized CD-covers from a database of 40000 images. The vocabulary tree showed good results when a large number of distinctive descriptors form a large vocabulary. Many different approaches to this solution have been developed in the last few years [12] and [13], showing its competency organizing several objects. Having regard to these good results, this solution will be tested in this paper with SIFT descriptors in a vocabulary tree.

In this context, the aim of this work is to propose, compare and evaluate a facial and head verification system for visible and thermal ranges, applying the SIFT algorithm and obtaining local distinctive descriptors from each image based on [14]. The construction of the vocabulary tree enables to have these descriptors hierarchically organized and ready to carry out a search to find a specific object.

This paper is organized as follows. The proposed system is presented in section 2. Description of experiments and results are outlined in section 4. Discussions are described in section 4. Finally, conclusions are given in section 5.

2   Facial Recognition Approach

In our system proposed, SIFT descriptors are used to extract information from thermal and visible images in order to verify the identity of a test subject. Local distinctive descriptors are obtained from each face in the database and are used to build a vocabulary tree, through the use of the k-means function. For each test image, only its new descriptors are calculated and used to search through the hierarchical tree in order to build a vote matrix, in which the most similar image of the database can be easily identified. This approach mixes the singularity of the SIFT descriptors to perform reliable matching between different views of a face and the efficiency of the vocabulary tree for building a high discriminative vocabulary. A description of the system is provided in the next subsections.

2.1   Approach Proposed

This approach is composed by four stages: face segmentation, SIFT descriptors calculator, vocabulary tree construction and matching module.

While face segmentation is executed manually, the matching module searches in the vocabulary tree the best correspondence between the test descriptors and those of the database. Therefore, firstly the explanation is focused on the SIFT parameters and tree classification, and secondly a brief description of the matching module is given.

A block diagram of the system is shown in figure 2.

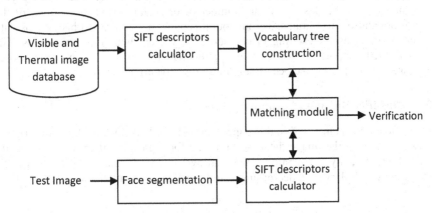

Fig. 2. Diagram of the proposed thermal face recognition system

2.2   Features Based on Scale-Invariant Feature Transform

The use of SIFT descriptors is applied in the most part of the results achieved by D. Lowe in [15] as a guideline, only determinant parameters are modified in order to adapt the algorithm to the system. Keypoints are detected using a cascade filtering, searching for stable features across all possible scales. The scale space of an image, $L(x,y,\sigma)$ is produced from the convolution of a variable-scale Gaussian, $G(x,y,\sigma)$ with an input image, $I(x,y)$;

$$L(x, y, \sigma) = G(x, y, \sigma) * I(x, y) \tag{1}$$

where * is the convolution operation in x and y, and

$$G(x, y, \sigma) = \frac{1}{2\pi\sigma^2} \cdot e^{\frac{-(x^2+y^2)}{2\sigma^2}} \tag{2}$$

Following (Lowe, 2004), scale-space in the Difference-of-Gaussian function (DoG) convolved with the image, $D(x, y, \sigma)$ can be computed as a difference of two nearby scales separated by a constant factor k:

$$D(x, y, \sigma) = \big(G(x, y, k\sigma) - G(x, y, \sigma)\big) * I(x, y) = L(x, y, k\sigma) - L(x, y, \sigma) \qquad (3)$$

From [16], it is stated that the maxima and minima of the scale-normalised Laplacian of Gaussian (LGN), $\sigma^2 \nabla^2 G$ produce the most stable image features in comparison with other functions, such as the gradient or Hessian. The relationship between D and $\sigma^2 \nabla^2 G$ is:

$$G(x, y, k\sigma) - G(x, y, \sigma) \approx (k-1)\sigma^2 \nabla^2 G \qquad (4)$$

The factor $(k - 1)$ is a constant over all scales and does not influence strong location. A significant difference in scales has been chosen, $k = \sqrt{2}$, which has almost no impact on the stability and the initial value of $\sigma = 1.6$ provides close to optimal repeatability according to [15].

After having located accurate keypoints and removed strong edge responses of the DoG function, orientation is assigned. There are two important parameters for varying the complexity of the descriptor: the number of orientations and the number of the array of orientation histograms. Throughout this paper a 4x4 array of histograms with 8 orientations is used, resulting in characteristic vectors with 128 dimensions. The results in [15] support the use of these parameters for object recognition purposes since larger descriptors have been found more sensitive to distortion.

2.3   Classifier Based on Vocabulary Tree

The verification scheme used in this paper is based on [11]. Once the SIFT descriptors are extracted from the image database, it's time for organizing them in a vocabulary tree. A hierarchically verification scheme allows to search selectively for a specific node in the vocabulary tree, decreasing search time and computational effort.

Fig. 3. Two levels of a vocabulary tree with branch factor 3

The k-means algorithm is used in the initial point cloud of descriptors for finding centroids through the minimum distance estimation so that a centroid represents a cluster of points.

The k-means algorithm is applied iteratively, since the calculation of the centroid location can vary the associated points. The algorithm converges if centroids location does not vary. Each tree level represents a node division of the nearby superior stage.

The initial number of clusters is defined by 10, with 5 tree levels. These values have shown good results, working with the actual database.

A model of a vocabulary tree with 2 levels and 3 initial clusters is shown in figure 3.

3   Experimental Settings

3.1   Databases Used

Authors have built an image database in order to develop this work. This database contains 738 images of 704×756 pixels and 24 bit per pixel using a SAT-S280 SATIR camera, which contains two sensors, one of them is a thermal sensor and another is a visible camera. An example of that image is observed on figure 4.

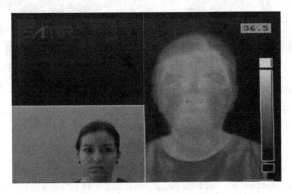

Fig. 4. Example of image from our visible and thermal camera

Our database is composed by 41 subjects, corresponding to 18 images per subject, acquired in 3 different sessions, 6 images per session, during one month. These images are composed by two images, visible and thermal image, as it is observed in figure 4. We have divided that image in two parts, and we have got 1476 images, 738 visible face images and 738 thermal facial images. In particular, false thermal colour is given for the sensor according to characteristics of each person. All images have been stored in PNG format. Besides, after segmentation process, we have equally split in two sets of 738 images with different modalities: face and head. Therefore, the images are divided into categories depending on the type of information they provide, having a total of 2952 images:

- Heads: Thermal images of full heads of subjects (738 images).
- Heads: Visible images of full heads of subjects (738 images).
- Faces: Thermal images of facial details. (738 images).
- Faces: Visible images of facial details. (738 images).

The following figures present some examples of thermal and visible images of heads (in figure 5) and faces (in figure 6) with the specified format.

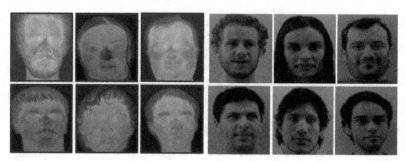

Fig. 5. 6 thermal and visible head images of the database. The examples show additional facial features such as head shape, hair and chin.

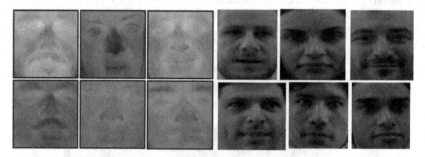

Fig. 6. 6 thermal and visible face images of the database from the same subjects in figure 4. The examples only show basic facial features such as eyes, lips and nose, representing the minimum information needed to verify a subject in the system.

The images were taken indoors by a SAT-S280 camera in 3 different sessions with different facial expressions such as happiness, sadness or anger, various facial orientations and distinctive changes in the haircut or facial hair of subjects.

In order to assure the independence of results, both sets of images are equally divided into 2 subsets in a second stage: test and training, under a 50% hold-out cross validation methodology. For each modality, 369 test images and 369 training images are available for the experiments.

The set of head images collects interesting details for recognition tasks, such as ear shape, haircut and chin. On the other hand, the set of facial images provides the minimum information that is nose, mouth and eyes areas.

Face detection and segmentation were manually realized in order to not to lose details in the process. One by one, from each image head or face was segmented and stored in separate files.

3.2   Experimental Methodology and Its Results

The aim of the experiments was to find how important is the extra information provided by head shape for human verification versus face information for thermal and visible ranges. Additionally, a comparison between head and facial verification results is carried out for both ranges. The proposed methodology consisting on the matching of facial features of thermal and visible images is compared with the matching of thermal and visible images of heads.

For each subject, an equally random division of the image database is made so that 9 images per individual are used for testing and the remaining 9 for training purposes (50% hold-out validation method). As previously commented, 369 test images and 369 training images randomly chosen are available for the experiments in each modality. This division is carried out 41 times that is subject by subject in 41 iterations.

The process of face/head verification for a subject is the following. Firstly, the previously stated division of the database is made. Secondly, each of the 9 images of the test subject is compared with the 369 training images and results are obtained. Once these 9 images are processed, the database is joined together again and the process restarts with the next subject until the 41 subjects of the database are processed.

The parameters that take part in the experiments are the False Rejection Rate (FRR), False Acceptance Rate (FAR) and Equal Error Rate (EER), commonly used in biometric studies. Mean times are also considered in this study. These parameters are collected in form of vectors depending on a variable, the histogram threshold.

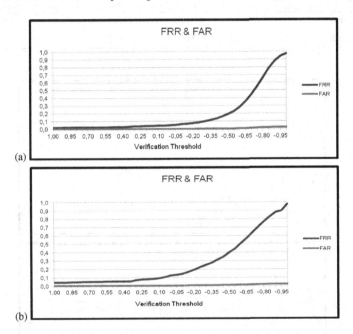

Fig. 7. FRR (blue line) and FAR (red line) in terms of the histogram threshold in (a), visible head verification, and in (b), thermal head verification.

Since the verification process finishes, a histogram with the contributions of each image in the database is obtained. The database image that best fits the test image shows the biggest value in the histogram. In a second stage, histogram values are normalized with regard to the biggest value, from 1 to -1. A threshold is used during the experiments for considering only the contributions of images that are above this limit; those below are discarded in that moment. The histogram threshold descends from value 1 to -1 in order to consider different samples each time. In figure 7 and 8 FRR and FAR are shown in relation to the histogram threshold. X-axis represents the threshold variation and Y-axis shows FRR and FAR values.

In practical terms, the threshold fall represents how the system becomes less demanding, taking more samples in account, increasing the FRR and FAR, since the additional samples do not belong to the test subject.

In table 1, it is shown average computational times. Although the verification time remains the same, the database updating (model building time) with head images is substantially higher as these images possess more information than facial images and therefore, consume more time and need more computational effort.

Table 1. Average Computational Times of head and face images verification during the experiment for thermal and visible ranges

MEAN TIMES HEAD AND FACE IMAGES				
Concept	Visible Head Time (sec.)	Visible Face Time (sec.)	Thermal Head Time (sec.)	Thermal Face Time (sec.)
Model Building	283.47	135.08	121.56	102.55
Test Verification	0.49	0.28	0.26	0.26

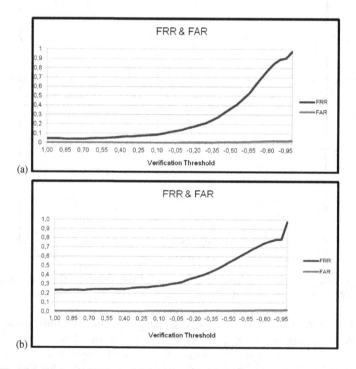

Fig. 8. FRR (blue line) and FAR (red line) in terms of the histogram threshold in (a), visible facial verification and in (b) thermal facial verification

The best result obtained in experiments with thermal head images is 97.60% in relation to 88.20% in thermal face verification. And the best result for visual head images is 99.05% in relation to 97.65% in visible face verification. It can be observed that the accuracy rate with head images is higher in comparison with facial images in both cases. Besides, we can observe that visible range has more discriminate information than thermal range.

4   Discussions

The four variants have been compared in this work under a performance for verification. The cause is the amount of information provided by each format. On the one hand, head images preserve important discriminative characteristics about the original visible and thermal images for identifying a subject that facial images do not include. This can be observed for both ranges. On the other hand, it becomes clear that in case of head images more SIFT descriptors are produced and therefore, more essential data for the verification process is extracted. Additionally, faces of different subjects have often common features that provide no discriminate information. In particular, for thermal images, this effect is bigger than visible images. In visible range, a similar accuracy is found in our experiments; only 2% is the difference between head and facial images, although head images give more discriminate information.

Between both ranges, visible images reach better accuracy for head and facial images. The difference is 2% for head images and 9% for facial images. Therefore, for this kind of camera and using SIFT keypoints; the visible range can be a better approach.

5   Conclusions

The main contribution of this work is the comparison of a head versus facial verification system based on SIFT descriptors with a vocabulary tree, applying on thermal and visible range, and using the accuracy rate, as comparison element between types of information. After the use of both ranges, it is observed good results, but better for visible range. Besides, we can conclude the head images for thermal and visible ranges are more discriminative than only the facial images.

As future work, this work will be a preliminary step in the development of face verification systems using SIFT descriptors in the fusion or combination of thermal and visible images. Too, we would like to increase considerably the size of database, and to include outdoor images. The proposed approach will be validated in this extended database. Too, we propose to apply fusion between both ranges, in order to improve the accuracy, and give more efficiency to this camera, which contains two different sensors.

Acknowledgements. This work was partially supported by "Cátedra Telefónica - ULPGC 2010/11", and partially supported by research Project TEC2012-38630-C04-02 from Ministry of Science and Innovation from Spanish Government.

Special thanks to Jaime Roberto Ticay-Rivas for their valuable help during the building of this database.

References

1. Soon-Won, J., Youngsung, K., Teoh, A.B.J., Kar-Ann, T.: Robust Identity Verification Based on Infrared Face Images. In: 2007 International Conference on Convergence Information Technology, ICCIT 2007, pp. 2066–2071 (2007)

2. Huang, H., He, H.: Super-Resolution Method for Face Recognition Using Nonlinear Mappings on Coherent Features. IEEE Transactions on Neural Networks 22(1), 121–130 (2011) ISSN: 1045-9227

3. Imtiaz, H., Fattah, S.A.: A wavelet-domain local feature selection scheme for face recognition. In: 2011 International Conference on Communications and Signal Processing, ICCSP 2011, Kerala, India, p. 448 (2011)

4. Socolinsky, D.A., Selinger, A.: Thermal Face Recognition in an Operational Scenario. In: CVPR 2004, Proceedings of the 2004 IEEE Computer Society Conference on Computer Vision and Pattern Recognition, vol. 2, pp. 1012–1019 (2004) ISSN: 1063-6919/04

5. Jiang, L., Yeo, A., Nursalim, J., Wu, S., Jiang, X., Lu, Z.: Frontal Infrared Human Face Detection by Distance From Centroide Method. In: Proceedings of 2004 International Symposium on Intelligent Multimedia, Video and Speech Processing, Hong Kong, pp. 41–44 (2004)

6. Heo, J., Savvides, M., Vijayakumar, B.V.K.: Performance Evaluation of Face Recognition using Visual and Thermal Imagery with Advanced Correlation Filters. In: Proceedings of the 2005 IEEE Computer Society Conference on Computer Vision and Pattern Recognition, CVPR 2005, pp. 9–15 (2005) ISSN: 1063-6919/05

7. Lowe, D.G.: Object recognition from local scale-invariant features. In: Proceedings of the Seventh IEEE International Conference on Computer Vision, ICIP 1999, vol. 2, pp. 1150–1157 (1999)

8. Soyel, H., Demirel, H.: Improved SIFT matching for pose robust facial expression recognition. In: 2011 IEEE International Conference on Automatic Face & Gesture Recognition and Workshops, FG 2011, pp. 585–590 (2011)

9. Jian-Gang, W., Jun, L., Wei-Yun, Y., Sung, E.: Boosting dense SIFT descriptors and shape contexts of face images for gender recognition. In: 2010 IEEE Computer Society Conference on Computer Vision and Pattern Recognition Workshops, CVPRW 2010, pp. 96–102 (2010)

10. Majumdar, A., Ward, R.K.: Discriminative SIFT features for face recognition. In: 2009 Canadian Conference on Electrical and Computer Engineering, CCECE 2009, pp. 27–30 (2009)

11. Nister, D., Stewenius, H.: Scalable Recognition with a Vocabulary Tree. In: 2006 IEEE Computer Society Conference on Computer Vision and Pattern Recognition, CVPR 2006, vol. 2, pp. 2161–2168 (2006)

12. Ober, S., Winter, M., Arth, C., Bischof, H.: Dual-Layer Visual Vocabulary Tree Hypotheses for Object Recognition. In: 2007 IEEE International Conference on Image Processing, ICIP 2007, vol. 6, pp. VI-345-VI-348 (2007)

13. Slobodan, I.: Object labeling for recognition using vocabulary trees. In: 19th International Conference on Pattern Recognition, ICPR 2008, pp. 1–4 (2008)

14. Crespo, D., Travieso, C.M., Alonso, J.B.: Thermal Face Verification based on Scale-Invariant Feature Transform and Vocabulary Tree - Application to Biometric Verification Systems. In: International Conference on Bio-inspired Systems and Signal Processing 2012, pp. 475–481 (2012)

15. Lowe, D.G.: Distinctive image features from scale-invariant keypoints. International Journal of Computer Vision 60(2), 91–110 (2004)

16. Mikolajczyk, K.: Detection of local features invariant to affine transformations, Ph.D. thesis. Institut National Polytechnique de Grenoble, France (2002)

Part IV

Health Informatics

Multiparameter Sleep Monitoring Using a Depth Camera

Meng-Chieh Yu[1,*], Huan Wu[2], Jia-Ling Liou[2], Ming-Sui Lee[1,2], and Yi-Ping Hung[1,2]

[1] Graduate Institute of Networking and Multimedia, National Taiwan University,
Roosevelt Road, Taipei, Taiwan
monjay.ntu@gmail.com, {mslee,hung}@csie.ntu.edu.tw
[2] Department of Computer Science and Information Engineering, National Taiwan University,
Roosevelt Road, Taipei, Taiwan
{spidey.wu,jalinliou}@gmail.com

Abstract. In this study, a depth analysis technique was developed to monitor user's breathing rate, sleep position, and body movement while sleeping without any physical contact. A cross-section method was proposed to detect user's head and torso from the sequence of depth images. In the experiment, eight participants were asked to change the sleep positions (supine and side-lying) every fifteen breathing cycles on the bed. The results showed that the proposed method is promising to detect the head and torso with various sleeping postures and body shapes. In addition, a realistic over-night sleep monitoring experiment was conducted. The results demonstrated that this system is promising to monitor the sleep conditions in realistic sleep conditions and the measurement accuracy was better than the first experiment. This study is important for providing a non-contact technology to measure multiple sleep conditions and assist users in better understanding of his sleep quality.

Keywords: Non-contact breath measurement, Sleep position, Sleep cycle, Head detection, Depth camera.

1 Introduction

Sleep is essential for a person's mental and physical health. Studies indicate that sleep plays a critical role in immune function [6], metabolism and endocrine function [25], memory, learning [17], and other vital functions. However, there are some sleep disorders, such as sleep apnea, insomnia, hypersomnia, circadian rhythm disorders, which might interfere with physical, mental and emotional functioning. For better understanding of the sleep problems, many sleep centers and research groups are devoted to the sleep study. Polysomnography (PSG) is a multi-parametric test used in the study of sleep and as a diagnostic tool in sleep medicine. It monitors many body functions including brain activity (EEG), eye movement, muscle activity, heart rhythm, and breathing while sleeping [9]. In this study, we focus on the research issues in sleep cycle, sleep breathing, and sleep positions. For sleep cycle measurement, EEG monitoring is the most accurate method to detect user's sleep cycle, including the period of non-rapid eye movement (NREM) and rapid eye movement (REM). However, it is not convenient to use. In recent years, the motion sensor and pressure sensor array are widely used to monitor user's sleep conditions and body movement

J. Gabriel et al. (Eds.): BIOSTEC 2012, CCIS 357, pp. 311–325, 2013.

while sleeping [11, 20, 30], as well as estimate the sleep cycle and evaluate the sleep quality. For breath measurement while sleeping, it is important and mainly used to detect the symptom of sleep apnea. Sleep apnea is one of the most important sleep disorder characterized by abnormal pauses in breathing or instances of abnormally low breathing during sleep. For decades, the breath measurement methods would direct contact to the user while monitoring, and it might interfere with the user and affect the sleep quality. Although some non-contact breath measurement methods are proposed in recent years, such as ultra wideband (UWB) and structured light plethysmography (SLP), these still have some measurement limitations. For sleep position, in order to prevent the sleep apnea, studies show that side-lying position is the best sleeping posture for individuals with sleep apnea [8, 12, 16, 26]. A study analyzed six common sleep positions, and concluded that supine positions were more likely to lead to snoring and a bad night's sleep [13]. However, to date, there has been relatively little research conducted on the measurement of sleep positions.

In this study, a sleep monitoring system using a depth camera was proposed to monitor users' breathing rate, body movement, and sleep position while sleeping. Moreover, we evaluated the measurement accuracy of the system, including the accuracy of head and torso detection, breath measurement (compared to the RIP), and sleep movement (compared to Actigraphy). Through the experimental results, we confirmed that the system could accurately monitor user's sleep conditions. This paper is structured as follows: The first section deals with the introduction of present sleep studies. The second section of the article is a review of several breath measurement methods and activity monitoring while sleeping. The proposed system design is described in the third section. The experimental results are demonstrated in section four followed by the discussion on some important findings. Finally, conclusion and suggestions are given for further research.

2   Related Work

Breathing is important while sleeping. There are many breathing-related sleep disorders, such as apnea and hyperventilation syndrome (HVS). Currently, many methods are proposed to monitor the breath conditions while sleeping. Most screening tools consist of an airflow measuring device, a blood oxygen monitoring device, and the respiratory inductance plethysmography (RIP). Thermistor (TH) measurements have been traditionally used to determine airflow during PSG studies. It is placed over the nose and mouth and infers airflow by sensing differences in the temperature of the warmer expired air and the cooler inhaled ambient air. However, low accuracy in detecting hypopneas is a major drawback [4]. The pulse oximeter is a medical device that monitors the oxygen saturation of user's blood, and changes in blood volume in the skin. Low oxygen levels in the blood often occur with sleep apnea and other respiratory problems [9]. Respiratory Inductance Plethysmography (RIP) measures the body movement of chest wall or abdominal wall caused by breathing exercise [7, 29], and then the breathing conditions can be estimated accurately. However, most of the breath measurement methods are essential to directly contact to the user while measuring, and it might affect the user and decrease the sleep quality.

In recent years, some non-contact breath measurement methods are developed. A study used a CCD video camera to detect the optical flow of the user in bed [19]. PneumaCare [21] developed a non-invasive method called Structured Light Plethysmography (SLP), which utilizes the distortion with movement of a structured pattern of light to calculate a volume or change in volume of a textured surface. Another study conducted an experiment and the results showed that SLP was comparable in performance to spirometer [22]. Moreover, slit lights projection [1, 2] is another non-invasive method which measures the breathing conditions by projecting the near-infrared multiple slit-light patterns on the user and measuring the breathing status. In addition to computer vision-based methods, there is a non-contact method which uses ultra wideband (UWB) to measure the breathing status. A study proposed an application of UWB radar-based heart and breathing activities for intensive care units and conventional hospital beds [24]. Another study used UWB to measure baby's breathing and heart rate especially in terms of opportune apnea detection and sudden infant death syndrome prevention [31].

For monitoring the sleep activity through movement, actigraphy has been used to study the sleep patterns for over 20 years. Actigraphy is a non-invasive method of monitoring human activity cycles [23]. It is useful for determining sleep patterns and circadian rhythms. The advantage of actigraphy over traditional PSG is that actigraphy can conveniently record the sleep activity [3]. In recent years, many commercial products were developed, such as Fitbit, WakeMate, and Actiwatch. In general, these products detect the information of time to fall asleep, time to wake up, and totally sleeping time. A study evaluated the measurement results of actigraphy and compared to PSG, and the experimental results showed that sleep parameters from actigraphy corresponded reasonably well to PSG [14]. In addition, there is a non-contact method which uses a microphone and an infrared sensor to monitor the sleep status [5]. Moreover, some studies utilize motion sensors (accelerometer, piezoelectric sensor) inside the pillow [27] or bed [10, 18] to monitor the sleep movement and sleep positions. However, none of related research in our survey has a complete study to provide a non-contact and multi-functioning sleep monitoring technique to monitor the sleep conditions. In this study, we developed a non-contact sleep monitoring system which can monitor user's sleep position, breathing condition, and body movement in the same time.

3   System Design

In this study, a cross-section object detection method is proposed to detect user's head and torso using a depth camera. The sleep position, body movement, and breathing condition are monitored once the head and torso is detected. The procedure of this method is as follows: First, the view transformation is estimated. Then, a median filter is adopted to reduce the image noise after view transformation. Next, a cross-section method is used to detect user's head and torso so that the sleep position and body movement can be measured. Besides, a breath measurement method is proposed to detect the breathing conditions through the movement of the torso.

3.1   System Setup

A depth camera [15] is used to capture the sequence of depth images of the user on the bed. The depth camera consists of an infrared laser projector combined with a monochrome CMOS sensor, which captures color images and a depth images under ambient light condition. In addition, the depth image can also be captured under the no-light condition. For the reason of easy setup and preventing the interference with the sight view of the user while sleeping, depth camera is placed on the wall behind the head instead of suspending from the ceiling. Besides, in order to ensure that the user's head and torso can be captured, and for the issues of breath measurement distance (the shorter the better), the limitation of sensing distance (larger than 0.8m), the depth camera is placed in the distance of 125 cm from the bed. The region between gray dotted lines indicates the sight view of the depth camera, and the region between yellow lines indicates the sight view of the user.

3.2   Depth Image Processing

Although the skeleton of the user body can be extracted easily through Microsoft Kinect SDK, the skeleton of body while lying on the bed cannot be extracted easily. It is because that the background is too close to the user, and the body might be covered by a quilt. In this study, a cross-section method is proposed to detect user's head and torso with a depth camera. We process the depth image signals at the resolution of 320 pixels in width and 240 pixels in height, and the frame rate is 30 frames per second.

View Transformation. In order to determine the cross-sections of the depth image, we would like to transform the camera view from the side view to the top view. To do

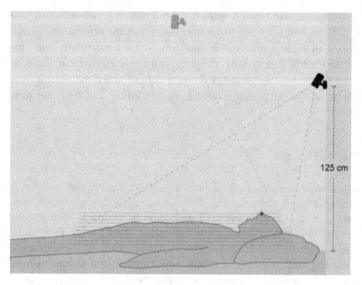

Fig. 1. Cross-sections of the lying user from top to bottom. Red point indicates the highest point of the user.

that, we need to calculate bed's normal vector first. In order to rotate the camera view to the top of the bed, three points and one rotate center point need to be specified manually. After taking three 3D-points on the bed and using cross product, the system could get the normal vector of the bed. Then, these 2D-points could project to 3D-points in the real world. Then, we proceed to calculate rotation matrix for the bed's normal vector. Once we have the rotation matrix, we can project all 2D points back to 3D point-cloud. Again, we project it back to 2D depth image. However, it will lose some information after rotating the camera view, so a median filter is used to fill empty holes. Fig. 2b shows the original depth image, and Fig. 2c shows the depth image after view transformation.

Cross-Section. We generate several binary images by setting different thresholds starts from the shallowest point of the depth image to the depth of the bed. We generate cross-sections every 2 cm from top to bottom. Generally, the distance between the highest point of the human body and the bed is around 18~28 cm, therefore, there would be 9~13 transverse sections of the person from top to bottom. Fig. 1 shows ten cross-sections (red line) from the highest point of the red point to the bed.

Head and Torso Detection. By using connect-component analysis, the components from each cross-section can be extracted. The concept of this method is to find out spheres in each cross-section. Once there is a circle growing larger from top section to bottom section, we assume that it might be a sphere there. So far, this algorithm might find other spheres. To decide the highest sphere, we collect each circle's contribution from each section. More circles at the same location means higher probability to have sphere there. If sphere candidates have n different locations, the probability that might be a sphere at location l is:

$$P(l) = \frac{\sum^{sections} \# \ of \ circles \ at \ l}{\sum_{i=0}^{n} \sum^{sections} \# \ of \ circles \ at \ i} \qquad (1)$$

In addition to detecting head from single depth image, we need to leverage the advantage of video sequence. Hence, we push every head location found by each frame into a queue. Then, we use the same idea to re-locate the highest probability head-like sphere. This will avoid some occasional misleading failed detection. Once the head is detected, the next step is to detect the torso's ROI (region-of-interest). We adopt almost the same way as detecting the head, but this time we track cuboids rather than spheres. However, there is a problem that the pillow might be recognized as a torso. Therefore, we reject cuboids if there is a head on it. Fig. 2 shows the processing procedure of head and torso detection in this system.

Breath Measurement. The breathing signal can be extracted from the torso ROI once we detect the head and torso. While the user is inhaling, his chest wall will expand, and the average depth value of the torso ROI will decrease; on the contrary, the average depth value of the torso ROI will increases while the user is exhaling. Therefore, the sequential of the average depth value of the torso ROI is considered as the breathing signal under the premise that the user is sleeping. For breath measurement, a turning point detection algorithm is proposed. At first, a mean filter is used for reducing the noises caused by the sensing deviation and body movements. Then, the turning points of the breathing signal are detected using the second derivative method. Finally, in order to eliminate redundant turning points, a dynamic threshold is applied to find the exact peak points and valley points. Fig. 3 shows a fragment of the breathing signal (blue line) and the measurement results (vertical gray line) during a realistic overnight sleep. While the turning points of the breathing signals are detected, the information of the breathing conditions can be figured out easily. The breathing conditions include the breathing rate, breathing depth, breathing stability, inhalation time, exhalation time, inhalation/exhalation ratio, and sleep apnea symptoms.

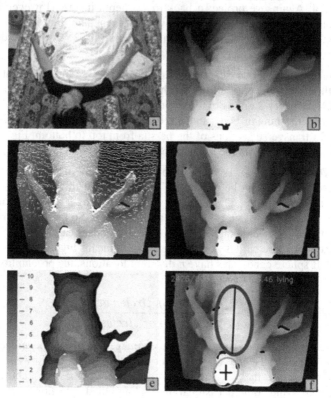

Fig. 2. Depth image Processing Procedure of our system. (a) Captured color image. (b) Captured depth image. (c) View transformed image. (d) Filtered image. (e) Cross-section image. (f) Final result of head/torso detection.

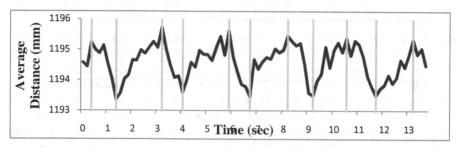

Fig. 3. Breath Measurement. The blue line indicates the raw breathing signals, and the gray lines indicate the turning points which we detected.

Body Movement. The body movement is defined as the sum of head movement and torso movement. The HR_t indicates the average depth value of the head ROI in time t, and the TR_t indicates the average depth value of the head ROI in time t. Then, the absolute difference value between two adjacent images frames could be calculated by Eq. 2. M_t indicates the movement value of the user in time t.

$$M_t = |HR_t - HR_{t-1}| + |TR_t - TR_{t-1}| \qquad (2)$$

Sleep Position. In this system, two main sleep positions (supine position and side-lying position) can be recognized. After the head and torso are detected, the highest point of head ROI and torso ROI can be found. Then, the ratio of the highest head point to torso point is calculated. Fig. 4 shows the highest point of head ROI (blue dot) and torso ROI (red dot) in the side-lying position and supine position.

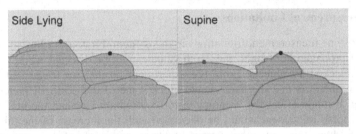

Fig. 4. Sleep Positions. Red dot indicates the highest point of torso ROI, and blue dot indicates the highest point of head ROI

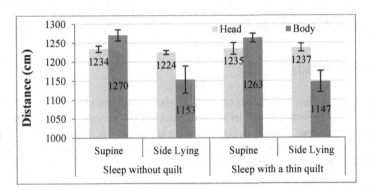

Fig. 5. The distance between the depth camera and the highest point of head ROI and torso ROI in two positions (supine and side-lying) and two conditions (sleep with no quilt and sleep with a thin quilt)

Next, the sleep position can be classified according to the ratio defined in equation 3. In order to find out the threshold to classify the sleep position, an experiment was conducted to record five participant's (two females and three males) highest points of head ROI and torso ROI in two sleep positions (side-lying and supine) and two conditions (sleep with no quilt and sleep with a thin quilt) (Fig. 5). The results revealed that the distance of the highest head point does not change significantly in different sleep positions and conditions. However, the distance of the highest torso point changed significantly in different sleep positions. The average ratio is -0.02633 in supine position, and it is 0.0652 in side-lying position. Therefore, the detection threshold of sleep position is set to the median value: 0.01. While the ratio is larger than the threshold value, the sleep position is defined as the supine position. Otherwise the sleep position is defined as the side-lying position, as shown in Eq. 4. We can observe that the standard deviation of the body distance is bigger in side-lying position than others. It is because the highest torso points are different for female and male. However, our method can also distinguish the sleep position accurately no matter the gender.

$$Ratio = \frac{D_{head} - D_{torso}}{D_{head}} \tag{3}$$

$$Posture = \begin{cases} Side\ lying\ if\ Ratio > 0.01 \\ Supine\ Otherwise \end{cases} \tag{4}$$

3.3   Measurement Limitations

There are some measurement limitations in this system. First, according to the law of rectilinear propagation of light, the depth value cannot be detected while the IR patterns are blocked by objects. From the experiment results, we found that the most common problem is that the hand would block some of the depth IR patterns while side-lying. It might affect the accuracy of torso detection. Second, the breathing amplitude of torso movement would be decreased with the increase of the thickness of quilt. According to our test, the average breathing amplitude of the torso movement with no quilt is 0.5 cm and it is 0.35 cm while sleeping with a thin quilt. The thickness of the thin quilt in our test is 0.6 cm. However, while sleeping with a thick quilt, such as thick silk-padding quilts, the system might not accurately detect the torso movement caused by breathing exercise.

4   Experiments

Two experiments were conducted to evaluate the measurement accuracy of head/torso detection, sleep position, body movement, and breath measurement of this system. First experiment was mainly designed to evaluate the measurement reliability for different users. Second experiment was designed to evaluate the measurement accuracy in realistic overnight-sleep condition.

4.1   Experiment I: Sleeping Simulation

Experimental Design. Eight participants volunteered to participate in this experiment (five males and three females). The average age is 33.8 years old (SD = 17.6), including two sixty-year old participants, five young participants (25~30 years old), and a ten-year old participant. The body mass index (BMI) of them is in the range between 18.6~29.75. In this experiment, participants were asked to lie down on the pillow, and a breathing sensor, RIP [28], was used to record the breathing conditions as the ground truth. During the experimental procedure, they were asked to change the sleep position every fifteen breathing cycles. The procedure of this experiment is in the sequence of supine, lying on the right side, supine, lying on the left side, supine, and lying on the right side. Totally, the participant needed to change the sleep position five times. Besides, the experimental procedure needed to be done twice, including a condition that the participants sleep with a thin quilt, and a condition that they sleep with no quilt. Before each task, participants were reminded not to breathe deliberately.

Experimental Results. The sleep measurement were divided into four different conditions in this experiment, including two sleep positions (side-lying and supine) and two circumstances (sleep with a thin quilt and sleep with no quilt). For each condition, the total numbers of correct head detection frames were calculated manually as well as the total numbers of correct torso detection frames. The average of accurate rate and standard deviation in each condition are listed below. The experimental results showed that while participants slept with no quilt, the measurement accuracy of head detection was 98% (SD = 0.036) while in the side-lying position, and it was 99.3% (SD = 0.018) in the supine position. Moreover, the measurement accuracy of torso detection was 91.5% (SD = 0.16) in the side-lying position, and it was 99.3% (SD = 0.01) in the supine position. Besides, while participants slept with a thin quilt, the measurement accuracy of head detection is 96.7% (SD = 0.11) in the side-lying position, and it was 99.5% (SD = 0.02) in the supine position. Moreover, the measurement accuracy of torso detection was 94.5% (SD = 0.1) in the side-lying position, and it was 99.5% (SD = 0.008) in the supine position. Overall, the average accurate rate was 98.4% in head detection and 96.4% in torso detection. The experimental results of head and torso detection are shown in Fig. 6.

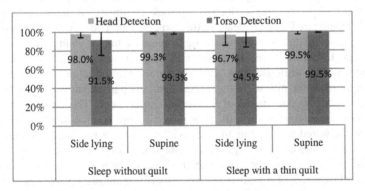

Fig. 6. Results of head and torso detection in experiment I

For breath measurement, the measurement accuracy is defined as the ratio of the totally breathing cycles we detected to the totally breathing cycles the RIP system detected. The measurement results of the RIP system was regarded as the ground truth of the breathing conditions. The experimental results show that while the user sleeps with no quilt, the measurement accuracy of breathing rate was 81.9% (SD = 0.11) in the side-lying position, and it was 90.4% (SD = 0.07) in the supine position. Moreover, while the user sleeps with a thin quilt, the measurement accuracy of breathing rate was 84.1% (SD = 0.05) in the side-lying position, and it was 88% (SD = 0.08) in the supine position. Overall, the average accurate rate of breath measurement was 86.3%. The experimental results of the breath measurement are shown in Fig. 7. For sleep position, the experimental results showed that in the circumstance of sleeping with a thin quilt, the detection accuracy was 100% (N=24) in the side-lying position, and it was 100% (N=24) in the supine position. Besides, while the user slept with no quilt, the detection accuracy was 95.8% (N=24) in the side-lying position, and it was 100% (N=24) in the supine position.

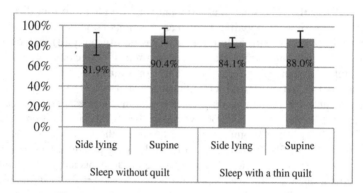

Fig. 7. Results of breath measurement in experiment I

4.2   Experiment I I: Realistic Overnight-Sleep Monitoring

Experimental Design. The experiment was conducted to ensure that the system could monitor the realistic overnight-sleep conditions accurately. A male participant (28 years-old) volunteered to participate in this experiment. The same as the first experiment, the breathing sensor (RIP) was used to measure the breathing conditions as the ground truth. In addition, an actigraphy was used to measure the movement of the non-dominant hand while sleeping. There was only one limitation that the participant was asked to lie on the pillow. In this experiment, the participant was asked to participate in a ten-day overnight-sleep monitoring experiment. The experiment did not specify the time to go to the bed, the time to getting up, and the totally sleeping time. Besides, we required participants to sleep with a thin quilt for five days, and to sleep with no quilt for another five days. Participant's breathing rate, body movement, and sleep position were monitored by our method and compared to the RIP and actigraphy. Fig. 8 shows one of a realistic overnight-sleep monitoring results in day 3. Fig. 8a shows the measurement results of breathing rate. Red curve indicates the measurement results of RIP system, and the blue curve indicates the measurement results of our system. Lower part of Fig. 8a shows the sleep positions we detected (blue) and real condition (red). Fig. 8b shows the movement level detected by an actigraphy, and Fig. 8c shows the movement level detected by our system. In this day, the participant slept with a thin quilt from 1:30 AM to 4:53 AM.

Experimental Results. The same with experiment I, the sleep measurement were divided into four different conditions, including two sleep positions (side-lying and supine) and two sleep circumstances (sleep with a thin quilt and sleep with no quilt). Totally, the participant slept 42 hours in ten nights. Following shows the experimental results. In the circumstance of sleeping with no quilt, the measurement accuracy of head detection was 89.4% (SD = 0.14) in the side-lying position and it was 99.9% (SD = 0.0007) in the supine position. Moreover, the measurement accuracy of torso detection was 89.3% (SD = 0.014) in the side-lying position and it was 89.3% (SD = 0.0003) in the supine position. Besides, in the circumstance of sleeping with a thin quilt, the measurement accuracy of head detection was 99.9% (SD = 0.007) in the side-lying position and it was 98.8% (SD = 0.17) in the supine position. Moreover, the measurement accuracy of torso detection is 99.4% (SD = 0.0003) in the side-lying position and it is 99.9% (SD = 0.003) in the supine position. The experimental results of head and torso detection are shown in Fig. 9. Overall, the average accurate rate of head detection was 96.7% (SD = 0.073), and the average accurate rate of torso detection was 96.8% (SD = 0.031).

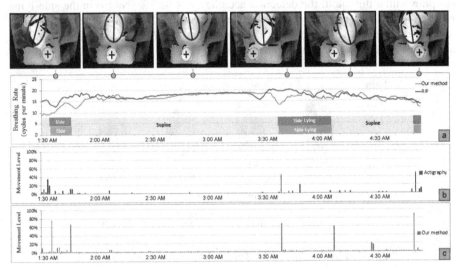

Fig. 8. A Realistic Overnight-Sleep Monitoring. Red color indicates the ground truth measured by RIP and actigraphy, and the blue color indicates the results of our system. (a) The results of breathing rate and sleep positions. (b) The movement level detected by an actigraphy. (c) The movement level detected by our system.

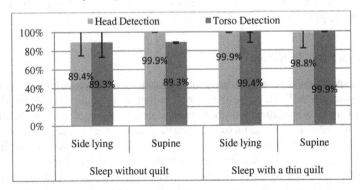

Fig. 9. Measurement results of head and torso detection in experiment II

For body movement, the times of the movement events in our method and actigraphy were compared. According to the observation, we observed that big body movement can be measured both in our system and actigraphy, such as the event of turning over the body. Besides, micro-movement could be measured, (see Fig. 8b and 8c). For breath measurement, the measurement accuracy of breathing rate was 89.7% (SD = 0.05) in the side-lying position and it was 92.8% (SD = 0.05) in the supine position in the circumstance of sleeping with no quilt Moreover, the measurement accuracy of breathing rate was 92.4% (SD = 0.07) in the side-lying position, and it was 92.7% (SD = 0.07) in the supine position. Overall, the average accurate rate of breath measurement was 92.03% (SD = 0.044). The comparison of the breath measurement in these different conditions is shown in Fig. 10.

For sleep position, the experimental results showed that in the circumstance of sleeping with a thin quilt, the detection accuracy was 94.7% (N=19) in the side-lying position, and it was 100% (N=21) in the supine position. Besides, while the user slept with no quilt, the detection accuracy was 88.23% (N=17) in the side-lying position, and it was 100% (N=20) in the supine position.

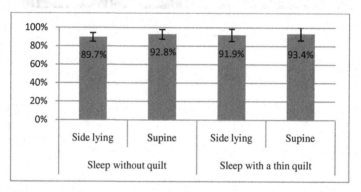

Fig. 10. Measurement results of breath measurement in experiment II

5   Discussions

The aim of this section is to summarize, analyze and discuss the results of experiments and give guidelines for the future developments.

5.1   Head/Torso Detection

From the experimental results of head and torso detection, we observed some issues worthy of discussion. First, while the user slept with a thin quilt, the overall detection accuracy of torso was better than uncovered. One reason might be that the thin quilt could smooth the shape of the torso, and enhance the measurement accuracy. Second, we found that the gesture might affect the head detection. In this system, the head and torso could be detected accurately under the premise that the shape of the head or torso is not overlapped by hand or other objects. Fig. 11a and 11b show two special sleep gestures that can be detected accurately. It is because that the shape of the head is not overlapped. However, there are some conditions that the head or torso could not be detected well. According to the observation, we found that the head could not be detected well while the user is scratching (Fig. 11c and 11d). In this condition, the shape of the head might be changed and no longer a sphere contour. In this case, the system could not recognize it as a head. One possible solution method is to detect the hand position, and then we can estimate the head position while the head is overlapping by hand.

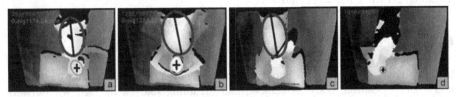

Fig. 11. Special Sleep Postures

In addition, there are some sleep conditions or sleep positions that we did not discuss. First, the algorithm of head and torso detection we proposed can be applied to detect multiple heads and torsos. Moreover, a shortest distance pairing procedure is used to pair the head and torso of specified sleeper. However, we still have the detection problem while the users are overlapping. Second, the head and torso can be detect in the prone position. Howerer, this system can not recogize whether the user is in the supine position or prone position now. Fig. 12 shows the detection result of the head and torso detection in the prone position and multiple users.

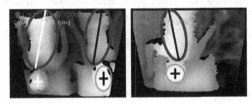

Fig. 12. Other Sleep Conditions. Left: multiple sleepers. Right: prone position

5.2   Breath Measurement

From the experimental results of breath measurement, we observed some phenomenon which was similar to the conditions of head and torso detection. First, the measurement accuracy of breathing is higher while sleeping in the supine position than in the side-lying position. We observed that there are more noises in the side-lying position than in the supine position. Besides, according to our measurement results, the average amplitude of the breathing signals is 0.8 cm in the supine position. However, the average amplitude of the breathing signals is about 0.5 cm in the side-lying position. Overall, less signal noises and more breathing amplitude would increase the measurement accuracy while sleeping in the supine position. Second, the overall measurement accuracy of breathing while sleeping with a thin quilt was better than the accuracy while sleeping with no quilt. We speculate that it might because the reason that the thin quilt reduces the wrinkles of the torso surface. Third, the movement of the torso is seen as the breathing signal, but the system cannot identify whether the movement is caused by breathing exercise or other exercises. According to the observations of the measurement results in experiment I, we found that there were more detection errors while the user is moving or turning around in bed. In experiment I, the participants were asked to turn over the body frequently, and the participant was almost in static condition in experiment II. That's the reason that the measurement accuracy of breathing is lower in experiment I than in experiment II. One possible solution method is to suspend the breathing measurement while the user is moving.

6   Conclusions

In this study, we proposed a depth image sequence analysis technique to monitor user's sleep position, body movement, and breathing rate on the bed without any physical contact. A depth image-based processing method is proposed to monitor the sleeping conditions. The results of experimental I showed that the proposed method is promising to detect the head and torso with various sleeping postures and body shapes. The results of experimental II showed that the system can accurately monitor the sleeping conditions. Therefore, we confirm that the system could provide relevant sleep information and sleep report to the user. Furthermore, the sleep parameters which we detected can provide to the sleep center to diagnose the sleep problems. This study is important for providing a non-contact technology to measure the sleep conditions and assist users in better understanding of his sleep quality. In the future, we expect to detect more sleeping conditions and solve some measurement limitations, such as the problems of overlapping. Besides, we will develop a multimedia feedback sleep-assisted system which can detect the breathing status and provide appropriate sleeping guidance in real time to help users shorten the time to fall asleep. In addition, a web-based browser will be developed to provide the personalized sleeping information to the user.

Acknowledgements. This work was supported in part by the National Science Council, Taiwan, under grants NSC 98-2221-E-002-128-MY3. And many thanks to HealthConn Corp. (http://hcc.healthconn.com/), who provided many assistances of the knowledge.

References

1. Aoki, H., Koshiji, K.: Non-contact Respiration monitoring method for screening sleep respiratory disturbance using slit light pattern projection. World Congress on Medical Physics and Biomedical Engineering 14(7), 680–683 (2006)
2. Aoki, H., Koshiji, K., Nakamura, H., Takemura, Y., Nakajima, M.: Study on respiration monitoring method using near-infrared multiple slit-lights projection. In: IEEE International Symposium on Micro-NanoMechatronics and Human Science, pp. 291–296 (2005)
3. Ancoli-Israel, S., Cole, R., Alessi, C., Chambers, M., Moorcroft, W., Pollak, C.P.: The role of actigraphy in the study of sleep and circadian rhythms. Sleep 26(3), 342–392 (2003)
4. BaHammam, A.: Comparison of nasal prong pressure and thermistor measurements for detecting respiratory events during sleep. Respiration 71(4), 385–390 (2004)
5. ZZZ checker, http://www.zzzchecker.com/ (retrieved on April 2011)
6. Born, J., Lange, T., Hansen, K., Molle, M., Fehm, H.L.: Effects of sleep and circadian rhythm on human circulating immune cells. The Journal of Immunology 158(9), 4454–4464 (1997)

7. Cantineau, J.P., Escourrou, P., Sartene, R., Gaultier, C., Goldman, M.: Accuracy of respiratory inductive plethysmography during wakefulness and sleep in patients with obstructive sleep apnea. Chest 102(4), 1145–1151 (1992)

8. Cartwright, R.D.: Effect of sleep position on sleep apnea severit. Sleep 7(2), 110–114 (1984)

9. Douglas, N., Thomas, S., Jan, M.: Clinical value of polysomnography. Lancet 339, 347–350 (1992)

10. Erina, K., Yousuke, K., Kajiro, W.: Sleep stage estimation by non-invasive biomeasurement. In: SICE-ICASE International Joint Conference, pp. 1494–1499 (2006)

11. Fitbit, http://www.fitbit.com/ (retrieved on May 2011)

12. Hoque, E., Dickerson, R.F., Stankovic, J.A.: Monitoring body positions and movements during sleep using WISPs. Wireless Health, 44–53 (2010)

13. Idzikowski, C.: Sleep position gives personality clue. BBC News (September 16, 2003)

14. Kushida, C.A., Chang, A., Gadkary, C., Guilleminault, C., Carrillo, O., Dement, W.C.: Comparison of actigraphic, polysomnographic, and subjective assessment of sleep parameters in sleep-disordered patients. Sleep Med. 2(5), 389–396 (2001)

15. Kinect, http://www.xbox.com/kinect (retrieved on March 2011)

16. Loord, H., Hultcrantz, E.: Positioner - a method for preventing sleep apnea. Acta Otolaryngologica 127(8), 861–868 (2007)

17. Maquet, P.: The role of sleep in learning and memory. Science 294, 1048–1052 (2001)

18. Malakuti, K., Albu, A.B.: Towards an intelligent bed sensor: Non-intrusive monitoring of sleep irregularities with computer vision techniques. In: International Conference on Pattern Recognition, pp. 4004–4007 (2010)

19. Nakajima, K., Matsumoto, Y., Tamura, T.: Development of real-time image sequence analysis for evaluating posture change and respiratory rate of a subject in bed. Physiological Measurement 22, N21–N28 (2001)

20. Philips Actiwatch, http://www.healthcare.philips.com/ (retrieved on August 2010)

21. PneumaCare, http://www.pneumacare.com/ (retrieved on June 2011)

22. Wareham, R., Lasenby, J., Cameron, J., Bridge, P.D., Iles, R.: Structured light plethysmography (SLP) compared to spirometry: a pilot study. In: European Respiratory Society Annual Congress, Vienna (2009)

23. Sadeh, A., Sharkey, K.M., Carskadon, M.A.: Activity-based sleep-wake identification: an empirical test of methodological issues. Sleep 17(3), 201–207 (1994)

24. Staderini, E.M.: UWB radars in medicine. Aerospace and Electronic Systems Magazine 17(1), 13–18 (2002)

25. Spiegel, K., Leproult, R., Cauter, D.V.: Impact of sleep debt on metabolic and endocrine function. The LANCET 354(9188), 1435–1439 (1999)

26. Szollosi, I., Roebuck, T., Thompson, B., Naughton, M.T.: Lateral sleeping position reduces severity of central sleep apnea/cheyne-stokes respiration. Sleep 29(8), 1045–1051 (2002)

27. Harada, T., Sakata, A., Mori, T., Sato, T.: Sensor pillow system: monitoring respiration and body movement in sleep. In: Intelligent Robots and Systems, IROS 2000, vol. 1, pp. 351–356 (2000)

28. Thought Technology Ltd., http://www.thoughttechnology.com/ (retrieved on June 2010)

29. Whyte, K.F., Gugger, M., Gould, G.A., Molloy, J., Wraith, P.K., Douglas, N.J.: Accuracy of respiratory inductive plethysmograph in measuring tidal volume during sleep. Journal of Applied Physiology 71(5), 1866–1871 (1991)

30. WakeMate, http://www.wakemate.com/ (retrieved on May 2011)

31. Ziganshin, E.G., Numerov, M.A., Vygolov, S.A.: UWB baby monitor. Ultra-wideband and Ultrashort Impulse Signals, 159–161 (2010)

Integration of a Heart Rate Prediction Model into a Personal Health Record to Support the Telerehabilitation Training of Cardiopulmonary Patients

Axel Helmer[1], Riana Deparade[2], Friedrich Kretschmer[3], Okko Lohmann[1],
Andreas Hein[1], Michael Marschollek[4], and Uwe Tegtbur[2]

[1] R&D Division Health, OFFIS Institute for Information Technology,
Escherweg 2, D-26121 Oldenburg, Germany
[2] Institute of Sports Medicine, Medical School Hannover,
Carl-Neuberg-Strasse 1, D-30625 Hannover, Germany
[3] Computational Neuroscience, University of Oldenburg,
Carl-von-Ossietzky-Strasse 9-11, D-26129 Oldenburg, Germany
[4] Peter L. Reichertz Institute for Medical Informatics, University of Braunschweig,
Institute of Technology and Hannover Medical School, Mühlenpfordtstr. 23,
D-38106 Braunschweig and Carl-Neuberg-Str. 1, D-30625 Hannover, Germany
{axel.helmer,okko.lohmann,andreas.hein}@offis.de,
{deparade.riana,tegtbur.uwe}@mh-hannover.de,
friedrich.kretschmer@uni-oldenburg.de,
michael.marschollek@plri.de

Abstract. Chronic obstructive pulmonary disease (COPD) and coronary artery disease are severe diseases with increasing prevalence. Studies show that regular endurance exercise training affects the health state of patients positively. Heart Rate (HR) is an important parameter that helps physicians and (tele-) rehabilitation systems to assess and control exercise training intensity and to ensure the patients' safety during the training. On the basis of 668 training sessions (325 F, 343 M), we created linear models predicting the training HR during five application scenarios. Personal Health Records (PHRs) are tools to support users to enter, manage and share their own health data, but usage of current products suffers under interoperability and acceptance problems. To overcome these problems, we implemented a PHR that is physically localized in the user's home environment and that uses the predictive linear models to support physicians during the training plan creation process. The prediction accuracy of the model varies with a median root mean square error (RMSE) of ≈ 11 during the training plan creation scenario up to ≈ 3.2 in the scenario where the prediction takes place at the beginning of a training phase.

Keywords: Modeling, Heart rate, Prediction, Personal health record, Cardiopulmonary rehabilitation.

J. Gabriel et al. (Eds.): BIOSTEC 2012, CCIS 357, pp. 326–338, 2013.
© Springer-Verlag Berlin Heidelberg 2013

1   Introduction

1.1   Background

Patients with chronic obstructive pulmonary disease (COPD) are suffering from the consequences of a chronic inflammation of their pulmonary system. This leads to an obstruction of the bronchi that causes airflow limitation and shortness of breath. Often, immobility and social isolation are the consequences, which in turn reinforce the degeneration of muscle mass and aggravate the symptoms. The Global Initiative for Chronic Obstructive Lung Disease (GOLD) summarizes: "COPD is the fourth leading cause of death in the world and further increases in its prevalence and mortality can be predicted in the coming decades" [1]. Just the direct medical costs attributable to COPD were estimated at $49.5 billion in the US [2].

Beside the pharmacological treatment, an important part of therapy is regular endurance training. Pulmonary rehabilitation training improves physical capacity, reduces breathlessness, reduces the number of hospitalizations and increases the quality of life [1]. The cost of continuous monitoring of these training sessions in clinics is high and additionally requires the patient to travel to a clinic for each single session. Performing the rehabilitation training at home can raise the patients' compliance and reduce costs.

Another unsolved problem in today's healthcare systems is the connection and data exchange between these different actors over the borders of institutions and health care sectors. Especially for COPD it is recommended to involve lounge specialists, nutritionists, psychologists, and family doctors to ensure an optimal treatment. In addition the patient should be involved in his own treatment and is although the only one who could provide data about the own sleeping behavior, nutrition, tobacco consume, and sport activities.

1.2   Related Work

To ensure a safe telerehabilitation at home, the detection of abnormal events during the training session and an autonomous training control are critical prerequisites. Nearly all the existing detection and control algorithms compare the patient's "is" state with the "should" state. Therefore different sets of vital signs are used as an indicator to derive the health state of the patient, which reflects the training intensity.

Achten and Jeukendrup summarized current research achievements in the field of heart rate monitoring in 2003 and state: "...the most important application of HR monitoring is to evaluate the intensity of the exercise performed" [3]. They conclude that the important influence factors on HR are age, gender, environmental temperature, hydration and altitude. They estimated the day-to-day variance under controlled conditions to be 2-4 beats per minute (bpm).

Different techniques have been proposed for controlling the training performance and to raise alarms on basis of HR. Velikic et al. used data from an accelerometer for a comparison of different models (linear, non-linear, Kalman filter) for HR prediction of healthy subjects and such with congestive heart failure [4]. The two linear models delivered the best results for a short term prediction of 20 minutes. Su et al. introduced a model to control HR during treadmill exercise [5]. Further approaches for the same application were provided by Cheng et al. [6] and Mazenc et al. [7]. Neither have any of these models been checked for their applicability to cardiopulmonary patients nor do specialized HR models exist for these.

Song et al. introduced a set of rules to control the training performance of COPD patients, Nee et al. used Bayesian networks for an adaptive alert system for patients with heart issues [8], Schulze et al. used Bayesian networks for the training control of COPD patients [9]. These machine learning based approaches showed to be well suited for adapting systems to the individual differences between users, but they have to be trained first. The rule based system is a more abstract definition that fits for most of the users, but uses hard coded thresholds which are not individualized. There exists no model that can be used to define individual reference values for the rule based thresholds or initial / start values for the machine learning based approaches.

Individualized models have to be able to take all values into account, which may be relevant for the prediction. Multiple professional maintained Electronic Patient Records (EPRs) containing data from one patient are typically maintained in several different health service institutions, storing different data in different formats and providing a variety of proprietary interfaces. The user controlled Personal Health Record (PHR) is meant to overcome these limitations by providing standardized communication interfaces, thus enabling data exchange in the course of medical treatments among various institutions.

Many PHR products with a wide range of features were introduced in the last ten years, but lately some of the big companies removed their products from the market (e.g. Google Health, ICW LifeSensor), because the usage fell short of the expectations. In [10] the Markle Foundation introduced PHR architectures. The typical architecture for PHRs is a web-based third-party tethered PHR, where a user enters and uploads his health related data to a web server, which is maintained by private companies like Microsoft or Google. A survey of the California HealthCare Foundation [11] revealed that 55 percent of patients with chronic conditions are concerned about the confidentially of their health data. As privacy concerns inhibit the adoption of PHRs, it is critical to obtain broader acceptance. Interoperability is a critical prerequisite to connect the users PHR with professional EPRs, where most of the user´s health data is stored. We compared web-based PHR products in [12] and found that only a minority of the existing PHR systems make use of existing standards that enable interoperability for data exchange.

1.3 Aim and Scope

HR is an important vital parameter and thereby an important indicator of a patients physical state during rehabilitation training [13]. The knowledge about factors that have an influence on the exercise physiology might help physicians and autonomous systems to take this information into account when deciding how much load a patient can undergo during a training session. Hence it could be used to support creation and optimization of training schedules and during the current training session itself to derive the future course.

A difference between the predicted trend of a normal training and a measured heart rate may give a hint on a potentially abnormal development and thereby help to detect critical states before they occur. This is especially important in telerehabilitation settings, where patient's train under unsupervised conditions at home (see [14, [15]).

The first aim of our research, which is presented in this paper, is to introduce a model which predicts the patients HR on basis of information about the patient and the environment. Secondly, we present a PHR system that uses a standardized data exchange to parameterize the HR model by combining professional and user generated information. The system also has the potential to overcome the acceptance problems of existing PHRs, by giving users the physical control about their health data.

2   Methods

2.1   Characteristics and Preparation of the Model Data

The data for the heart rate prediction model was obtained during outpatient rehabilitation from cardiopulmonary patients with NYHA 1-2 and COPD level 2-3. The only exclusion criterion was the inability to perform training.

We started with an original dataset of 164 patients (82 W, 82 M) and 1201 training sessions, which were collected between July and September 2009 in the exercise training center of the Medical School Hannover in addition to regular ambulatory training sessions. Patients performed their sessions twice a week, whereas in mean each patient performed ~8 training sessions (± 7.7). HR was obtained on basis of electrocardiogram (ECG) data. The following additional data was available:

- Patient demographics: age, sex
- Training data: date and time, duration, load
- Vital signs data: resting HR before training, recovery HR after training, blood pressure (BP) (rest, load, recovery – systolic and diastolic), Borg value [16] (used scale 6-20), HR during the whole training (sample rate ≈ 1 Hz.).

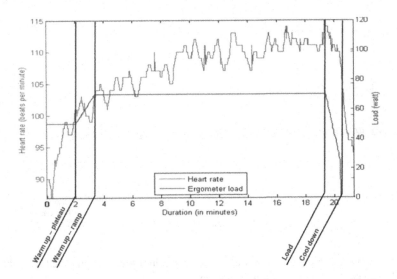

Fig. 1. Sample training session with heart rate, training load and distinction into the four training phases

We also included environmental variables, which could have a possible influence. For this purpose we procured data from the German weather service that was recorded by a weather station in Hannover (station id: 2014), where the training took place. We chose temperature, humidity and air pressure as main descriptors for weather and added them to the training data.

Only fully completed training sessions with a specific phase, and a typical load course showing the characteristics of a successful four-phase rehabilitation training (see Fig. 1) were included into the dataset. The first phase (warm up) consists of a load plateau at a certain level. This load increases stepwise over time in the second phase. The third phase (load phase) shows a constant load for at least 10 minutes. In the fourth phase (cool down) the load is reduced stepwise until it reaches null. We also excluded sessions, where a monitoring physician interfered by de- or increasing the training load, because the reasons of such changes were not documented in our data. This could also be an indicator for training under suboptimal conditions, e.g. the training load was too high or low for the patient due to an inadequately adapted training schedule.

To automatically and robustly extract the sessions from the dataset matching the previously mentioned criteria, we calculated the difference derivate of the load values over time to detect whether the load was in- or decreasing during a training session. Additionally, a session had to last from 12 to 26 minutes in total. After discarding all training sessions not fulfilling these criteria, we reduced the above mentioned number of 1201 to 668 (325 F, 343 M) training sessions from 115 patients (in mean 5.8± 4.5 trainings per patient).

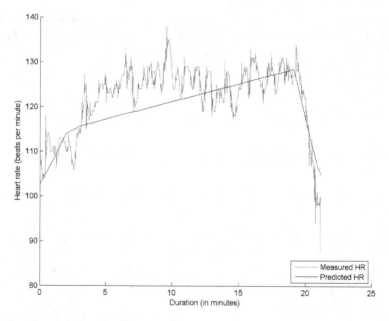

Fig. 2. Sample training session for scenario four with measured and predicted heart rate. The RMSE is calculated for the four training phases.

2.2 Model Creation

For the integration of the predictive model into existing training systems the set of potential predictors (input variables which explain a significant part of the response variable) varies depending on the point in time and the use case.

We designed five different scenarios with expanding/extending datasets, which represent typical settings of telerehabilitation training and ive training in clinics. The first scenario describes a situation before training when the schedule is created, but no reliable weather forecast is available (approximately three days before the training day). The second scenario includes the weather forecast. In the third scenario the patient already wears the sensors, but the training has not yet been started. The fourth scenario depicts an ongoing training and the prediction includes data that was gathered during previous completed training phases. To provide an example, the average heart rate of the warm up plateau phase can be included into the dataset for the load phase. The fifth scenario describes the situation after the training and does also include data like the subjective perceived exertion of the patient expressed on the Borg scale.

The following list is sorted in ascending order by the number of predictors available and the time in relation to the training session. Each scenario expands the predictor set of the previous one:

- *Scenario S1 (training plan creation)*: patient demographics and training plan data (load, duration of each phase)
- *Scenario S2 (training plan creation few days before the training day):* weather data
- *Scenario S3 (at the beginning of the training)*: resting HR, resting BP
- *Scenario S4 (during the training)*: average HR of the former phase, HR at the end of the former phase, BP during the load phase (phase three)
- *Scenario S5 (after the training)*: average HR of current training phase, average HR of load phase, average HR of all phases, recovery pulse, recovery BP, average of all BP values, Borg value

The final list of predictors for scenario five included 24 items (see table 1).

To build a hypothesis about which values have a relevant influence on the HR, we used a stepwise regression analysis [17]. This algorithmic approach performs a multi-linear regression and determines a model, by adding or removing the variable with the highest or lowest correlation of the model's F-statistics stepwisely.

So the variable with the highest chance of explaining the variance of the given normally distributed data set is added to the model, when the correlation is big enough to reject the null hypothesis. This is done until all variables with significant influence (predictors) have been added and all variables with non-significant influence have been removed from the final model. We used the standard entrance and exit tolerances of $p \leq 0.05$ and $p \geq 0.10$ for the model. Additionally, we performed chi-square tests to confirm the normal distribution of the HR dataset.

The stepwise regression determines a set of coefficients (B_i) and an intercept (also called constant term) (c) as result. Together with a number of given predictor values (X_i) it yields a linear combination of the following form to calculate the response variable (Y):

$$Y = c + b_1\, x_1 + b_2\, x_2 \ldots + b_i\, x_i \tag{1}$$

We created such a submodel for each training phase (warm up plateau, warm up ramp, training and cool down) to reflect the different physiological targets. These four submodels were then concatenated to a complete model for one training session (see Fig. 2). This also simplified the comparison to the real HR of the training sessions used for validation.

2.3   Model Evaluation

To determine the quality of our model and to prevent overfitting, we performed a 2-fold cross-validation. We divided the dataset into two parts d_0 and d_1. Both parts were of the same size and contained randomly selected training sessions ($n=334$) from the dataset. First, we used d_0 to train the model and validated it against the d_1 dataset then we performed this procedure vice versa.

We calculated the root mean square error (RMSE) which quantifies the deviation between measured and predicted heart rate over a whole training.

It is not easy to determine, which predictor of the resulting model explains which part of the response variable, as each added predictor depends on the former one. To uncover which predictors are of and have to be stored in the PHR, we measured the percental improvement of the RMSE when a predictor is added to the model in relation to the former one.

Table 1. Mean contribution of the predictors on the scenario (S1-S5) model. All values represent the improvement of the former RMSE in percent by addition of a predictor during stepwise regression. The "-" symbol denotes that a predictor is not available in the given scenario. The calculated average influence of a predictor is shown in column "Overall". The order of these values is additionally illustrated by a rank order in the last column.

Predictor	S1	S2	S3	S4	S5	Overall	Rank
Age	11.032	11.032	11.112	9.002	8.018	10.040	3
Gender	0.754	0.754	0.745	0.156	0	0.482	8
Load	0.368	0.368	5.556	0.646	0.078	1.403	6
Overall training duration	0.065	0.065	0	0	0	0.026	12
Duration of current training phase	0.040	0.040	0.015	0.019	0.011	0.025	13
Air pressure	-	0.023	0.013	0.141	0.136	0.079	11
Temperature	-	0	0	0	0	0	-
Humidity	-	0	0	0.059	0	0.015	14
Resting HR	-	-	40.990	7.154	5.268	17.804	2
Resting BP systolic	-	-	0.494	0.118	0.119	0.244	9
Resting BP diastolic	-	-	0	0	0.086	0.029	11
Average HR of former phase	-	-	-	57.064	54.3	55.682	1
Load phase BP systolic	-	-	-	0	0.005	0.003	16
HR at the end of former phase	-	-	-	0	0.027	0.013	15
Load phase BP diastolic	-	-	-	0	0	0	-
Average HR of current phase	-	-	-	-	5.648	5.648	4
Average HR of load phase	-	-	-	-	3.548	3.548	5
Recovery pulse	-	-	-	-	0.683	0.683	7
Recovery BP diastolic	-	-	-	-	0.122	0.122	10
Borg value	-	-	-	-	0	0	-
Average HR of all phases	-	-	-	-	0	0	-
Average of all BP values systolic	-	-	-	-	0	0	-
Average of all BP values diastolic	-	-	-	-	0	0	-
Recovery BP systolic	-	-	-	-	0	0	-
Total number of predictors	5	8	11	15	24		

2.4   Personal Health Record System

To provide a data source that is both highly available for the patient and always under his full control, our approach physically locates the PHR in the user's home environment where it can be installed on normal PCs as well as on selected set top boxes. It is important to allow the user to disable any access to the PHR from outside (e.g. by appropriately configuring the system or by simply keeping it off the Internet connection) to achieve higher acceptance, than a fully Internet-based system with remote data storage at an "unknown'" location outside the control of the user would.

Because the medical models such as the HR model requiring interoperability with different systems, it is important to choose appropriate standards for data storage and communication. The "Integrating the Healthcare Enterprise" (IHE) initiative works on improving the way computer systems in healthcare share information [18]. IHE does not define new standards but uses and combines existing ones in support of specific use cases. IHE is mainly involved with the professional side, but has defined an integration profile based on the Clinical Document Architecture (CDA) that describes the information exchange between EPRs and PHRs, named Exchange of Personal Health Record Content (XPHR) [19].

XPHR documents can be bound to one of IHE's transport profiles describing how a standards-based communication between EPR and PHR is implemented. There are three such IHE profiles for the exchange of documents: Cross-enterprise Document Sharing (XDS) uses a central registry and repository infrastructure, which is not available yet in most countries. Cross-enterprise Document Reliable Interchange (XDR) describes a point-to-point document exchange using secure Web Services and E-Mail. Cross-Enterprise Document Media Interchange (XDM) allows users to exchange data through media such as USB sticks or CD-Rs (see [20]).

We have implemented XPHR as content profile and XDM as transport profile for the interoperable document exchange. This strengthens the approach to give the user the physical control about the own health data. Relevant predictors have been predetermined in the model creation process and can be either transmitted through standardized medical documents or entered directly in the PHR by physicians and patients.

3   Results

We modeled the four stages of a training session (one for each training phase) for the five different scenarios and determined the weighted RMSE to quantify the error of each model (see Fig. 2). As a proof of concept we implemented a training schedule definition in the PHR that uses the S1 and S2 models to predict the patient's HR during the creation of the schedule (see Fig. 3).

Table 1 shows the contribution of different predictors to the model and their effect on the RMSE. Because of their naturally high correlation (also known as multicollinearity) it is no surprise that four of the first five predictors have a high impact on the model. Important other predictors are age and load (Overall>1.4). The only predictor from the weather data is the air pressure with just a very small influence of ≈0.08%. Most of the different blood pressure values and the Borg value have no impact on the model.

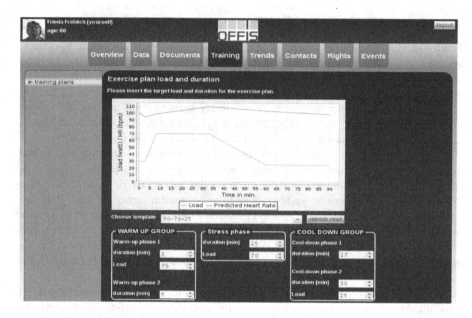

Fig. 3. Training schedule definition with integrated heart rate prediction in the PHR

Table 2 shows the accuracy of the prediction. For the calculation of average and median over the complete training the phases are weighted by their duration.

The RMSE for scenario S1 and S2 is similar (mean ≈12.3 and median ≈11.1). This also shows that the available weather data has nearly no effect on HR prediction. With an average HR of ≈98.4 bpm over all training sessions, this is equivalent to a relative mean error of ≈12.5%. The third scenario shows an average and median error of ≈8.5 and ≈6.1 which corresponds to a relative mean error of ≈8.6%. The difference between the average and mean error suggests that there are some outlier trainings that have a strong influence on the average error.

Due to the additional predictors in S3, the median error is almost reduced to 50% compared to S2. The main reason for this strong improvement is one dominating predictor: the resting heart rate (see table 1). The overall ranking of this predictor is dominated by its S3 value of ≈41%. This strongly increases the average value where the values are much lower in S4 (≈7.2%) and S5 (≈5.3%). This might be caused by the dependence between resting HR and the average HR of the former phase. The latter seems to be the better predictor.

S3 is also the scenario in which the training load has by far the highest influence (≈5.6%) with a distance of 5% to the next smaller value in S4 (≈0.6%). A plausible explanation for this value might be that training sessions with cardiopulmonary patients are generally conducted at a very low load of ≈35 watt on cycle ergometers. Therefore the leg movement might have a stronger influence on the real training load, than the selected load of the bicycle ergometer.

S4 / S5 are further increasing the precision of the prediction (mean ≈4.7 / ≈4.9, median ≈3.2 / ≈3.5 in table 2) with an average relative error of ≈4.8% / ≈5%. This is mainly caused by time-near HR-based values (average HR of former ≈55.7% and current phase ≈5.6%).

Although more predictors contribute to scenario S5 a higher prediction error is calculated compared to S4, whereas it was vice versa during the model building process (≈1.56 improvement for the mean and ≈0.93 for the median RMSE). This is an indicator for overfitting of the S5 model, which might occur due to the usage of too many explanatory variables.

Table 2. Mean error of the prediction. All values refer to the RMSE of the model in relation to the real HR.

Scenario	Phase 1	Phase 2	Phase 3	Phase 4	Average	Median
S1	11.448	10.267	12.079	12.954	12.254	11.069
S2	11.443	10.267	12.079	12.986	12.260	11.084
S3	5.528	6.514	8.528	9.561	8.498	6.068
S4	4.347	3.636	4.637	6.572	4.733	3.266
S5	4.762	2.940	4.734	5.281	4.906	3.542

4   Discussion

The stepwise regression algorithm leads to a local optimum which is not necessarily the global optimum. A stepwise addition of variables decreases the models' RMSE. When using only the RMSE as an indicator for the degree of influence for each individual predictor this has the disadvantage, that a later added predictor may have less influence, because a part of his improvement is already explained by the previously added variable. Thereby the result depends on the order of the steps and could lead to a suboptimal model when applied to highly correlated variables (like systolic and diastolic BP).

Therefore a stepwise regression can never replace expert knowledge. On a statistical level, we want to improve the model by performing a factor analysis that will reduce the number of predictors and provide a better knowledge about their correlation to each other. This might also eliminate the potential overfitting of S5 and enable the transfer to other training modalities.

The accuracy of our model strongly depends on the scenario and the associated data items. The first scenario takes place during the training plan creation and the calculated model shows the highest error. This result might still be good enough to gain an impression about HR development of a common cardiopulmonary patient during training time. We believe that the error of this scenario can be improved by adding further predictors related to the patients metabolic response like weight, medication and information about the current training state.

The available weather data only had a minor influence and lowers the precision of the model. This may reflect the fact that weather has no direct influence on the patient when he trains in a tempered environment. However, that does not mean that the direct environment has no influence at all. We want to examine this by the measurement of the conditions inside the training area. Furthermore we are going to examine if the weather indirectly affects the Borg value, another very important value to control the intensity of the rehabilitation training.

The influence of the resting HR at the beginning of a training in S3 leads to a good precision of the model during the training itself. This predictor is probably influenced by many other, hard to measure variables like medical treatment, stress, dehydration and coffee consumption, which might have a strong impact on the metabolic system. This leads to the unexpected observation that the given blood pressure values show only a very small effect on the HR. Blood pressure kinetics are in close relationship to HR, but not to absolute values, due to antihypertensive treatment in most patient's.

The prediction can be used to estimate the patient's physical state on the day of testing and thereby help to define an appropriate training intensity before the training starts.

The phase-wise prediction in S4 during the runtime of the training shows a relative error below 5%. This should be precise enough to robustly detect abnormal HR developments and calculate the optimal load for the next phase. In future we will focus on the analysis of other time dynamic predictors that might increase the model accuracy and also facilitate high refresh rates without the abstract distinction between training phases.

Integration of the first two HR models into the PHR showed to be a promising way to bring them into application. Key challenges for the adoption have been addressed by the implementation of standards and by localizing the system in the user's own home to avoid acceptance problems. Whether this concept is successful also depends on future backup solutions of the PHR's health data and the access models that have to be implemented, when such a system will be integrated into existing health infrastructures.

5   Conclusions

We created a statistical model to predict HR as an important vital parameter for the rehabilitation training of cardiopulmonary patients and integrated it into a PHR that is localized in the user's home to overcome acceptance and interoperability problems. We considered demographic data, training plan information, vital parameters and weather information as potential predictors and classified them into five aim-specific scenarios where they can be used as individualized initial or reference values to parameterize alert or training control algorithms. The implementation of the first two application scenarios into the training plan creation of the PHR was presented as a proof of concept for the integration of the model. The validation of the model revealed that weather and the measured blood pressure have nearly no direct influence on HR. Age and previously measured HR based variables like the resting HR strongly influence the responding HR.

The models prediction results in an overall low error of ≈11 bpm in median, when used for the creation of a training schedule (scenario 1). The error is reduced by about 50%, when the model is used for prediction at the beginning of a training session. The error decreases to less thanthe significance level when the model is used during a training to predict HR at the beginning of each of the four training phases. This makes it potentially suitable to detect critical situations before they appear.

The precision of the prediction might be improved by additionally including expert knowledge and further statistical methods, but it already serves as a good basis for the integration of HR predictive mechanisms into training related systems such as the training plan creation in the PHR and might potentially increase the safety and efficiency during the rehabilitation training of cardiopulmonary patients.

Acknowledgements. This work was funded in part by the Ministry for Science and Culture of Lower Saxony within the Research Network "Design of Environments for Ageing" (grant VWZN 2420/2524).

References

1. Rodriguez-Roisin, R., Vestbo, J.: Global strategy for the diagnosis, management, and prevention of chronic obstructive pulmonary disease. Report (February 2011),
   http://www.goldcopd.org/uploads/users/files/
   GOLD_Report_2011_Feb21.pdf
2. National Heart Lung and Blood Institute: Morbidity & Mortality: 2009 Chart book on Cardiovascular Lung and Blood Diseases. U.S. Department of Health and Human Services National Institutes of Health (2009)
3. Achten, J., Jeukendrup, A.E.: Heart rate monitoring: applications and limitations. Sports Med. 33(7), 517–538 (2003)
4. Velikic, G., Modayil, J., Thomsen, M., Bocko, M., Pentland, A.: Predicting heart rate from activity using linear and non-linear models. In: Proceedings of 2011 IEEE 13th International Conference on e-Health Networking, Applications and Services (2010)
5. Su, S.W., Wang, L., Celler, B.G., Savkin, A.V., Guo, Y.: Identification and control for heart rate regulation during treadmill exercise. IEEE J BME 54(7), 1238–1246 (2007)
6. Cheng, T.M., Savkin, A.V., Celler, B.G., Su, S.W., Wang, L.: Nonlinear modeling and control of human heart rate response during exercise with various work load intensities. IEEE J BME 55(11), 2499–2508 (2008)
7. Mazenc, F., Malisoff, M., de Queiroz, M.: Model-based nonlinear control of the human heart rate during treadmill exercising. In: Proc. 49th IEEE Conf. Decision and Control, CDC, pp. 1674–1678 (2010)
8. Nee, O.: Clinical decision support with guidelines and bayesian networks. Ph.D. dissertation, University of Oldenburg (2010)
9. Schulze, M., Song, B., Gietzelt, M., Wolf, K.H., Kayser, R., Tegtbur, U., Marschollek, M.: Supporting rehabilitation training of copd patients through multivariate sensor-based monitoring and autonomous control using a bayesian network: prototype and results of a feasibility study. Inform. Health Soc. Care 35(3-4), 144–156 (2010)
10. Markle Foundation: Connecting Americans to their Health Care: Final Report – Working Group on Policies for Electronic Information Sharing Between Doctors and Patients. Markle Foundation (2004)
11. The California HealthCare Foundation: Consumers in Health Care: The Burden of Choice (2005) ISBN 1-932064-97-4
A. Helmer, Lipprandt, M., Frenken, T., Eichelberg, M., Hein, A.: Empowering Patients through Personal Health Records: A Survey of Existing Third-Party Web-Based PHR Products. Elctronic Journal of Health Informatics 6(3), 1–19 (2011) ISSN 1446-4381
B. Song, Wolf, K., Gietzelt, M., Al Scharaa, O., Tegtbur, U., Haux, R., Marschollek, M.: Decision support for teletraining of copd patients. Methods Inf. Med. 49(1), 96–102 (2010)

12. Helmer, Song, B., Ludwig, W., Schulze, M., Eichelberg, M., Hein, A., Tegtbur, U., Kayser, R., Haux, R., Marschollek, M.: A sensor-enhanced health information system to support automatically controlled exercise training of COPD patients. In: 2010 4th International Conference on Pervasive Computing Technologies for Healthcare, PervasiveHealth, pp. 1–6 (2010),
    `http://ieeexplore.ieee.org/xpls/`
    `abs_all.jsp?arnumber=5482235&tag=1`
13. Lipprandt, M., Eichelberg, M., Thronicke, W., Kruger, J., Druke, I., Willemsen, D., Busch, C., Fiehe, C., Zeeb, E., Hein, A.: Osami-d: An open service platform for healthcare monitoring applications. In: Proc. 2nd Conference on Human System Interactions, HSI 2009, pp. 139–145 (2009)
14. Borg, G.: Perceived exertion as an indicator of somatic stress. Scandinavian Journal of Rehabilitation Medicine 2(2), 92–98 (1970)
15. Hair, J.F., Tatham, R.L., Anderson, R.E., Black, B.: Multivariate Data Analysis. Cram101 Textbook Outlines. Cram101 Incorporated (2006) ISBN: 9781428813755,
    `http://books.google.de/books?id=ZWrWAAAACAAJ`
16. Integrating the Healthcare Enterprise International. IHE International (January 2012),
    `http://www.ihe.net/`
17. Integrating the Healthcare Enterprise: Patient Care Coordination (PCC) Technical Framework Volume I Revision 7.0. Tech. Rep. (2011)
18. Integrating the Healthcare Enterprise: IT Infrastructure Technical Framework Volume 2b Revision 7.0. Integrating the Healthcare Enterprise, Tech. Rep. (2011)

Exploiting Cloud-Based Personal Health Information Systems in Practicing Patient Centered Care Model

Juha Puustjärvi[1] and Leena Puustjärvi[2]

[1] Department of Computer Science, University of Helsinki, P.O. Box 68, Helsinki, Finland
juha.puustjarvi@cs.helsinki.fi
[2] The Pharmacy of Kaivopuisto, Neitsytpolku 10, Helsinki, Finland
leena.puustjarvi@kolumbus.fi

Abstract. The introduction of new emerging healthcare models, such as patient-centered care, pharmaceutical care, and chronic care model, are changing how people think about health and of patients themselves These healthcare models need technology solutions that support the co-operation within patient's healthcare team, provide a platform for sharing patient's healthcare data among the healthcare team, and provide a mechanism for disseminating relevant educational material for the patient and the healthcare team. Unfortunately current health information technology solutions only provide the connection between patients and healthcare providers, and thus do not support the new emerging healthcare models. Instead, cloud-based healthcare delivery models will potentially have more impact on developing appropriate technology for the new healthcare models. In this paper, we describe our work on designing a personal health information system, which supports patient remote monitoring and the new emerging healthcare models as well. The key idea is to develop the system by integrating relevant e-health tools through a shared ontology and to exploit the flexibility of cloud computing in its implementation. In developing the ontology we have used semantic web technologies such as OWL and RDF.

Keywords: Remote patient monitoring, Personal health systems, Healthcare models, Cloud computing, Semantic web, Ontologies, OWL, RDF.

1 Introduction

Cloud computing represents new way of delivering information technology: anyone with a suitable Internet connection and a standard browser can access an application in a cloud. In addition, cloud computing allows for more efficient computing by centralizing storage, memory, processing and bandwidth.

However, in spite of the widespread adoption of cloud computing by most industries, the healthcare sector has been rather slow in adopting cloud-based solutions. Slow adoption in healthcare sector is partially due to concerns about data security and compliance with key regulations, which defines numerous offenses relating to health care and sets civil and criminal penalties for them.

Making sure that data security and compliance with key regulations are met, cloud computing will provide significant benefits to healthcare organizations and help them

J. Gabriel et al. (Eds.): BIOSTEC 2012, CCIS 357, pp. 339–352, 2013.

improve patient care. It also promotes the introduction of new healthcare delivery models that will make healthcare more efficient and effective.

The introduction of new emerging healthcare models, such as patient-centered care, pharmaceutical care, and chronic care model, are changing how people think about health and of patients themselves.

Patient-centered care [1, 2, 3] emphasizes the coordination and integration of care, and the use of appropriate information, communication, and education technologies in connecting patients, caregivers, physicians, nurses, and others into a healthcare team where health system supports and encourages cooperation among team members. It is based on the assumption that physicians, patients and their families have the ability to obtain and understand health information and services, and make appropriate health decisions [4]. This in turn requires that health information should be presented according to individuals understanding and abilities [5].

Pharmaceutical care emphasizes the movement of pharmacy practice away from its original role on drug supply towards a more inclusive focus on patient care [6, 7, 8]. It emphasizes the responsible provision of drug therapy for the purpose of achieving definite outcomes that improve patient's quality of life [9, 10].

Chronic care model [11, 12] emphasizes patients' long-term healthcare needs as a counterweight to the attention typically paid to acute short-term, and emergency care. In this sense, the traditional care models are not appropriate as the patients with chronic illness do not receive enough information about their condition, and they are not supported in caring themselves after they leave the doctor's office or hospital.

Patient remote monitoring and home telehealth technologies provide a variety of tools for patients to take an active role in the management of their chronic diseases. Especially, the ability to monitor and interact with patient from a distance by exploiting electronic devices to record and send the measurements of patients' vital signs to a caregiver has been a key technology in fostering patients' ability to receive care at home. Earlier the only reliable means of controlling such measurements has been for a medical professional to take them directly, or for a patient to be constantly monitored in hospital, which would normally only happen once the patient has become seriously ill.

The new technologies and principle of practicing medicine holds significant promise of improving on major health care delivery problems. However, there are many functions in the patient-centered care, pharmaceutical care and chronic care models that the home telehealth devices and the e-heath tools such as personal health records do not support.

For example, patients that are out of hospital or who are left hospital often have concerns about their medicines, and so there is strong demand for extending the functionalities of home telehealth devices by the functions of pharmaceutical care. Neither the current e-heath tools support the coordination of the care, nor the social connections among the members of patient's healthcare team. They also fail in providing comprehensive access to patient's health data and in promoting patient's medical education.

Inspired by the (semantic) web technologies and the flexibility of cloud computing, we have studied their suitability for supporting the emerging healthcare models. Our studies have indicated that Personal Health Information Systems (PHIS) should support the functionalities of many traditional e-health tools such as remote patient monitors, personal health records, health-oriented blogs, and health-oriented information servers. It is also turned out that by gathering these functionalities into one system we can achieve synergy, i.e., achieve functionalities that would not be obtainable by any of the e-health tools independently.

In gathering the functionalities we have adapted the ideas of knowledge centric organizations to PHIS, i.e., we have revolved the e-health tools around a health oriented knowledge base. So, all the e-heath tools share patient's health data. Further by exploiting the characteristics provided by cloud computing we can easily ensure the interoperation of patient's healthcare team: accessing the PHIS requires only internet connection. Instead of the prevailing systems provided by healthcare organizations do not provide appropriate technology for co-operation as their use is devoted to organization's healthcare personnel only.

The rest of the paper is organized as follows. First, in Section 2, we motivate our work by considering the recent advances in patient remote monitoring. Then, in Section 3, we present the requirements of PHIS that we have derived from emerging healthcare models, and then, in Section 4, we analyze the suitability of cloud computing for satisfying these requirements. In Section 5, we present the architecture of the knowledge oriented PHIS and the PHIS-ontology that is shared by the e-health tools. In Section 6 we describe how the PHIS-ontology can be exploited in promoting patient's medical education and in delivering relevant information within patient's healthcare team. In Section 7, we illustrate how XSLT-transformations is used in transforming XML-coded medical data in the format that is compliant with the PHIS-ontology. Finally Section 8 concludes the paper by discussing the challenges of our solutions as well as our future research.

2   Remote Patient Monitoring

Telemedicine is the use of medical information exchanged from one site to another via communications to improve a patient's [13]. Telemedicine is viewed as a cost-effective alternative to the more traditional face-to-face way of providing medical care [14].

Telemedicine can be broken into three main categories: store-and-forward, interactive services and remote patient monitoring.

• *Store-and-forward telemedicine* involves acquiring medical data and then transmitting this data to the system that is accessible to patient's physician. So it does not require the presence of patient and physician at the same time.

• *Interactive services* provide real-time interactions between patient and physician. It includes phone conversations, online communication and home visits.

• *Remote monitoring* enables medical professionals to monitor a patient remotely using various technological devices. Remote monitoring is above all used for managing chronic diseases such as heart disease, diabetes and asthma.

Nowadays remote patient monitoring technologies are becoming a more sophisticated, integrated, and systematic approach to healthcare that can be personalized to each patient's medical needs. In particular, Personal Health Systems (PHSs) go beyond the simple remote patient monitoring systems in that they enable the communication between patients and healthcare professionals and provide clinicians with access to current patient data. They also provide interactive tools for personalized care management including vital sign collection, patient reminders and communication tools such as video conferencing capabilities, allowing remotely located health care professionals to interview, observe and educate the patient, as well as assist in the use of the peripherals or other medical devices. Some devices also have the ability to show video, which can be used for educating the patient.

From technology point of view personal health systems consist of a hub and wireless peripheral devices that collects physiologic data. Typical peripheral devices include blood pressure cuffs, pulse oximeters, weight scales blood glucose meter. The data gathered from peripheral devices are transmitted by the hub to a clinical database for later analysis (Figure 1).

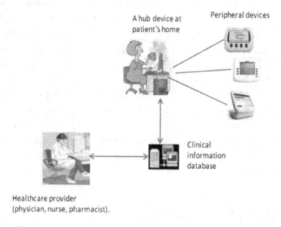

Fig. 1. Remote monitoring through a clinical information database

3   Information Flows in PHIS

The technology that supports both patient centered healthcare and pharmaceutical care of the patients with chronic conditions have to coordinate the flows of information that are coming from a variety of sources. These information flows include:

• Vital sign information from peripheral devices (that collect physiologic data) to PHIS.

• Health regimen information between healthcare providers (physicians, nurses and pharmacists).

• Information between patient and healthcare providers.

• Healthcare information between healthcare providers and patient's family members.

• Relevant educational health information from healthcare providers to patient.

Supporting these information flows is much more challenging as the simple vital sign information flow that characterized an earlier generation of remote patient monitoring. In particular the traditional remote monitoring model (illustrated in Figure 1) supports only partially these requirements as the usage of the clinical information system is isolated from third parties such as from patient's family members.

Apart from the co-operation support, the member's of patient's healthcare team should have a seamless access to patient's health data, which is usually stored in electronic health record (EHR) [15] or personal health records (PHR) [16]. The former is managed by medical authorities while the latter managed by the patient and all that are authorized by the patient are allowed to access it [17]. Hence patient's PHR, which in our architecture is a component of the PHIS, has a central role to support emerging healthcare models.

4   Cloud-Based PHIS

Cloud computing is a technology that uses the Internet and central remote servers to maintain data and applications [18]. It is an evaluation of the widespread adoption of virtualization, service oriented architecture and utility computing. The name cloud computing was originally inspired by the cloud symbol that's often used to represent the internet in diagrams.

Cloud computing allows consumers and businesses to use applications without installation, and they can access their personal files at any computer with internet access. This technology allows for more efficient computing by centralizing storage, memory, processing and bandwidth. Further, unlike traditional hosting it provides the following useful characteristics:

• The resources of the cloud can be used on demand, typically by the minutes.

• The used resources are easily scalable in the sense that users can have as much or as little of a service as they want at any given time.

• The resources are fully managed by the provider. The consumer does not need any complex resource, only a personal computer with internet access.

Software as a service (SaaS), is a type of cloud computing. In this service model, a service provider licenses an application to customers either as a service on demand, through a subscription, in a "pay-as-you-go" model, or at no charge [19]. The SaaS model to application delivery is part of the utility computing model where all of the technology is in the "cloud" accessed over the internet as a service.

There are various architectural ways for implementing the SaaS model including the followings [18]:

• Each customer has a customized version of the hosted application that runs as its own instance on the host's servers.

• Many customers use separate instances of the same application code.

• A single program instance serves all customers.

In the case of PHISs the required computation is rather small compared to traditional business applications and thus the last mentioned architecture is appropriate for the implementation of the PHIS, i.e., a single PHIS serves all patients. However, patient specific data can only be accessed by the patient and those that are authorized by the patient.

The SaaS-based PHIS and its users are presented in Figure 2.

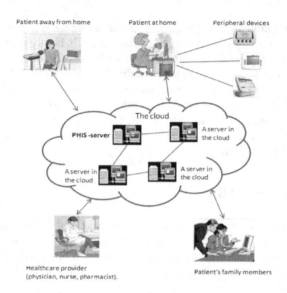

Fig. 2. The users of the cloud-based PHIS

We next itemize some clarifying aspects of the figure:

• The cloud takes the advantages of SOA (Service Oriented Architecture) in the interoperation of the services, e.g., in importing patient's health data the PHIS-server interoperates with the servers of other healthcare organizations including hospitals, physicians' offices and health centers.

• As the figure illustrates the peripheral devices that the patient has at home are connected to patient's PC, and so the vital signs collected by the devices are transmitted via the PC to the cloud, i.e., to the PHIS.

• The patient accesses his or her health data stored in PHIS through the browser. As the patient needs nothing but an internet access, the patient can easily connect to the PHIS at home, as well as being away from home.

• Healthcare providers and patient's family members that are authorized by the patient can access patient's health data as well as communicate through their browsers.

Next, we consider the internal structure of the PHIS-server.

5   PHIS-Ontology

The architecture of the PHIS and its connections in the cloud are presented in Figure 3. As the figure illustrates patient and the members of his or her healthcare team access the PHIS-server through the personalized health portal. It is a site on WWW that provides personalized capabilities for its users and links to other relevant servers.

In designing the PHIS we have followed the idea of knowledge oriented organizations [20], where the key idea is to revolve all applications around a shared ontology (stored in a knowledge base), which we call *PHIS-ontology*. It is developed by integrating the ontologies of the e-health tools supported by the PHIS. For now we have

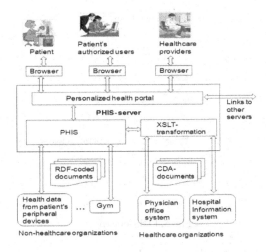

Fig. 3. The components of the PHIS-server and its external connections

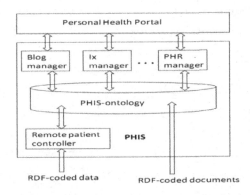

Fig. 4. e-Health tools accessing the PHIS-ontology

integrated the ontologies of the Blog manager, Information therapy (Ix) manager, Remote manager, and PHR manager. Such an internal architecture of the PHIS is presented in Figure 4.

Figure 5 illustrates the idea of the knowledge base and the case where PHIS-ontology is developed by integrating the Blog-ontology, Ix-ontology, PHR-ontology and RM-ontology (Remote Monitoring ontology). In the figure ellipses represent OWL´s classes, rectangles represent OWL's data properties and the lines between ellipses represent OWL's object properties. Accordingly class A is shared by all the four ontologies.

In order to illustrate shared classes, A could be class *Disease*, B class *Patient*, and C class *Informal_entity*. Further assume that object property A-B is *suffer_from*, object property *A-E* is *deals*, data property *b1* is *patient_name*, and data property *e1* is a *url*. In such as setting we could specify by RDF (Resource Description Framework) that John Smith suffers from diabetes and the educational material dealing diabetes is stored in a specific *url*.

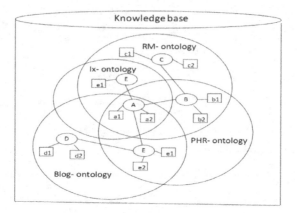

Fig. 5. PHIS-ontology

A portion of the PHIS-ontology is graphically presented in Figure 6. In this graphical representation ellipses represent classes and subclasses, and rectangles represent data and object properties.

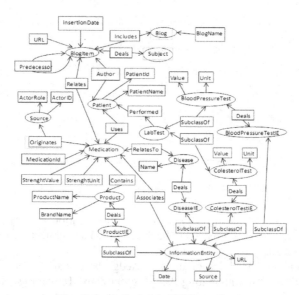

Fig. 6. A portion of the PHIS-ontology

A portion of the graphical ontology of Figure 6 is presented in OWL in Figure 7.

In order to understand the relationship of XML, OWL and RDF note that XML (Extensible Mark-up Language) [21] is just a meta language for defining markup languages. By a meta language we refer to a language used to make statements about statements in another language, which is called the object language. Accordingly RDF [22] and OWL [23] are object languages. Instead, XML says nothing about the semantics of the used tags. It just provides a means for structuring documents. Due to

```
<rdf:RDF
      xmlns:rdf=http://www.w3.org/1999/02/22-rdf-syntax-ns#
      xmlns:rdfs=http://www.w3.org/2000/01/rdf-schema#
      xmlns:owl=http://www.w3.org/2002/07/owl#>

      <owl:Ontology rdf:about=""PHA/>
      <owl:Class rdf:ID="Blog/">
      <owl:Class rdf:ID="BlogItem/">
      <owl:Class rdf:ID="Patient/">
      <owl:Class rdf:ID="Medication/">
      <owl:Class rdf:ID="Source/">
      <owl:Class rdf:ID="Product/">
      <owl:Class rdf:ID="LabTest/">
      <owl:Class rdf:ID="BloodPressureTest">
                  <rdfs:subClassOf rdf:resource="#LabTest"/>
      </owl:Class>
      <owl:Class rdf:ID="ColesterolTest">
                  <rdfs:subClassOf rdf:resource="#LabTest"/>
      </owl:Class>

      <owl:ObjectProperty rdf:ID="Relates">
                  <rdfs:domain rdf:resource="#BlogItem"/>
                  <rdfs:range rdf:resource="#Medication"/>
      </owl:ObjectProperty>

      <owl:ObjectProperty rdf:ID="Uses">
                  <rdfs:domain rdf:resource="#Patient"/>
                  <rdfs:range rdf:resource="#Medication"/>
      </owl:ObjectProperty>

                  .
                  .
                  .
</rdf:RDF>
```

Fig. 7. A portion of the PHIS-ontology in OWL

the lack of semantics we do not use XML for representing PHIS-ontology but instead we use ontology languages RDF and OWL.

6   Providing Information Flows and Medical Education

As we have already stated the technology that supports emerging healthcare models should coordinate the flows of information within patient's healthcare team as well as provide educational relevant material. In our developed solutions these functionalities are carried out by the information stored in the PHIS–ontology. In particular these functionalities exploit the instances of the classes BlogItem and InformationEntity.

Each instance of the class BlogItem represents an entry in patient blog. By its object property Predecessor the entries are presented in chronological order with the latest entry listed first, and by using the object property Deals blog's entries can be classified into different subjects. Patient and patient's healthcare team are allowed to access and create new entries for the blog through the functionalities provided by the Blog manager.

Each instance of the class InformationEntity represents an educational material. Its data property url specifies the location of the actual content of the material, i.e., the instance. As illustrated in Figure 6, the PHIS-ontology also specifies the relationships of the class InformationEntity to other relevant classes such as Medication and Disease. Thus, based on these relationships relevant educational material can be automatically delivered to the patient. For example, we can query the information entities that deal the diseases that patient John Smith suffers from. Further, by activating such queries (by the Ix-manager) when new disease is inserted for a patient, we can automate information therapy, i.e., prescribe the right information to right patient at right time.

7   Transforming XML-coded Documents

As we have already illustrated in Figure 3, the PHIS-server does not only import data from patients home telehealth devices but it also imports data from other information sources such as a hospital laboratory, gym and physician office.

If the format of the imported data does not coincide with the PHIS-ontology, then the stylesheet engine [20] is required for transforming the imported data before its insertion into the PHIS-ontology. Such a transformation is illustrated in Figure 8.

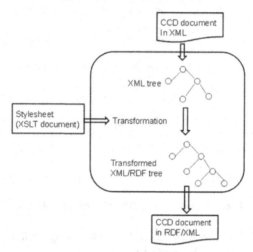

Fig. 8. Transforming CDA-documents into semantic documents

CDA-documents are typical XML-based documents, and so they are not compliant with the PHIS-ontology. By the CDA-documents we refer to documents that are based on the Clinical Document Architecture (CDA), which is an ANSI approved HL7 standard [24]. It is proven to be a valuable and powerful standard for a structured exchange of persistent clinical documents between different software systems 17.

However, in the case of non-persistent documents with CDA we encounter many problems. The main reason for this is that the semantics of the CDA-documents is bound to the shared HL7 Reference Information Model (RIM) [24]. Thereby introducing new document types would require to extending the RIM, which is a long lasting standardization process. So this approach contradicts with our requirement of flexibility in introducing new document types in importing health information into the PHIS-ontology. Therefore in importing data from XML-based data sources (e.g., from HL7 CDA compliant systems) requires that the XML-formatted data is first translated (by an XSLT-based style sheet engine [21] into RDF and then inserted into the PHIS-ontology.

In order to illustrate this transformation consider the CDA document of Figure 9.

Figure 9 represents a CCR file that has a medication list (element Medications), which is comprised of one medication (element Medication) that is source stamped by the Pharmacy of Kaivopuisto. The CCR file is based on the CCR standard [25].

The CCR standard as well as the CCD standard [26] are originally a patient health summary standards, and later on these standards are commonly exploited in structuring the data in personal health records. From technology point of view these standards represent two different XML schemas designed to store patient clinical summaries [27]. However, both schemas are identical in their scope in the sense that they contain the same data elements.

```
<ContinityOfCareRecord>
    <Patient>
        <ActorID>AB-12345</ActorID>
        <ActorName> Susan Taylor</ActorName>
    </Patient>
    <Medications>
        <Medication>
            <Source>
                <ActorID>Pharmacy of Kaivopuisto</ActorID>
                <ActorRole>Pharmacy</ActorRole>
            </Source>
            <Description>
                <Text>One tablet ones a day</Text>
                <GenericSubstitutionInfo>
                    http://www.../medicalinfo/SubstitutionInfo
                </GenericSubstitutionInfo>
            </Description>
            <Product>
                <ProductName>Valsartan</ProductName>
                    <ProductInfo>
                        http://www.../medicalinfo/ValsartanInfo
                    </ProductInfo>
            </Product>
            <Strenght>
                <Value>50</Value>
                <Unit>milligram</Unit>
            </Strenght>
            <Quantity>
                <Value>30</Value><Unit>Tabs</Unit>
            </Quantity>
        </Medication>
    </Medications>
</ContinityOfCareRecord>
```

Fig. 9. A CCR document

After the XSLT transformation the CCR document of Figure 9 is in the RDF/XML format presented in Figure 10. In this format the document is compliant with PHIS-ontology and can be inserted into the PHIS-ontology.

The RDF/XML-formatted document of Figure 10 is comprised of four RDF-descriptions. Further, the first RDF-description is comprised of three RDF-statements. The first statement states that the type of the instance identified by "AB-12345" is Patient in the PHIS-ontology. The second RDF-statement states that the name of the instance identified by "AB-12345" is Susan Taylor.

In developing a stylesheet for a document, say for a CCR document, we have first developed an OWL-ontology for the document. In this stage we have used the XML-schema of the CDA document as follows:

1. The complex elements of the XML-schema are transformed into OWL classes.

2. The simple elements of the XML-schema are transformed into OWL data properties such that the complex element is the domain of the data properties.

3. The attribute of the XML-schema are transformed into OWL data properties.

4. The relationships between complex elements are be named and transformed to OWL object properties.

```
<rdf:RDF
  xmlns:rdf=http://www.w3.org/1999/02/22-rdf-syntax-ns#
  xmlns:po=http://www.lut.fi/ontologies/PHIS-ontology#>
  <rdf:Description rdf:about="AB-12345">
        <rdf:type rdf:resource="&po;Patient"/>
        <po:PatientName>Susan Taylor</po:PatientName>
        <po:Uses rdf:resource="&po;Med-07092010"/>
  </rdf:Description>
  <rdf:Description rdf:about="Med-07092010">
        <rdf:type rdf:resource="&po;Medication"/>
        <po:Contains rdf:resource="&po;Valsartan"/>
            <po:StrenghtValue rdf:datatype=
                "&xsd;integer">30</po:StrenghtValue>
            <po:StrenghtUnit>Tabs</po:StrenghtUnit>
  </rdf:Description>
  <rdf:Description rdf:about="Valsartan">
            <rdf:type rdf:resource="&po;Product"/>
        <po:Deals rdf:resource="&info;ValsartanInfo"/>
  </rdf:Description>
  <rdf:Description rdf:about="Pharmacy of Kaivopuisto">
            <rdf:type rdf:resource="&po;Source/>
        <po:ActorRole>Pharmacy</po:ActorRole>
  </rdf:Description>
</rdf:RDF>
```

Fig. 10. Transformed CDA document in RDF/XML format

Note that as the OWL does not support structured attributes we have not transformed all complex elements to classes but rather the complex elements that do not have identification have been transformed to a set of properties. For example the following complex element:

```
<Strenght>
<Value>50</Value>
<Unit>milligram</Unit>
</Strenght>
```

is first transformed into data properties StrenghtValue and StrenghtUnit, and then connected to the OWL class Medication.

The ontology developed from the CCR document is called the CCR-ontology. The stylesheet for the CCR document is then specified in the way that the resulted document is an instance of the CCR-ontology. So, for example, the document of Figure 10 is an instance of the CCR-ontology, which in turn is a portion of the PHIS-ontology.

8   Conclusions

Monitoring a patient's vital signs provides an important source of information to the physician that treats the patient. Nowadays information and communication technology provides the possibility of a new generation of lightweight monitoring systems which a patient can wear while being at home or while going about their daily business. Formerly the only reliable means of such monitoring has been for a medical professional to take them directly, or for a patient to be constantly monitored in hospital.

The new patient remote monitoring technology holds significant promise of improving on major health care delivery problems. However, there are many functions in the emerging healthcare models (including patient centered care, pharmaceutical care and chronic care models) that the modern monitoring devices and systems do not support as they only provide the communication between patient and healthcare provider. Instead the emerging healthcare models require Information and Communication Technology (ICT) support for the co-operation of patient's healthcare team and the support in delivering relevant educational material for the patient and the members of the healthcare team.

Our studies have shown that the ICT-support of these requirements requires the integration of patient's e-health tools as it significantly simplifies patients' interaction with the services, enables the co-operation within the healthcare team and the development of new services such as automated information therapy.

From technology point of view we have integrated e-heath tools through the shared PHIS-ontology that is stored in the knowledge base, which exploits semantic web technologies such as OWL and RDF. The management of the shared ontology requires that in importing data the documents that are not compliant with the ontology have to be transformed by XSLT transformation into the RDF-format that is compliant with the ontology, i.e., a stylesheet has to be defined for each non-compliant document type.

In our future work we will study the effects of introducing cloud-based health information systems on the mind-set of patient and healthcare personnel as the introduction of these technologies also changes the daily duties of the patient and many healthcare employees. Therefore we assume the most challenging aspect will not be the technology but rather the changing the mind-set of patient's healthcare team.

References

1. Bauman, A., Fardy, H., Harris, H.: Getting it right; why bother with patient centred care? Medical Journal of Australia 179(5), 253–256 (2003)
2. Gillespie, R., Florin, D., Gillam, S.: How is patient-centred care understood by the clinical, managerial and lay stakeholders responsible for promoting this agenda? Health Expectations 7(2), 142–148 (2004)
3. Little, P., Everitt, H., Williamson, I.: Observational study of effect of patient centredness and positive approach on outcomes of general practice consultations. British Medical Journal, 908–911 (2001)
4. Michie, S., Miles, J., Weinman, J.: Patient-centredness in chronic illness: what is it and does it matter? Patient Education and Counselling, 197–206 (2003)
5. Stewart, M.: Towards a global definition of patient centred care: The patient should be the judge of patient centred care. British Medical Journal 322, 444–445 (2004)
6. Wiedenmayer, K., Summers, R., Mackie, C., Gous, A., Everard, M., Tromp, D.: Developing pharmacy practice. World Health Organization and International Pharmaceutical Federation (2006)
7. Mil, I., Schulz, J., Tromp, M.: Pharmaceutical care, European developments in concepts, implementation, teaching, and research: a review. Pharm World Sci. 26(6), 303–311 (2004)
8. Hepler, C.D., Strand, L.M.: Opportunities and responsibilities in pharmaceutical care. Am. J. Hosp. Pharm. 47, 533–543 (1990)

9. WHO: The role of the pharmacist in the health care system. Preparing the future pharmacist: Curricular development. Report of a third WHO Consultative Group on the role of the pharmacist, Vancouver, Canada, 27-29. World Health Organization, Geneva, WHO/PHARM/97/599, http://www.who.int/medicinedocs/

10. Hepler, C.D.: Clinical pharmacy, pharmaceutical care, and the quality of drug therapy. Pharmacotherapy 24(11), 1491–1498 (2004)

11. Fiandt, K.: The Chronic Care Model: Description and Application for Practice, http://www.medscape.com/viewarticle/549040

12. Boult, C., Karm, L., Groves, C.: Improving Chronic Care: The "Guided Care" Model, http://www.guidedcare.org/pdf/Guided%20Care%20model_Permanente%20Journal_Winter%202008.pdf

13. Angaran, D.M.: Telemedicine and Telepharmacy: current status and future implications. American Journal of Health System Pharmacy 56, 1405–1426 (1999)

14. Kontaxakis, G., Visvikis, D., Ohl, R., Sachpazidis, I., Suarez, J., Selby, B., Peter, et al.: Integrated telemedicine applications and services for oncological positron emission tomography. Oncology Reports 15, 1091–1100 (2006)

15. EHR:Electronic Health Record, http://en.wikipedia.org/wiki/Electronic_health_record

16. Raisinghani, M.S., Young, E.: Personal health records: key adoption issues and implications for management. International Journal of Electronic Healthcare 4(1), 67–77 (2008)

17. Puustjärvi, J., Puustjärvi, L.: Designing and Implementing an Active Personal Health Record System. In: The Proc. of the International Conference on eHealth, Telemedicine, and Social Medicine (2011)

18. Chappel, D.: A Short Introduction to Cloud Computing: An Enterprise-Oriented View, http://www.davidchappell.com/CloudPlatforms-Chappell.pdf

19. Khajeh-Hosseini, A., Sommerville, I., Sriram, I.: Research Challenges for Enterprise Cloud Computing, http://arxiv.org/ftp/arxiv/papers/1001/1001.3257.pdf

20. Daconta, M., Obrst, L., Smith, K.: The semantic web: A Guide to the Future of XML, Web Services, and Knowledge Management. John Wiley & Sons (2003)

21. Harold, E., Scott Means, W.: XML in a Nutshell. O'Reilly & Associates (2002)

22. RDF Resource Description Language, http://www.w3.org/RDF/

23. OWL WEB OntologyLanguage, http://www.w3.org/TR/owl-features

24. Dolin, R., Alschuler, L., Beerb, C., Biron, P., Boyer, S., Essin, E., Kimber, T., Lincoln, J.E.: The HL7 Clinical Document Architecture. J. Am. Med. Inform. Assoc. 8(6), 552–569 (2011)

25. CCR Continuity of Care Record, http://www.ccrstandard.com/

26. HL7, What is the HL7 Continuity of Care Document?, http://www.neotool.com/blog/2007/02/15/what-is-hl7-continuity-of-care-document/

27. Puustjärvi, J., Puustjärvi, L.: The role of medicinal ontologies in querying and exchanging pharmaceutical information. International Journal of Electronic Healthcare 5(1), 1–13 (2009)

Online Social Networks Flu Trend Tracker: A Novel Sensory Approach to Predict Flu Trends

Harshavardhan Achrekar[1], Avinash Gandhe[2], Ross Lazarus[3],
Ssu-Hsin Yu[2], and Benyuan Liu[1]

[1] Department of Computer Science, University of Massachusetts Lowell, Massachusetts, U.S.A.
[2] Scientific Systems Company Inc, 500 West Cummings Park, Woburn, Massachusetts, U.S.A.
[3] Department of Population Medicine, Harvard Medical School, Boston, Massachusetts, U.S.A.

Abstract. Seasonal influenza epidemics cause several million cases of illnesses cases and about 250,000 to 500,000 deaths worldwide each year. Other pandemics like the 1918 "Spanish Flu" may change into devastating event. Reducing the impact of these threats is of paramount importance for health authorities, and studies have shown that effective interventions can be taken to contain the epidemics, if early detection can be made. In this paper, we introduce Social Network Enabled Flu Trends (SNEFT), a continuous data collection framework which monitors flu related messages on online social networks such as Twitter and Facebook and track the emergence and spread of an influenza. We show that text mining significantly enhances the correlation between online social network(OSN) data and the Influenza like Illness (ILI) rates provided by Centers for Disease Control and Prevention (CDC). For accurate prediction, we implemented an auto-regression with exogenous input (ARX) model which uses current OSN data and CDC ILI rates from previous weeks to predict current influenza statistics. Our results show that, while previous ILI data from the CDC offer a true (but delayed) assessment of a flu epidemic, OSN data provides a real-time assessment of the current epidemic condition and can be used to compensate for the lack of current ILI data. We observe that the OSN data is highly correlated with the ILI rates across different regions within USA and can be used to effectively improve the accuracy of our prediction. Therefore, OSN data can act as supplementary indicator to gauge influenza within a population and helps to discover flu trends ahead of CDC.

1 Introduction

Seasonal influenza epidemics result in about three to five million cases of severe illness and about 250,000 to 500,000 deaths worldwide each year [11]. In 1918, the so-called "Spanish flu" killed an estimated 20-40 million people worldwide, and since then, human-to-human transmission capable influenza virus has resurfaced in a variety of particularly virulent forms much like "SARS" and "H1N1" against which no prior immunity exists, resulting in a devastating situation with severe casualties. Reducing the impact of seasonal epidemics and pandemics such as the H1N1 influenza is of paramount importance for public health authorities. Studies have shown that preventive measures can be taken to contain epidemics, if an early detection is made or if we have some form of an early warning system during the germination of an

J. Gabriel et al. (Eds.): BIOSTEC 2012, CCIS 357, pp. 353–368, 2013.
© Springer-Verlag Berlin Heidelberg 2013

epidemic [7,14]. Therefore, it is important to be able to track and predict the emergence and spread of flu in the population.

The Center for Disease Control and Prevention (CDC) [3] monitors influenza-like illness (ILI) cases by collecting data from sentinel medical practices, collating reports and publishing them on a weekly basis. It is highly authoritative in the medical field but as diagnoses are made and reported by doctors, the system is almost entirely manual, resulting in a 1-2 weeks delay between the time a patient is diagnosed and the moment that data point becomes available in aggregate ILI reports. Public health authorities need to be forewarned at the earliest to ensure effective preventive intervention, and this leads to the critical need of more efficient and timely methods of estimating influenza incidences.

Several innovative surveillance systems have been proposed to capture the health seeking behaviour and transform them into influenza activity. These include monitoring call volumes to telephone triage advice lines [6], over the counter drug sales [15], and patients visit logs on Physicians for flu shots. Google Flu Trends uses aggregated historical log on online web search queries pertaining to influenza to build a comprehensive model that can estimate nationwide ILI activity [9].

In this paper, we investigate the use of a novel data source, OSN data, which takes advantage of the timeliness of early detection to provide a snapshot of the current epidemic conditions and makes influenza related predictions on what may lie ahead, on a daily or even hourly basis. We sought to develop a model which estimates the number of physician visits per week related to ILI as reported by CDC.

Our approach treats OSN users within United States as "sensors" and collective message exchanges showing flu symptoms like "I have Flu", "down with swine flu",etc. - as early indicators and robust predictors of influenza. We expect these posts on OSN's to be highly correlated to the number of ILI cases in the population. We analyze messages, build prediction models and discover trends within data to study the characteristics and dynamics of disease outbreak. We validate our model by measuring how well it fits the CDC ILI rates over the course of two years from 2009 to 2011. We are interested in looking at how the seasonal flu spreads within the population across different regions of USA and among different age groups.

In this paper, we extend our preliminary analysis [1,2], and provide a continuing study of using OSN's to track the emergence and spread of seasonal flu in the year 2010-2011. OSN data which demonstrated high correlation with CDC ILI rate for the year 2009-2010, was affected by spurious messages and so text mining techniques were applied. We show that text mining can significantly enhance the correlation between the OSN data and the ILI data from CDC, providing a strong base for accurate prediction of ILI rate.

For prediction, we build an auto-regression with exogenous input (ARX) model where ILI rates of previous weeks from CDC form the autoregressive component of the model, and the OSN data serve as exogenous input. Our results show that while previous ILI data from CDC offer a realistic (but delayed) measure of a flu epidemic, OSN data provides a real-time assessment of the current epidemic condition and can be used to compensate for the lack of current ILI data. We observe that the OSN data are in fact highly correlated with the ILI data across the different regions within United

States. Using fine-grained analysis on user demographics and geographical locations along with prediction capabilities will provide public health authorities an insight into current seasonal flu activities.

This paper is organized as follows: Section 2 describes applications that harness the collective intelligence of online social network (OSN) users, to predict real-world outcomes. In Section 3, we give a brief introduction to our data collection and modeling framework. In Section 4, we introduce our data filtering technique for extracting relevant information from the Twitter and Facebook datasets. Detailed data analysis is performed to establish correlation with CDC reports on ILI rates. Then we go one step further and introduce our influenza prediction model in Section 5. In Section 6, we perform region-wise analysis of flu activities in the population based on the Twitter and Facebook. Finally we conclude in Section 7 and acknowledgements are provided in Section 8.

2   Related Work

A number of measurement related studies have been conducted on different forms of social networks like Del.icio.us, Facebook and Wikipedia etc [8,22]. Sitaram et al. demonstrated how social media content like chatter from Twitter can be used to predict real-world outcomes of forecasting box-office revenues for movies [21]. Sakaki et al. used a probabilistic spatio-temporal model to build an autonomous earthquake reporting system in Japan using twitter users as sensors and applying Kalman filtering and particle filtering for location estimation [19]. Meme Tracking in news cycles as explained by Leskovec et al. was an attempt to model information diffusion in social media like blogs and tracking handoff from professional news media to social networks [13].

Ginsberg et al. in his paper discussing his approach for estimating Flu trends proposed that the relative frequency of certain search terms are good indicators of the percentage of physician visits and established a linear correlation to weekly published ILI percentages between 2003 and 2007 for all nine regions identified by CDC [9]. Culotta used a document classification component to filter misleading messages out of Twitter and showed that a small number of flu-related keywords can forecast future influenza rates [5].

OSN data has been used for real-time notifications such as large-scale fire emergencies, downtime on services provided by content providers [17] and live traffic updates. There have been efforts in utilizing twitter data for measuring public interest/concern about health-related events [18,20], predicting national mood [16], currency tracing and performing market and risk analysis [10] . Tweetminster, a media utility tool design to make UK politics open and social, analyses political tweets, to establish the correlations between buzz on Twitter and election results. In June 2010, we introduced the SNEFT architecture as a continuous data collection engine which combines the detection and prediction capability on social networks in discovering real world flu trends [1,2,4].

3   Data Collection

In this section we describe our data collection methodology by introducing the SNEFT architecture, provide a description of our dataset, explore strategies for data cleaning, and apply filtering techniques in order to perform quantitative spatio-temporal analysis.

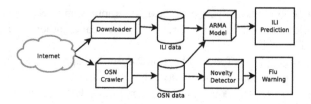

Fig. 1. The system architecture of SNEFT

3.1 SNEFT Architecture

We propose the Social Network Enabled Flu Trends (SNEFT) architecture along with its crawler, predictor and detector components, as our solution to predict flu activity ahead of time with a certain accuracy.

CDC ILI reports and other influenza related data are downloaded into the "ILI Data" database from their corresponding websites (e.g., CDC [3]). A list of flu related keywords ("Flu" , "H1N1" and "Swine Flu") that are likely to be of significance are used by the OSN Crawler as inputs into public search interfaces to retrieve publicly available posts mentioning those keywords. Relevant information about the posts such as time,location and other demographic information is collected along with the relative keyword frequency and stored in a spatio-temporal "OSN Data" database for further data analysis.

An Autoregressive with Exogenous input (ARX) model is used to predict ILI incidence as a linear function of current and past OSN data and past ILI data thus providing a valuable "preview" of ILI cases well ahead of CDC reports. Novelty detection techniques can be used to continuously monitor OSN data, and detect transition in real time from a "normal" baseline situation to a pandemic using the volume and content of OSN data enabling SNEFT to provide a timely warning to public health authorities for further investigation and response.

3.2 OSN Crawler

Based on the search API provided by Twitter and Facebook, we have developed crawlers to fetch data at regular time intervals.

The Twitter search service accepts single or multiple keywords using conjunctions ("flu" OR "h1n1" OR "#swineflu") to search for relevant tweets. Search results are typically 15 tweets (maximum 50) per page up to 1,500 tweets arranged in chronologically decreasing order, obtained from a real time stream known as the public timeline. The tweet has the User Name, the Post with status id and the Timestamp attached with each post. From the twitter username, we can get the number of followers, number of friends, his/her profile creation date, location and status update count for every user.

The "Post by everyone" option allows us to search public posts for given keywords in Facebook. All results that show up are available to the public for a limited time period. We are interested in getting useful information (profile ID, time stamp of the post, and the post content) out of posts. Given a profile ID, we will retrieve the detailed information of the profile, which typically includes, among other things, name, gender, age, affiliations (school, work, region), birthday, location, education history, and friends.

The location field helps us in tracking the current/default location of a user. Geo location codes are present in a location enabled mobile tweet/post. For all other purposes, we assume the location attribute within the profile page to be his/her current location and pass it as an input to Google's location based web services to fetch geo-location codes (i.e., latitude and longitude) along with the country, state, city with a certain accuracy scale. All the data extracted from posts and profile page are stored in a spatio-temporal "OSN data" Database.

We apply filters to get quantitative data within Unites States and exclude organizations and users who posts multiple times during a certain period of time on flu related activities. This data is fed into the Analysis Engine which has a detector and ARX predictor model. The visualization tools and reporting services generate timely visual and data centric reports on the ILI situation. The CDC monitors Influenza-like illness cases within USA by collecting data about number of Hospitalizations, percentages weighted ILI visits to physicians, etc, and publishes it online. We download the CDC data into "ILI data" database to compare with our results.

4 Data Set

In this section we briefly describe our datasets used for influenza prediction. OSN has emerged as a primary source of user interactions on daily events, health status updates, entertainment, etc. At any given time, tens of millions of users are logged onto OSN's, with each user spending an average of tens of minutes daily. Since Oct 18, 2009, we have searched and collected tweets and profile details of Twitter users who mentioned flu descriptors in their tweets. Facebook opened their Search functionality in early February 2010 and since then we have been fetching status updates and wall posts of Facebook users with mention of flu descriptors. The preliminary Twitter analysis for the year 2009-2010 is documented in [1]. For 2010-2011, we have 4.5 million tweets from 1.9 million unique users and 2.0 million facebook posts from 1.5 million unique facebook users. Twitter allows its users to set their location details to public or private from the profile page or mobile client. So far our analysis on location details of the Twitter dataset suggest that 22% users on Twitter are within USA, 46% users are outside USA and 32% users have not published their location details. Analysis on location details of Facebook dataset suggest that 22% users are within USA, 17% users are outside USA and 61% users have not published their location details.

Initial analysis for the period 2009-2010 indicated a strong correlation between CDC and Twitter data on the flu incidences [1]. However results for the year 2010-2011 showed a significant drop in the correlation coefficient from 0.98 to 0.47. In an attempt to investigate such a drastic drop in correlation we looked at data samples and found spurious messages which suppressed the actual data. To list a few, tweets like "I got flu shot today.", "#nowplaying Vado - Slime Flu..i got one recently!" (Slime flu is the name of a debut mixtape from an artist V.A.D.O. released in 2010) are false alarms. In the year 2009-2010, the Swine Flu event was so evident that the noise did not significantly affect the correlation that existed then. To mitigate this problem, we removed the spurious tweets using a filtering technique that trains a document classifier to label whether a message is indicative of a flu event or not.

4.1 Text Classification

In an information retrieval scenario, text mining seeks to extract useful information from unstructured textual data. Using a simple "bag-of-words" text representations technique based on a vector space, our algorithm classifies messages wherein user mentions having contracted the flu himself or has observed the flu among his friends, family, relatives, etc. Accuracy of such a model is highly dependent on how well trained our model is, in terms of precision, recall and F-measure.

The set of possible labels for a given instance can be divided into two subsets, one of which is considered "relevant". To create such an annotated dataset which demands human intelligence, we use Amazon Mechanical Turks to manually classify a sample of 25,000 tweets and 10,000 status updates. Every message is classified by exactly three Turks and the majority classified result is attached as the final class for that message.

Table 1. Twitter Text Classification 10 fold cross validation results (left) followed by Facebook's 10 fold cross validation results (right)

| Classifier | Class | Twitter | | | Facebook | | |
		Precision	Recall	F-value	Precision	Recall	F-value
J48	Yes	0.801	0.791	0.796	0.684	0.785	0.731
	No	0.813	0.704	0.755	0.629	0.501	0.557
Naive Bayesian	Yes	0.725	0.829	0.773	0.688	0.847	0.759
	No	0.813	0.704	0.755	0.69	0.47	0.559
SVM	Yes	**0.807**	**0.822**	**0.814**	**0.696**	**0.857**	**0.768**
	No	**0.829**	**0.814**	**0.822**	**0.71**	**0.485**	**0.576**

The training dataset is fed as an input to different classifiers namely decision tree (J48), Support Vector Machines (SVM) and Naive Bayesian. For efficient learning, some configurations that we incorporated within our text classification algorithm include setting term frequency and inverse document frequency (tf-idf) weighting, stemming, using a stopwords list, limiting the number of words to keep (feature vector set) and reordering class. Based on the results shown in Table 1, we conclude that SVM classifier with highest precision and recall rate outperforms other classifiers when it comes to text classification for our data set. Application of SVM on unclassified data originating from within the United States resulted in a Twitter dataset with 280K positively classified tweets from 187K unique twitter users and 185K positively classified facebook posts from 164K unique Facebook users. In order to gauge if the number of unique twitter users mentioning the flu per week is a good measure of the CDC's ILI reported data, we plot (in Figure 2) the number of Twitter users/week against the percentage of weighted ILI visits, which yields a high Pearson correlation coefficient of 0.8907. A similar plot was generated for the number of unique Facebook users mentioning about flu per week against the percentage of weighted ILI visits resulting in Pearson correlation coefficient of 0.8728.

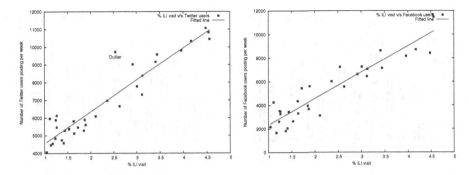

Fig. 2. Number of OSN users per week versus percentage of weighted ILI visit by CDC. (Twitter on left and Facebook on right)

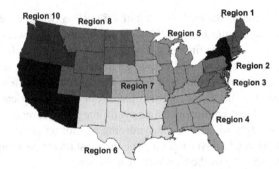

Fig. 3. Regionwise division of USA into ten regions by United States Health and Human Services

This increase in the number of users posting about the flu is accompanied by an increase in the percentage of weighted ILI visits reported by CDC in the same week. The marked outlier present in the Twitter data as identified in Figure 2 is consistent with Google Flu Trends data when high tweet volumes were witnessed in the week starting January 2, 2011. The CDC has divided the United States into 10 regions as shown in Figure 3. The CDC publishes their weekly reports on percentage weighted ILI visits collated from its ten regions and aggregates then for United States. Figure 4 compares the OSN dataset with CDC reports with and without text classification for each of the ten regions defined by the CDC and for the entire United States as a whole. We observe that the correlation coefficients have significantly improved with text classification, across all the regions and USA overall. Thus our text classification techniques play a vital role in improving the overall prediction performance.

4.2 Data Cleaning

The OSN dataset required data cleaning to discount retweets and successive posts from the same users within a certain period of time.

– *Retweets*: A retweet in Twitter is a post originally made by one user that is forwarded by another user. For flu tracking, a retweet does not indicate a new ILI case, and thus

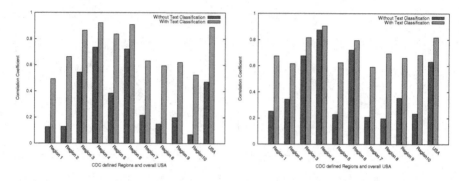

Fig. 4. Classified OSN (Twitter on left and Facebook on right) dataset achieves higher correlation with CDC reports on Nationwide and Regional levels

should not be counted in the analysis. Out of 4.5 million tweets we collected, there are 541K retweets, accounting for 12% of the total number of tweets.

- *Syndrome Elapsed Time*: An individual patient may have multiple encounters associated with a single episode of illness (e.g., initial consultation, consultation 1–2 days later for laboratory results, and follow-up consultation a few weeks later). To avoid double counting from common pattern of ambulatory care, the first encounter for each patient within any single syndrome group is reported to CDC, but subsequent encounters with the same syndrome are not reported as new episodes until more than six weeks have elapsed since the most recent encounter in the same syndrome [12]. We call this the Syndrome Elapse time.

Hence, we created different datasets namely: Twitter dataset with No Retweets (Tweets starting with RT) and Twitter dataset without Retweets and with no tweets from same user within certain syndrome elapsed time. For Facebook we create dataset namely Facebook dataset with no posts from same user within certain syndrome elapsed time.

When we compared the different datasets mentioned in Table 2 with CDC data, we found that Twitter dataset without Retweets showed a high correlation (0.8907) with CDC Data. Similarly Facebook data with Syndrome elapse time of zero showed a high correlation of 0.8728. As opposed to a common practice in public health safety, where medical examiners within U.S. observe a syndrome elapse time period of six weeks [12], user behaviour on Twitter and Facebook follows a trend wherein we do not ignore successive posts from same user. Thus Twitter dataset without Retweets is our choice of dataset for all subsequent experiments. Similarly Facebook data within same week becomes our choice of dataset for all subsequent analysis.

From Figure 5, we observe that the Complementary Cumulative Distribution Function (CCDF) of the number of tweets posted by same individual on Twitter can be fitted by a power law function of exponent -2.6429 and coefficient of determination (R-square) 0.9978 with a RMSE of 0.1076 using Maximum likelihood estimation. Most people tweet very few times (e.g., 82.5% of people only tweet once and only 6% of people tweet more than two times). However, we do not observe the power-law behavior in the CCDF of number of posts per user on Facebook, as shown in plot on the right hand side of Figure 5.

Table 2. Correlation between OSN Datasets and CDC along with its Root Mean Square Errors (RMSE)

		Twitter		Facebook	
Syndrome Elapse Time	Retweets	Correlation coefficient	RMSE errors	Correlation coefficient	RMSE errors
0 week	**No**	**0.8907**	**0.3796**	**0.8728**	**0.4287**
1 week	No	0.8895	0.3818	0.8709	0.4314
2 week	No	0.8886	0.3834	0.8698	0.4332
3 week	No	0.886	0.3878	0.8689	0.4346
4 week	No	0.8814	0.3955	0.8681	0.4357

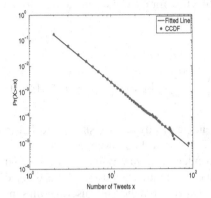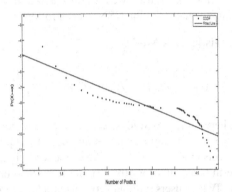

Fig. 5. Complementary Cumulative Distribution function (CCDF) of the number of tweets/posts on Twitter/Facebook by same users

Most of these high-volume tweets in Twitter are created by health related organization, who tweet multiple time during a day and users who subscribe to flu related RSS feeds published by these organizations. "Flu_alert","swine_flu_pro", "live_h1n1", "How_To_Tips", "MedicalNews4U" are examples of such agencies on Twitter. Similarly one can identify agencies like "Flu Trackers", "Influenza Flu" and specific users that actively post on Facebook.

5   Prediction Model

The correlation between OSN activity and CDC reports can change due to a number of factors. Annual or seasonal changes in flu-related trends, for instance vaccination rates that are affected by health cares, result in the need to constantly update parameters relating OSN activity and flu activity. However, particularly at the beginning of the influenza season, when prediction is of most significance, enough data may not be available to accurately perform these updates. Additionally predicting changes in ILI rates simply due to changes in flu-related OSN activity can be risky due to transient changes, such as changes in OSN activity due to flu-related news.

In order to establish a baseline for the ILI activity and to smooth out any undesired transients, we propose the use of Logistic Autoregression with exogenous inputs (ARX). Effectively, we attempt to predict a CDC ILI statistic during a certain week by using current and past OSN activity, and CDC data from previous weeks. The prediction of current ILI activity using ILI activity from previous weeks forms the autoregressive component of the model, while the OSN data from previous weeks serve as exogenous inputs. By CDC data, we refer to the percentage of visits to a physician for Influenza-Like Illness (also called ILI rate).

5.1 Influenza Model Structure

Although the percentage of physician visits is between 0% and 100%, the number of OSN users is bounded below by 0. Simple Linear ARX neglects this fact in the model structure. Therefore, we introduce a logit link function for CDC data and a logarithmic transformation of the OSN data as follows:

Logistic ARX Model

$$
\log \left(\frac{y(t)}{1 - y(t)} \right) = \sum_{i=1}^{m} a_i \log \left(\frac{y(t - i)}{1 - y(t - i)} \right) + \sum_{j=0}^{n-1} b_j \log(u(t - j)) + c + e(t) \quad (1)
$$

where t indexes weeks, $y(t)$ denotes the percentage of physician visits due to ILI in week t, $u(t)$ represents the number of unique Twitter/Facebook users with flu related tweets in week t, and $e(t)$ is a sequence of independent random variables. c is a constant term to account for offset. In our tests, the number of unique OSN users $u(t)$ is defined as Twitter users without retweets and having no tweets from the same user within syndrome elapsed time of 0 week or Facebook users having no posts from the same user within syndrome elapsed time of 0 week. The flu related messages are defined as posts with keywords "flu", "H1N1" and "swine flu". The rationale for the model structure in Eq. (1) is that OSN data provides real-time assessment of the flu epidemic. However, the OSN data may be disturbed at times by events related to flu, such as news reports of flu in other parts of the world, but not necessarily to local people actually getting sick due to ILI. On the other hand, the CDC data provides a true, albeit delayed, assessment of a flu epidemic. Hence, by using the CDC data along with the OSN data, we may be able to take advantage of the timeliness of the OSN data while overcoming the disturbance that may be present in the OSN data.

The objective of the model is to provide timely updates of the percentage of physician visits. To predict such percentage in week t, we assume that only the CDC data with at least 2 weeks of lag is available for the prediction, if past CDC data is present in a model. The 2-week lag is to simulate the typical delay in CDC data reporting and aggregation. For the OSN data, we assume that the most recent data is always available, if a model includes the OSN data terms. In other words, the most current CDC or OSN data that can be used to predict the percentage of physician visits in week t is week t-2 for the CDC data and week t for the OSN data.

In order to predict ILI rates in a particular week given current OSN data and the most recent ILI data from the CDC we must estimates the coefficients, a_i , b_j and c in Eq. (1). Also, in practice, the model orders m and n are unknown and must be estimated. In our experiment, we vary m from 0 to 2 and n from 0 to 3 in Eq. (1) in order to obtain the best values of m and n to use for prediction. Intuitively, this answers the question of how many weeks of OSN and ILI data should be used to predict the ILI activity in the current week. Within the ranges examined, $m = 0$ or $n = 0$ represent models where there are no CDC data, y, or OSN data, u, terms present. Also, if $m = 0$ and $n = 1$, we have a linear regression between OSN data and CDC data. If $n = 0$, we have standard auto-regressive (AR) models. Since the AR models utilize past CDC data, they serve as baselines to validate whether OSN data provides additional predictive power beyond historical CDC data.

Prediction with Logistic ARX Model. To predict the flu cases in week t using the Logistic ARX model in Eq. (1) based on the CDC data with 2 weeks of delay and/or the up-to-date OSN data, we apply the following relationship:

$$\log\left(\frac{\hat{y}(t)}{1 - \hat{y}(t)}\right) = a_i \log\left(\frac{\hat{y}(t-1)}{1 - \hat{y}(t-1)}\right) + \sum_{i=2}^{m} a_i \log\left(\frac{y(t-i)}{1 - y(t-i)}\right)$$

$$+ \sum_{j=0}^{n-1} b_j \log(u(t-j)) \tag{2}$$

$$\log\left(\frac{\hat{y}(t-1)}{1 - \hat{y}(t-1)}\right) = \sum_{i=1}^{m} a_i \log\left(\frac{y(t-i-1)}{1 - y(t-i-1)}\right) + \sum_{j=0}^{n-1} b_j \log(u(t-j-1)) \tag{3}$$

where $\hat{y}(t)$ represents predicted CDC data in week t. It can be verified from the above equations that to predict the CDC data in week t, the most recent CDC data is from week $t - 2$. If the CDC data lag is more or less than two weeks, the above equations can be easily adjusted accordingly.

5.2 Cross Validation Test Description

Based on ARX model structure in Eq. (1), we conducted tests using different combinations of m and n values. We currently have 33 weeks with both Twitter activity and CDC data available (10/3/2010–05/15/2011). Due to limited data samples, we adopted the K-fold cross validation approach to test the prediction performance of the models.

In a typical K-fold cross validation scheme, the dataset is divided into K (approximately) equally sized subsets. At each step in the scheme, one such subset is used as the test set while all other subsets are used as training samples in order to estimate the model coefficients. Therefore, in a simple case of a 30-sample dataset, 10-fold cross-validation would involve testing 3-samples in each step, while using the other 27 samples to estimate the model parameters.

In our case, the cross-validation scheme is somewhat complicated by the dependency of the sample $y(t)$ on the previous samples, $y(t-1), \ldots, y(t-m)$ and $u(t), \ldots, u(t-n+1)$ (see Eq. (1)). Therefore, the first sample that can be predicted is $y(\max(m + 1, n))$ not $y(1)$. In fact, since we are predicting "two weeks ahead" of the available CDC data, the first sample that can be estimated is actually $y(\max(m + 2, n + 1))$. Since, prediction equations cannot be formed for $y(1), \ldots, y(\max(m+2, n+1)-1)$, those samples were not considered in any of the K subsets during our experiment to be evaluated for prediction performance. However, they were still used in the training set to estimate the values of the coefficients a_i and b_j in Eq. (1).

Considering the above constraints, our K-fold validation testing procedure is as follows:

1. For each (m, n) pair from $m = 0, 1, 2$ and $n = 0, 1, 2, 3$, repeat the following:
   (a) Identify F, the index of first data sample that can actually be predicted. $F = max(m + 1, n)$
   (b) Represent the available data indices as $t = 1, \ldots, T$. Then divide the dataset into K approximately equally sized subsets $\{S_1, S_2, \ldots, S_K\}$, with each subset comprising members that have an approximately equal time interval between them. For example, the first set would be $S_1 = \{y(F), y(F+K), y(F+2K), \ldots\}$, the second would be $S_2 = \{y(F+1), y(F+K+1), y(F+2K+1), \ldots\}$ and so on.
   (c) For each $S_k, k = 1, \ldots, K$, obtain the values of the model parameters a_i and b_j using all the other subsets with the least squares estimation technique. Based on the estimated model parameter values and the associated prediction equations in Eq. (2), predict the value of each member of S_k.
2. For each (m, n) pair, we have obtained a prediction of the CDC time-series, $y(t)$ for $t = F_{mn}, \ldots, T$. Note that F still represents the first time index that can be predicted. However, we use the subscript mn to emphasize the fact that F varies depending on the values of m and n. By comparing the prediction with the true CDC data, we calculate the root mean-squared error (RMSE) as follows:

$$\epsilon = \sqrt{\frac{1}{T - F_{\max} + 1} \sum_t (y(t) - \hat{y}(t))^2} \tag{4}$$

The RMSE is computed over $t = F_{\max}, \ldots, T$, regardless of techniques and model orders to ensure fairness in comparison.

5.3   Cross Validation Results

We fit our model with Twitter data, Facebook data, and the combination of Twitter and Facebook data. According to the cross validation results in Table 3[1], the models corresponding to $m = 2$ and $n = 0$ have the lowest RMSE for both Twitter and Facebook. This indicates that two most recent data points are required to perform accurate prediction of influenza rates using Twitter or Facebook data. However the model corresponding to $m = 1$ and $n = 2$ for the combination of Twitter and Facebook data has the lowest RMSE among all models. Thus the model corresponding to $m = 1$ and $n = 2$ is used for accurate prediction of influenza rates and it uses most recent CDC ILI data, in addition to the two most recent OSN data points. In general, the addition of OSN data improves the prediction with past CDC data alone. For the 10-fold cross validation results presented in Table 3, for example, the AR model ($m = 1, n = 0$)

---

[1] Cross Validation Results presented for Twitter dataset differs from our previous work [2] as we disregard the scaling effect caused by creation of new Twitter accounts over time.

Table 3. Root mean squared errors from 10-fold cross validation applied to Twitter Dataset ,Facebook Dataset and combination of Twitter and Facebook Dataset. The m and n values in the table specify the model that results in the RMSE in the corresponding row and column respectively. The lowest RMSE in the table is highlighted.

	TWITTER				FACEBOOK				TWITTER + FACEBOOK			
	$n=0$	$n=1$	$n=2$	$n=3$	$n=0$	$n=1$	$n=2$	$n=3$	$n=0$	$n=1$	$n=2$	$n=3$
$m = 0$		0.3491	**0.3355**	0.3765		0.4077	**0.3651**	0.3812		0.5190	0.3449	0.4297
$m = 1$	0.6465	0.3708	0.3884	0.4175	0.6465	0.4088	0.4111	0.4061	0.6465	0.3553	**0.3108**	0.3398
$m = 2$	0.5527	0.3532	0.3665	0.4016	0.5527	0.3976	0.4015	0.4101	0.5527	0.4121	0.3675	0.3608

comprising of the $y(t - 2)$ term and the constant term for the prediction of $y(t)$ has a RMSE of 0.6465. For the same $m = 1$, the model with additional Twitter data (i.e. $n = 1$) has a lower RMSE of 0.3708, Facebook data (i.e. $n = 1$) has a lower RMSE of 0.4088 and combination of Twitter and Facebook data (i.e. $n = 1$) has a lower RMSE of 0.3553. We observe that the combination of Twitter and Facebook data provides further improvement for prediction accuracy over Twitter and Facebook alone. In this model, using OSN data ($m = 0$) alone is insufficient for prediction and the past ILI rates are critical in predicting future values, as is evident from our results. Therefore, the OSN data provides a real-time assessment of the flu epidemic (i.e. the availability of Twitter data in week t in the prediction of physician visits also in week t as shown in Eq. (2)), while the past CDC data provides the recent ILI rates in the prediction model. As shown earlier in the paper, there is strong correlation between the OSN data and the CDC data. Hence, the more timely OSN data can compensate for the lack of current CDC data and help capture the current flu trend.

Finally in Figure 6, we provide the plots for percentage weighted ILI visits, positively classified Twitter (left) and Facebook (right) users and predicted ILI rate using CDC and Twitter and Facebook for the year 2010-2011.

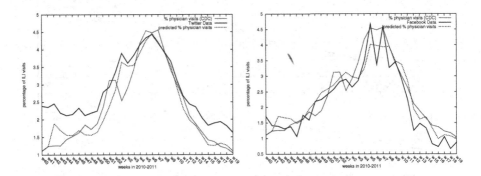

Fig. 6. Weekly plot of percentage weighted ILI visits, positively classified OSN (Twitter (Left) and Facebook (Right)) datasets and predicted ILI rate using CDC and OSN

6 Flu Prediction within Regions

We analyzed the relationship between the OSN activity and ILI rates across all geographic regions defined by the Health and Human Services (HHS) regions. For reference, the regions are shown on the USA map in Figure 3.

In studying the regional statistics, we would like to make some comparisons across regions. For instance (i) when the ILI rate peaks later in a particular region than the rest of country, do the Twitter reports also peak later, (ii) is there in relationship between the decay in ILI rates and the decay in Twitter reports.

Figure 7 shows, for both ILI (left) and Twitter (right), the relative intensity across the ten Health and Human Services (HHS) regions (columns) during successive weeks (rows) in the year 2009-2010 during which the H1N1/Swine Flu was evident.

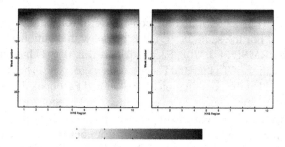

Fig. 7. Heatmap of CDC's Regionwise ILI data (left) and Twitter data (right). Colormap scale included (below).

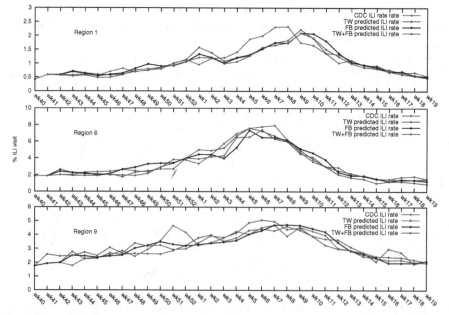

Fig. 8. Comparision between actual and predicted ILI rates for Region 1, Region 6 and Region 9

The colormap used is a scale with white representing low intensity and black, high intensity. We are comparing "trends" among the ILI and Twitter data.

Regional analysis shows that ILI seems to peak later in the Northeast (Regions 1 and 2) than in the rest of the country by at least week. The Twitter reports also follow this trend. In Region 9, Region 4 and the Northeast, the ILI rates seem to drop off fairly slowly in the weeks immediately following the peaks. This is also reflected in the Twitter reports. Approximately 20-25 weeks after the peak ILI, the northern regions have lower levels relative to the peaks in the southern regions. This is also true of the Twitter reports. The decline in ILI rates is slowest in Region 9.

Figure 8 depicts regionwise ILI prediction performance for the year 2010-2011 using our logit model. We select region 1, region 6 and region 9 to represent the regions, one each from the East, South and Western U.S. and plot the actual and predicted ILI values for each of these regions using Twitter data, Facebook data, and the combination of Twitter and Facebook data. We observe that the OSN reports and ILI rates are in fact correlated across regions and therefore corroborate our earlier findings that OSN can improve ILI rate prediction.

7   Conclusions

In this paper, we have described our approach to achieve faster, near real time prediction of the emergence and spread of influenza epidemic, through continuous tracking of flu related OSN messages originating within United States. We showed that applying text classification on the flu related messages significantly enhances the correlation between the Twitter and Facebook data and the ILI rates from CDC.

For prediction, we build an auto-regression with exogenous input (ARX) model where the ILI rate of previous weeks from CDC formed the autoregressive portion of the model, and the OSN data served as an exogenous input. Our results indicated that while previous ILI rates from CDC offered a realistic (but delayed) measure of a flu epidemic, OSN data provided a real-time assessment of the current epidemic condition and can be used to compensate for the lack of current ILI data.

We observed that the OSN data was highly correlated with the ILI rates across different HHS regions. Therefore, flu trend tracking using OSN's significantly enhances public health preparedness against the influenza epidemic and other large scale pandemics.

Acknowledgements. This research is supported in parts by the National Institutes of Health under grant 1R43LM010766-01 and National Science Foundation under grant CNS-0953620.

References

1. Achrekar, H., Gandhe, A., Lazarus, R., Yu, S.H., Liu, B.: Predicting flu trends using twitter data. In: International Workshop on Cyber-Physical Networking Systems (June 2011)
2. Achrekar, H., Gandhe, A., Lazarus, R., Yu, S.H., Liu, B.: Twitter improves seasonal influenza prediction. In: Fifth Annual International Conference on Health Informatics (February 2012)
3. Centers for Disease Control and Prevention: FluView, a weekly influenza surveillance report (2009), http://www.cdc.gov/flu/weekly

4. Chen, L., Achrekar, H., Liu, B., Lazarus, R.: Vision: towards real time epidemic vigilance through online social networks: introducing sneft – social network enabled flu trends. In: ACM Mobile Cloud Computing and Services, San Francisco, California (June 2010)
5. Culotta, A.: Detecting influenza outbreaks by analyzing twitter messages. In: Knowledge Discovery and Data Mining Workshop on Social Media Analytics (2010)
6. Espino, J., Hogan, W., Wagner, M.: Telephone triage: A timely data source for surveillance of influenza-like diseases. In: AMIA: Annual Symposium Proceedings (2003)
7. Ferguson, N.M., Cummings, D.A., Cauchemez, S., Fraser, C., Riley, S., Meeyai, A., Iamsirithaworn, S., Burke, D.S.: Strategies for containing an emerging influenza pandemic in southeast asia. Nature 437, 209–214 (2005)
8. Gauvin, W., Ribeiro, B., Towsley, D., Liu, B., Wang, J.: Measurement and gender-specific analysis of user publishing characteristics on myspace. IEEE Networks (September 2010)
9. Ginsberg, J., Mohebbi, M.H., Patel, R.S., Brammer, L., Smolinski, M.S., Brilliant, L.: Detecting influenza epidemics using search engine query data. Nature 457, 1012–1014 (2009)
10. Jansen, B., Zhang, M., Sobel, K., Chowdury, A.: Twitter power:tweets as electronic word of mouth. Journal of the American Society for Information Science and Technology 60(1532), 2169–2188 (2009)
11. Jordans, F.: WHO working on formulas to model swine flu spread (2009), http://www.physorg.com/news165686771.html
12. Lazarus, R., Kleinman, K., Dashevsky, I., Adams, C., Kludt, P., DeMaria Jr., A., Platt, R.: Use of automated ambulatory-care encounter records for detection of acute illness clusters, including potential bioterrorism events (2002), http://www.cdc.gov/ncidod/EID/vol8no8/02-0239.html
13. Leskovec, J., Backstrom, L., Kleinberg, J.: Meme-tracking and the dynamics of the news cycle. In: International Conference on Knowledge Discovery and Data Mining, Paris, France, vol. 495(978) (2009)
14. Longini, I., Nizam, A., Xu, S., Ungchusak, K., Hanshaoworakul, W., Cummings, D., Halloran, M.: Containing pandemic influenza at the source. Science 309(5737), 1083–1087 (2005)
15. Magruder, S.: Evaluation of over-the-counter pharmaceutical sales as a possible early warning indicator of human disease. Johns Hopkins University APL Technical Digest (2003)
16. Mislove, A.: Pulse of the nation: U.S. mood throughout the day inferred from twitter (2010), http://www.ccs.neu.edu/home/amislove/twittermood/
17. Motoyama, M., Meeder, B., Levchenko, K., Voelker, G.M., Savage, S.: Measuring online service availability using twitter. In: Workshop on Online Social Networks, Boston, Massachusetts, USA (2010)
18. Paul, M., Dredze, M.: You are what you tweet:analyzing twitter for public health. Association for the Advancement of Artificial Intelligence (2011)
19. Sakaki, T., Okazaki, M., Matsuo, Y.: Earthquake shakes twitter users: real-time event detection by social sensors. In: 19th International Conference on World Wide Web, Raleigh, North Carolina, USA (2010)
20. Signorini, A., Segre, A.M., Polgreen, P.M.: The use of twitter to track levels of disease activity and public concern in the U.S. during the influenza a h1n1 pandemic. PLoS ONE 6(5) (May 2011)
21. Sitaram, A., Huberman, B.A.: Predicting the future with social media. Social Computing Lab, HP Labs, Palo Alto, California, USA (2010)
22. Webb, S., Caverlee, J.: A large-scale study of myspace: Observations and implications for online social networks. Association for the Advancement for Artificial Intelligence (2008)

Clustering of Human Sleep Recordings Using a Quantile Representation of Stage Bout Durations

Chiying Wang[1], Francis W. Usher[1], Sergio A. Alvarez[2],
Carolina Ruiz[1], and Majaz Moonis[3]

[1] Department of Computer Science, Worcester Polytechnic Institute,
100 Institute Road, Worcester, MA 01609, U.S.A.
[2] Department of Computer Science, Boston College,
140 Commonwealth Avenue, Chestnut Hill, MA 02467, U.S.A.
[3] Department of Neurology, University of Massachusetts Medical School,
55 Lake Avenue North, Worcester, MA 01655, U.S.A.
alvarez@cs.bc.edu, ruiz@cs.wpi.edu

Abstract. In this paper, a condensed representation of stage bout durations based on the q-quantiles of the duration distributions is used as a basis for the discovery of duration-related patterns in human sleep data. A collection of 244 all-night hypnograms is studied. Quartiles ($q = 4$) provide a good tradeoff between representational detail and sample variation. 15 descriptive variables are obtained that correspond to the bout duration quartiles of wake after sleep onset, NREM stage 1, NREM stage 2, slow wave sleep, and REM sleep. EM clustering is used to identify distinct groups of hypnograms based on stage bout durations. Each group is shown to be characterized by bout duration quartiles of specific sleep stages, with statistically significant differences among groups ($p < 0.05$). Several sleep-related and health-related variables are shown to be significantly different among the bout duration groups found through clustering. In contrast, multivariate linear regression fails to yield good predictive models based on the same bout duration variables used in the clustering analysis. This work demonstrates that machine learning techniques are capable of uncovering naturally occurring dynamical patterns in sleep data that also provide sleep-based indicators of health.

Keywords: Sleep, Bout duration, Sleep dynamics, Data mining, Clustering, Machine learning.

1 Introduction

Sleep has been a source of fascination since antiquity. Despite much scientific attention, the process of sleep is not yet fully understood. Sleep in mammals has been thought to be controlled by body-wide mechanisms, in order to ensure energy conservation and recovery [3]. It has also been proposed that sleep may be an emergent property of the networks of neurons in the brain [22]. Sleep is known to play a key role in memory consolidation [13].

Sleep Staging. The scientific study of sleep traditionally uses a subdivision of the sleep process into distinct stages through polysomnography, which relies on the measurement of brain electrical activity through electroencephalography (EEG), supplemented

J. Gabriel et al. (Eds.): BIOSTEC 2012, CCIS 357, pp. 369–384, 2013.

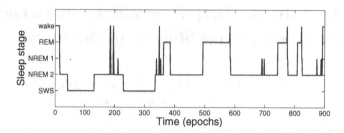

Fig. 1. Sample hypnogram from the present study

by other physiological signals [24], including heart rhythms via electrocardiography (ECG), and muscle movements via electromyography (EMG). A stage of sleep associated with dreaming, the Rapid Eye Movement (REM) stage, was subsequently identified [1], [11], leading to currently used staging standards [31], [18] that comprise the light sleep non-REM (NREM) stages NREM 1 and NREM 2, a deep sleep (slow-wave sleep (SWS)) stage or stages NREM 3/4, and REM sleep. Neuroimaging techniques, including fMRI and PET, have yielded specific information about brain activity in different regions of the brain during each of these sleep stages [9].

Sleep Architecture. Sleep normally progresses through the various stages during the course of a full night, albeit in a manner that is not predictable in detail. A sample diagram of the temporal progression of human sleep stages during the night, known as a hypnogram, is shown in Fig. 1. This particular diagram was generated from one of the 244 polysomnographic recordings used in the present paper. Some typical features to note are: most SWS (stages 3 and 4) occurs earlier in the night, a greater amount of REM sleep occurs later in the night, REM and stage 2 alternate semi-cyclically, and there are brief periods of wakefulness throughout the night, after the initial onset of sleep. Despite some common features across individuals, the detailed structure of sleep is known to vary from person to person, and is affected by a variety of factors, from such fundamental physical attributes as body composition [30] and handedness [28], to behaviors such as smoking [35] and the practice of yoga [33].

Description of Sleep Structure. Sleep structure is frequently described in terms of sleep stage composition, that is, in terms of the fraction of time accounted for by various sleep stages within a night of sleep (e.g., [10] and [20]). While sleep stage composition provides a good global summary of overall sleep stage content, it does not capture any information about the duration and ordering of uninterrupted episodes of different sleep stages during the night, which are important features of sleep architecture.

Sleep stage durations have previously been used to describe alterations in sleep dynamics due to health status or ingested neuroactive substances. For example, in [7], it is shown that sleep stage durations are affected in specific ways by age, caffeine, and hypnotic drug withdrawal. Obstructive Sleep Apnea (OSA) is shown to alter sleep stage dynamics in [27] and [4]. Differences in mean duration of stage 2 bouts between patients with fibromyalgia and normal control subjects have been described in [6]. Exponential and power-law functions have been proposed as models for the stage duration distributions [23], and the parameter values in these models have been shown to be affected by health conditions such as chronic fatigue syndrome [21].

Scope of the Present Paper. The work reported in the present paper uses a description of sleep dynamics based on the durations of continuous, uninterrupted bouts in the different sleep stages, as well as in wakefulness episodes after sleep onset. This representation captures temporal features of sleep that are not considered by standard sleep composition variables alone. The present paper is a revised and extended version of [34]. In addition to the durations of bouts in various sleep stages, it would be desirable to account for the specific stage to which a transition occurs at the end of each stage bout. However, the information in a full night hypnogram appears to be insufficient to adequately model such stage transitions [4].

In the present paper, the machine learning technique of Expectation-Maximization (EM) clustering is used to group hypnograms into families based on the distributions of their stage bout durations. Hypnograms within each family are more similar to one another, in terms of their bout duration statistics, than are hypnograms from different families. The prior work [20] also uses clustering to study sleep data, but considers only stage composition, not bout durations nor other aspects of sleep dynamics.

Each of the families found through clustering is shown to be characterized by bout duration statistics for specific sleep stages, the values of which are shown to be statistically significantly different from those of other families at the level $p < 0.05$, even after a suitable correction has been made for the magnification of type I error due to multiple statistical comparisons. The Benjamini-Hochberg framework [2] is used to bound the overall false discovery rate in a rigorous manner.

Several potentially health-related variables not involved in defining the bout duration families, such as a sensation of muscle weakness or paralysis that occurs in emotional situations, are also shown to differ significantly among the bout duration families identified through machine learning ($p < 0.05$). This is particularly noteworthy because, in contrast to machine learning, the widely used statistical technique of multivariate linear regression does not provide a good predictive model of this muscle paralysis variable based on the same bout duration variables. Our results show that machine learning can uncover interesting dynamical patterns in sleep data, and that such patterns may be used to predict selected aspects of individual patient health based on an all-night sleep study.

2 Methods

2.1 Human Sleep Data

Fully anonymized human polysomnographic recordings were obtained from the Sleep Clinic at Day Kimball Hospital in Putnam, Connecticut, USA. 244 recordings were used. In addition to the polysomnographic recordings and sleep stage information (section 2.2), health-related patient information was available through responses to a patient questionnaire. Summary statistics for the collection of sleep data are as in Table 1. The acronyms that appear in the header row of Table 1 have the following meanings. BMI: Body-Mass Index, the ratio of body weight to height-squared; ESS: Epworth Sleepiness

Table 1. summary statistics of sleep dataset

	Age (years)	BMI (kg m^{-2})	ESS (score)	BDI (score)	Mean SaO$_2$ (%)	Heart rate (bpm)
Male (n=122) $\mu \pm \sigma$	47.4±15.1	33.7±8.1	7.6±5.4	11.5±8.8	93.5±2.9	68.5±11.3
Female (n=122) $\mu \pm \sigma$	48.4±14.5	33.7±8.3	7.1±4.8	13.0±7.8	94.6±1.9	70.8±9.6
Overall (n=244) $\mu \pm \sigma$	47.9±14.8	33.7±8.2	7.4±5.1	12.2±8.3	94.1±2.5	69.7±10.5
min–max	20–85	19.2–64.6	0–23	0–48	70.2–97.9	46–99

Scale [19], a measure of daytime sleepiness based on responses to a questionnaire; BDI: Beck Depression Inventory [32], a questionnaire-based measure of affective depression; Mean SaO$_2$: mean level of oxygen-saturated hemoglobin in the blood.

2.2   Descriptive Data Features

The choice of data representation has a crucial impact on data analysis (e.g., [15]). The durations of continuous uninterrupted bouts in individual stages are natural candidates for representation of sleep stage sequences. However, their high dimensionality presents challenges. A dimensionality reduction technique based on the quantiles of the the bout duration distributions is described below. The resulting compressed data representation is used as the input to the clustering procedure (section 2.3).

Staging. The polysomnographic recordings (see section 2.1) were staged in 30-second epochs by expert sleep technicians following the Rechtschaffen and Kales (R & K) standard [31]. R & K NREM stages 3 and 4 were then combined to obtain a single slow wave sleep (SWS) stage. This combination procedure results in stage labels that provide [25] a good approximation to the more recent AASM standard [18].

Sleep Stage Bouts and Bout Durations. Next, stage bout durations in epochs were extracted from each hypnogram. A stage bout is defined to be a maximal uninterrupted segment of the given stage within a given hypnogram. For example, two distinct SWS bouts and four distinct REM bouts are visible in Fig. 1, together with a greater number of stage 2 and wake bouts. A stage bout that begins in epoch t has duration $T - t$, where T is the first epoch after t such that the sleep stages of the given hypnogram in epochs t and T are not the same. The distribution (probability mass function, or discrete probability density function (PDF)) of NREM stage 2 bout durations over the population considered in the present paper is shown in Fig. 2 (left).

Cumulative Distribution Function. Empirical probability density functions such as that shown in Fig. 2 (right) are subject to substantial sampling variation. Thus, the cumulative distribution function (CDF) is often preferred, for example in the Kolmogorov-Smirnov statistic for comparison of distributions. The CDF of stage X is the function F_X defined for each duration, d, as follows (the letter P denotes probability):

$$F_X(d) = P(\text{a bout of stage } X \text{ has duration } \leq d)$$

With an average of 250 possible bout durations per stage, this process yields a feature vector of length approximately 1250 for each data instance. The dimensionality of this raw feature space therefore exceeds the number of available instances by a factor of approximately 6, which leads to sparsely populated feature vectors.

Bout Duration Quantile Features. Selected features of the duration distributions were used to reduce the dimensionality of the data representation. Specifically, only selected quantiles (e.g., quartiles, deciles) were used to describe each stage. For a given integer $q \geq 2$, and each stage X, the q–quantiles are defined as follows. Let i be an integer in the range $i = 1, 2, \cdots q - 1$. The i-th q–quantile $X.Q^{(q)}i$ of stage X is:

$$X.Q^{(q)}i = \operatorname*{argmin}_{d}\{F_X(d) \geq \frac{1}{q}i\},$$

where F_X is the CDF of stage X. In words, the value of $X.Q^{(q)}i$ for a given set of hypnograms is the smallest d for which at least a fraction i/q of the stage X bouts in the input set have a duration of d or less. As an illustration in the case $q = 4$, the CDF of NREM stage 2 bout durations for the entire set of 244 hypnograms is shown in Fig. 2 (right), together with the compressed quartile representation, visualized as a piecewise constant approximation with jumps at the quartile durations.

Selection of the Number of Quantiles. As Fig. 2 (right) suggests, the CDF approximation error decreases as the number of quantiles, q, increases. However, the variance of the quantile estimates themselves will increase, as the number of samples available per quantile decreases with increasing q. Thus, one expects that there may be an optimal range of values for q. Experiments were therefore performed to determine how the results of the clustering technique (section 2.3) depend on q.

The mean Rand index stability value (see section 2.3) was observed to attain a maximum value at or near $q = 4$, for a number of clusters between 2 and 5. The value $q = 4$ is appealing because quartiles are easily understood. Hence, 4 quantiles were used in all subsequent work reported in the present paper. The three bout duration quartile values $X.Q^{(4)}1$, $X.Q^{(4)}2$, $X.Q^{(4)}3$ were used to describe each of the five stages, X (wake, N1, N2, SWS, REM), yielding a 15-dimensional feature vector for each instance. This data representation was used for the clustering analysis (section 2.3).

2.3 Clustering

Clustering was applied to the 15-dimensional feature vectors (section 2.2) to seek objectively defined groups of hypnograms with distinct bout duration characteristics.

Clustering Technique. The technique of Expectation-Maximization (EM) clustering was selected after an experimental comparison with k-means clustering showed higher stability of the EM clustering results with respect to pseudorandom initial parameter variation (see section 3.1). EM performs iterative maximum likelihood estimation of the cluster parameters [12,26]. Clustering experiments were carried out using the Weka data mining toolkit [16]. A mixture of Gaussians is used as the cluster model, and initial parameter values are found through k-means clustering.

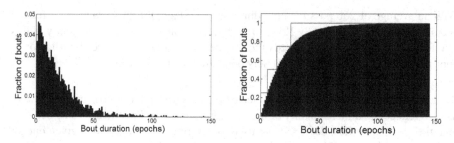

Fig. 2. Left: stage 2 bout duration PDF; right: CDF with quartiles as approximation

Measuring Clustering Stability. Since clustering parameters are initialized pseudo-randomly, the results may vary across runs. It is therefore important to gauge the variation of clustering results for different starting conditions. Stability of clustering results was assessed by comparing the clusters resulting from all pairs of 50 seed values for a given value of k. A measure of agreement of two clusterings based on the fraction of pairs of instances that are grouped together in the same cluster by each of the two clusterings, the adjusted Rand Index [17], was computed for all pairs of seed values. This index has a maximum value of 1, attained only for two identical clusterings. The adjusted Rand index of a randomly selected pair of clusterings is 0 on average. As compared with the standard Rand Index [29], the adjusted Rand Index is therefore much stricter, as it accounts for the degree of matching expected by chance. Subsequent experiments were performed with a clustering of maximum mean adjusted Rand Index.

2.4   Statistical Significance

Multiway and Pairwise Comparisons. When comparing means or medians of several populations (e.g., clusters), ANOVA or a Kruskal-Wallis test are used. Likewise, statistical significance of differences of means or medians between pairs of populations is tested by using either a t-test or Wilcoxon rank sum test, respectively. ANOVA and t-tests presuppose normality of the distribution of the means, a condition that may not hold exactly in all cases. Nearly all of the comparisons performed in the present paper involve populations with several dozen members, and the normality condition is satisfied approximately. In any case, the Kruskal-Wallis and Wilcoxon rank sum tests do not presuppose normality, and provide additional confidence regarding statistical validity. A two-sample Kolmogorov-Smirnov test is used to compare probabililty distributions without any assumptions of a particular functional form, and without targeting any particular statistic such as the mean or median.

Correction for Increased Type I Error due to Multiple Comparisons. Several of the results described are obtained through exploratory data analysis, involving the simultaneous testing of multiple statistical hypotheses. In any such situation, the risk of a type I inference error – incorrectly rejecting a null hypothesis – increases due to the accumulation of error over multiple comparisons. This issue is addressed in the present paper using the method of [2]. Given n prospective individual findings with associated

p-values $p_1 < p_2 < \cdots < p_n$, and a desired overall level of significance (p-value) p, the Benjamini-Hochberg procedure declares as significant the first k findings, where k is the largest index i, $1 \leq i \leq n$, for which $p_i i/n < p$. This approach provides rigorous control of the false discovery rate, the expected proportion of multiple null hypotheses that are incorrectly rejected due to multiple comparisons. In the present paper, control of the false discovery rate is performed at the significance level $p < 0.05$.

3   Results

This section describes the results of clustering that were obtained using the hypnogram data of section 2.1 represented in terms of the quartiles of the stage bout duration distributions (section 2.2, $q = 4$), utilizing EM clustering (section 2.3). In passing, we note that variants of the bout duration quartile data representation that use more than 4 quantiles were also considered for the present work. The advantage of using a greater number of quantiles is the ability to describe finer details in the bout duration distributions. However, clustering stability was considerably lower with such representations, and so the decision was made to use quartiles only (see section 2.2).

3.1   Clustering Stability

The mean observed value of the adjusted Rand Index (section 2.3) for EM is at least 0.87 for the values $k = 2, 3, 4$. The high values of the adjusted Rand Index show that the EM clustering is only slightly influenced by the initial parameter values, and represents a stable grouping of the hypnograms. In contrast, the adjusted Rand index of k-means stays between 0.36 and 0.53 over the range $k = 2, 3, 4$. For this reason, EM was selected as the clustering algorithm for the work discussed in the present paper. The seed value 8 was found to provide an EM clustering of maximum mean adjusted Rand Index as compared to the other 49 seed values considered, for each $k = 2, 3, 4$. All results discussed subsequently in this paper utilize the EM clustering resulting from the seed value 8.

3.2   Cluster Separation

Visualization of Cluster Separation. The visualization technique of multidimensional scaling (MDS) provides a low-dimensional nonlinear projection of a dataset in a way that minimizes distortion of the distances between pairs of data instances [5]. Fig. 3 shows a two-dimensional MDS projection of the set of data instances used in the present paper. The results of EM clustering did not enter into the generation of the MDS projection itself. The EM cluster labels for $k = 3$ were used to determine the glyph (marker) for each instance in the visualization shown. The MDS results show only moderate separation among the EM clusters in two dimensions, which indicates that more than two variables are likely to be needed in order to achieve high separation. See section 3.2.

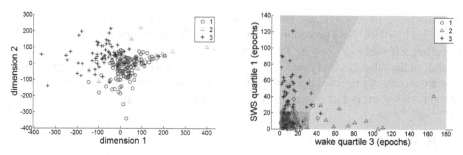

Fig. 3. Left: MDS view of clusters; right: LDA boundaries in W.Q3, SWS.Q1 space

Measurement of Cluster Separation via Classification

Classification Based on All Duration Quartile Variables. Separation among clusters was further assessed quantitatively by performing a classification task in which the EM cluster labels are viewed as the target class attribute, with the variables used for clustering used as predictive attributes. Classification accuracy, the fraction of instances for which the cluster label is correctly predicted, and the area under the Receiving Operating Characteristic (ROC) plot [14], remain consistently above 0.80 in the cases $k = 2, 3, 4$ for widely used classification techniques including C4.5 (J48) decision tree learning, naïve Bayes, and multilayer artificial neural networks (ANN). The area under the ROC plot accounts for prediction errors on a per-class basis, and is a better measure of classification performance in this context because the class (cluster) sizes are very dissimilar. Accuracy can produce overly optimistic results in such situations. Mean values of the area under the ROC plot for selected classifiers appear in Table 2. A 4-fold cross-validation protocol was employed to control variance due to data sampling.

Table 2. AUC selected classifiers

classifier	$k = 2$	$k = 3$	$k = 4$
ANN	0.94	0.97	0.91
J48	0.88	0.90	0.89
naive Bayes	0.99	0.98	0.98

Table 3. Bout duration cluster sizes

$k = 2$	$k = 3$	$k = 4$
$\{211, 33\}$	$\{148, 19, 77\}$	$\{127, 15, 48, 54\}$

Classification Based on a Single Pair of Duration Quartile Variables. Observed cluster separation is fair in two-dimensional projections of the bout duration dataset in terms of the bout duration clustering variables, as expected based on the MDS visualization in Fig. 3. An example involves the wake.Q3 and SWS.Q1 bout duration quartile variables. As observed in Fig. 3, there is considerable overlap among the clusters near the bottom left corner. Fig. 3 also shows sample decision boundaries in this reduced two-dimensional feature space using a linear discriminant analysis (LDA) classifier.

Classification Rule Description of Clusters. Use of the rule induction algorithm RIPPER [8] (JRIP) over the wake.Q3 and SWS.Q1 predictive variables alone, with the $k = 3$ cluster label as the class, yields, after pruning and simplification, the classification rules shown in Fig. 4. The final rule is a default rule that is used when the other

```
(wake.Q3 >= 50) => cluster=cluster2 (10.0/0.0)
(SWS.Q1 >= 30) => cluster=cluster3 (41.0/2.0)
(SWS.Q1 >= 17) and (6 >= wake.Q3 >= 4) => cluster=cluster3 (13.0/3.0)
=> cluster=cluster1 (180.0/37.0)
```

Fig. 4. JRIP rules for 3 clusters using wake.Q3 and SWS.Q1 only. Coverage/errors in parentheses

rules do not apply. This particular model attains an accuracy of 0.77 and a mean area under the ROC plot of 0.76. Although the classification performance of the model in Fig. 4 is unremarkable, it provides an easily understood rough description of the clusters in the case $k = 3$. In particular, it suggests that cluster 2 is associated with high wake bout durations. This is consistent with Fig. 3. More detailed characterizations of the various clusters are discussed in section 3.3.

3.3   Statistical Properties of the Bout Duration Clusters

Cluster Sizes and Membership. The sizes of the EM bout duration clusters appear in Table 3. Relationships among the three families of clusterings $k = 2, 3, 4$ are derived from the detailed membership lists of the various clusters. A simplified description of the relationships among clusterings for different values of k is the following. Additional characteristics of individual clusters are given in Table 4 and section 3.3.

Relationships between $k = 2$ and $k = 3$ Clusters. Cluster 1 in the $k = 2$ family splits into the two $k = 3$ clusters labeled 1 and 3. As discussed in section 3.3 below, the $k = 3$ cluster 3 portion is characterized by higher SWS bout duration quartiles than the $k = 3$ cluster 1 portion. Two-thirds of the $k = 2$ cluster 2 – with higher wake and lower SWS and REM bout duration quartiles – retains its identity in the $k = 3$ family; the remaining one-third of the $k = 2$ cluster 2 joins the $k = 3$ cluster 3. The only inaccuracy in this description is that 3 of the 33 instances in the $k = 2$ cluster 2 join the $k = 3$ cluster 1.

Relationships between $k = 3$ and $k = 4$ Clusters. In the transition between $k = 3$ and $k = 4$, cluster 1 remains largely unchanged (with only 12 of 148 instances leaving cluster 1 and joining cluster 4). Cluster 2 remains mainly within cluster 2 (with 4 of 19 instances joining cluster 4, which has higher mean REM bout duration quartiles than cluster 2; see section 3.3). Two-thirds of the $k = 3$ cluster 3 joins the $k = 4$ cluster 2, and the remaining one-third of $k = 3$ cluster 3 remains within the $k = 4$ cluster 3 (8 instances join $k = 4$ cluster 4). However, the $k = 4$ cluster 3 retains the characteristic, shared with $k = 3$ cluster 3, of having the highest SWS bout duration quartiles among clusters.

Cluster Bout Duration Summary Statistics. The mean, standard deviation, median, and mean absolute deviation of the 15 descriptive variables were computed for each of the EM clusters, with a view toward establishing statistical differences among clusters. Table 4 provides numerical values of the bout duration quartile means of the different clusters for $k = 2, 3, 4$. Fig. 5, 6, 7 show the mean values of the 15 clustering variables in the cases $k = 2, 3, 4$, respectively. These figures suggest that each cluster is characterized by different bout duration quartiles for one or more of the sleep stages than the other clusters (e.g., cluster 2 by higher wake duration quartiles).

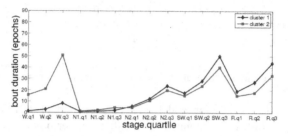

Fig. 5. Mean values of clustering variables, $k = 2$

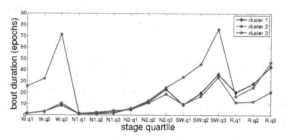

Fig. 6. Mean values of clustering variables, $k = 3$

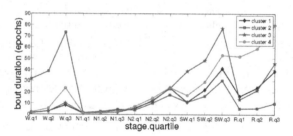

Fig. 7. Mean values of clustering variables, $k = 4$

Multiway Cluster Comparisons. Statistical significance of the observed differences in the means among clusters was assessed by ANOVA and Kruskal-Wallis tests for multiway comparisons, and by t, and Wilcoxon rank sum tests for pairwise comparisons. A nonparametric two-sample Kolmogorov-Smirnov test was also used to determine differences between pairs of clusters in the overall distributions of the bout duration quartile variables. All p-values were corrected for multiple comparisons using the Benjamini-Hochberg method on a per clustering basis, so that reported p-values are upper bounds on the false discovery rate relative to all findings over the given family of k clusters.

Stage Bout Duration Quartile Means. The results are as follows. For all clustering families, $k = 2, 3, 4$, the mean values of the wake stage and stage NREM1 duration quartile variables differ significantly among clusters in a multiway comparison using the Kruskal-Wallis test ($p < 0.05$). ANOVA results are in agreement with Kruskal-Wallis, with the exception that the inter-cluster difference in the quartile variable REM.Q1 is

Table 4. mean bout duration quartiles of different clusters, in epochs

	wake			N1			N2			SWS			REM		
	Q1	Q2	Q3	Q1	Q2	Q3	Q1	Q2	Q3	Q1	Q2	Q3	Q1	Q2	Q3
$k = 2$ cluster 1 (n=211)	1.4	2.9	8.4	1.1	1.4	2.2	5.9	12	24	17	28	50	19	27	44
$k = 2$ cluster 2 (n=33)	16	21	51	1.5	2.5	4.6	4.6	11	20	15	24	40	15	18	33
$k = 3$ cluster 1 (n=148)	1.5	3.1	8.6	1.0	1.3	2.0	5.8	12	23	9.4	20	36	20	28	43
$k = 3$ cluster 2 (n=19)	26	32	71	1.5	2.5	4.1	4.8	11	19	9.7	17	34	12	12	21
$k = 3$ cluster 3 (n=77)	1.4	3	10	1.2	1.8	3.3	5.8	13	25	34	45	76	17	25	47
$k = 4$ cluster 1 (n=157)	1.4	3.0	8.2	1.1	1.4	2.1	5.8	12.0	24	11	22	41	16	24	39
$k = 4$ cluster 2 (n=15)	32	39	73	1.6	2.7	4.6	4.1	10	18	11	16	30	5.3	5.6	9.9
$k = 4$ cluster 3 (n=48)	1.3	2.8	10	1.2	1.9	3.7	5.1	13	23	39	48	77	13	22	46
$k = 4$ cluster 4 (n=24)	2.2	5.4	23	1.1	1.5	2.3	7.2	15	24	17	29	53	52	59	80

not found to be significant for $k = 2, 3$. Additionally, both Kruskal-Wallis and ANOVA find highly significant ($p < 10^{-6}$) differences among clusters in the SWS bout duration quartile variables for $k = 3, 4$. In contrast, the differences in the stage NREM2 bout duration quartiles among clusters are not found to be significant for any of the clustering families, $k = 2, 3, 4$. Pairwise statistical comparisons provide additional information, and are discussed in section 3.3 below.

Pairwise Comparisons. Bout Duration Characteristics of Individual Clusters. The following are a few noteworthy statistically significant differences in bout durations. The reader is also referred to Table 4, and Fig. 5, 6, 7 in conjunction with this discussion. Below, the precise family (value of k) is omitted when bout duration characteristics of a given cluster number are qualitatively similar for different values of k.

Cluster 1. Clusters 1 and 3 share the property that their median wake bout duration quartiles are significantly lower than for clusters 2 and 4 (Wilcoxon $p < 0.05$). On the other hand, cluster 1 has significantly lower SWS bout duration quartiles than cluster 3. See Fig. 5, Fig. 6, and Fig. 7. The bout duration characteristics of cluster 1 are remarkably stable across values of k.

Cluster 2. Cluster 2 consistently has significantly higher wake bout duration quartiles than any other cluster, for $k = 2, 3, 4$ (Wilcoxon $p < 0.02$). The single exception is the variable wake.Q1 in the case $k = 4$. Low sample sizes for $k = 4$ clusters 2 and 4 (15 and 24, respectively) likely contribute to the latter isolated nonsignificance finding. One also observes that, in the progression from $k = 2$ to $k = 3$ to $k = 4$, cluster 2 has monotonically decreasing REM bout duration quartiles.

Cluster 3. As observed in section 3.3, 145 of the 148 instances (approximately 98%) in the $k = 3$ version of cluster 3 belong to the $k = 2$ version of cluster 1. The remainder of the $k = 2$ cluster 1 instances form the majority of the $k = 3$ cluster 3. Therefore, it is not surprising that many of the bout duration quartiles for the $k = 3$ version of cluster 3 are similar to those for cluster 1. See Fig. 6 and Table 4. However, there is an immediately noticeable difference between clusters 1 and 3 for $k = 3$, namely the fact that cluster 3 has visibly higher SWS bout duration quartiles than all other clusters, including cluster

```
(wake.Q3 >= 44) => cluster=cluster2 (12.0/1.0)
(wake.Q2 >= 6) and (SWS.Q1 <= 5) => cluster=cluster2 (5.0/1.0)
(NREM1.Q2 >= 3) and (NREM2.Q2 <= 9) => cluster=cluster2 (5.0/1.0)
(SWS.Q1 >= 30) => cluster=cluster3 (41.0/2.0)
(NREM1.Q3 >= 4) => cluster=cluster3 (22.0/1.0)
(SWS.Q3 >= 79) => cluster=cluster3 (15.0/3.0)
(SWS.Q2 >= 49) => cluster=cluster3 (5.0/2.0)
=> cluster=cluster1 (139.0/0.0)
```

Fig. 8. JRIP conjunctive rule model of the clusters for $k = 3$

1. In other words, cluster 3 for $k = 3$ consists mainly of those $k = 2$ cluster 1 instances with higher SWS bout duration quartiles. This high SWS bout duration description of cluster 3 persists for $k = 4$. However, the observed SWS quartile bout durations for cluster 3, though highest among all clusters, are not significantly higher than those of clusters 2 and 4, again due likely to the small sizes of the latter clusters.

Cluster 4. Cluster 4 is characterized by significantly higher REM bout quartile durations than any other cluster (Wilcoxon $p < 10^{-3}$).

Clustering Description via Classification Rules. One can compare the characterizations of the clusters described in the preceding paragraphs with the model constructed by the JRIP conjunctive rule classifier in the case $k = 3$. The model is as shown in Fig. 8, and achieves a classification accuracy of 0.86 and mean ROC area of 0.88. The rules of this model closely agree with the descriptions provided above.

3.4   Health-Related Cluster Differences

Comparisons of Sleep-Related and Health-Related Variables. The bout duration clusters identified by the EM procedure were examined to determine differences among them in the values of sleep-related and health-related variables not used in the clustering procedure itself. Group comparisons of means and medians were performed using ANOVA and Kruskal-Wallis tests, respectively. Pairwise comparisons of means and medians used a t-test and Wilcoxon rank sum test.

Sleep Latency. For all values of $k = 2, 3, 4$, Kruskal-Wallis and ANOVA determined that mean sleep latency (time elapsed from getting in bed until first non-wake epoch) differs significantly among bout duration clusters ($p < 0.05$). The highest mean value of sleep latency occurs in cluster 2. The pairwise difference in mean and median sleep latency between cluster 2 and all other clusters is also significant ($p < 0.05$). As observed in Table 4 and discussed in section 3.3, cluster 2 has the highest mean wake bout duration quartiles of all of the clusters. It is entirely possible that the high sleep latency contributes to the increased wake bout durations in cluster 2.

Sleep Questionnaire Variables. Certain variables that correspond to items in the Epworth Daytime Sleepiness questionnaire are significantly different among clusters, and are significantly different in pairwise comparisons between cluster 2 and the others:

```
paralysis =
  -0.015 wake.Q1 + 0.012 wake.Q2 + 0.0037 wake.Q3
  -0.22 NREM1.Q1 + 0.07 NREM1.Q3
  +0.03 NREM2.Q1 + 0.018 NREM2.Q2
  -0.0045 REM.Q1 - 0.012
```

Fig. 9. Least squares linear regression model of paralysis ($r^2 < 0.01$)

a sensation of muscular weakness or paralysis during laughter, anger, or emotional situations, and the recollection of vivid dreams and nightmares, differ significantly among clusters for $k = 2, 3$, and are highest in cluster 2 for $k = 2, 3$ ($p < 0.05$); an uncomfortable crawly sensation in the legs that is relieved by walking differs significantly ($p < 0.05$) among clusters for $k = 3, 4$, and is lowest in cluster 2.

Comparison with Multivariate Linear Regression. Given the significant differences in health variables in section 3.4, it is natural to ask if linear regression can provide good predictions of one of these variables, such as a muscle weakness or paralysis in emotional situations, based on bout duration statistics. For $k = 3$, least squares linear regression yields the model in Fig. 9 (coefficients to two significant digits).

Although terms involving wake bout duration quartiles, which as discussed in section 3.3 differentiate cluster 2 from the others, and in which paralysis attains its maximum value as discussed in section 3.4, appear in the regression model of Fig. 9, the linear correlation between paralysis and the predictions of the least squares linear regression model is less than 0.06. Thus, this model explains a fraction that is less than 0.06^2, much less than 1%, of the variance in paralysis. Nonlinear predictive models obtained through regression based on the machine learning technique of Support Vector Machines (SVM) provide slightly improved performance here. In any case, the fact that paralysis differs significantly among the bout duration-based groupings found through clustering, shows that machine learning can uncover structure in health-related data that is not clearly identified by traditional statistical techniques such as linear regression.

4   Conclusions and Future Work

The durations of maximal uninterrupted periods in a given sleep stage are important in the description of sleep structure. This paper has applied unsupervised machine learning to the discovery of patterns in human sleep data based on stage bout durations, utilizing a compressed representation in terms of the quantiles of the stage bout duration distributions. The results identify groups of hypnograms with statistically distinct differences in bout durations among groups ($p < 0.05$), even after a Benjamini-Hochberg correction for increased type I error due to multiple comparisons. Each group is characterized by bout duration features for specific sleep stages.

Sleep latency, a variable not among those used for clustering, is also shown to differ significantly among the bout duration groups. Significant differences are also found for several variables corresponding to items on the Epworth Daytime Sleepiness questionnaire, such as muscular weakness or paralysis associated with emotional situations, the recollection of vivid dreams or nightmares after waking, and an uncomfortable "crawly" sensation in the legs that is relieved by walking. It is found that these variables are significantly different in the bout duration group characterized by the highest mean duration of wake bouts. This finding provides a specific manner in which sleep dynamics reflects the values of variables that are not specific to sleep. It is noted that multivariate linear regression captures less than 1% of the variance associated with the muscular paralysis variable. Thus, machine learning provides access to findings beyond those available with more traditional statistical tools in this case.

The results presented in this paper are based on a highly compressed representation of the bout duration distributions, utilizing only the three quartile values of the cumulative bout duration distribution for each sleep stage. It is possible that this compression limits the capacity of the clustering technique to identify important dynamical features. Increasing the number of quantiles provides greater representational accuracy, but was found to also reduce stability of the clustering results. Future work should investigate alternative representations of sleep dynamical information that simultaneously provide important detail in the distributions and stability of the machine learning results.

A limitation of the current work is that it only considers the duration of each stage bout, without regard for what stage occurs immediately afterwards. It would be desirable to consider stage transitions. Work in progress by the authors examines the use of Markov-type variants for this purpose. A major obstacle in this direction is the sparsity of stage transitions available within a single night of sleep. More accurate modeling of the sleep stage transition statistics will require the use of multiple nights' sleep data, or ambulatory monitoring of physiological signals over extended periods of time.

References

1. Aserinsky, E., Kleitman, N.: Regularly occurring periods of eye motility, and concomitant phenomena, during sleep. Science 118(3062), 273–274 (1953)
2. Benjamini, Y., Hochberg, Y.: Controlling the false discovery rate: a practical and powerful approach to multiple testing. J. Royal Statistical Soc., Series B 57(1), 289–300 (1995)
3. Berger, R.J., Phillips, N.H.: Energy conservation and sleep. Behav. Brain Res. 69(1-2), 65–73 (1995)
4. Bianchi, M.T., Cash, S.S., Mietus, J., Peng, C.-K., Thomas, R.: Obstructive sleep apnea alters sleep stage transition dynamics. PLoS ONE 5(6), e11356 (2010)
5. Borg, I., Groenen, P.J.F.: Modern Multidimensional Scaling: Theory and Applications, 2nd edn. Springer Series in Statistics. Springer, Berlin (2005)
6. Burns, J.W., Crofford, L.J., Chervin, R.D.: Sleep stage dynamics in fibromyalgia patients and controls. Sleep Medicine 9(6), 689–696 (2008)
7. Březinová, V.: Duration of EEG sleep stages in different types of disturbed night sleep. Postgrad. Med. J. 52(603), 34–36 (1976)
8. Cohen, W.W.: Fast effective rule induction. In: Twelfth International Conference on Machine Learning, pp. 115–123. Morgan Kaufmann (1995)
9. Dang-Vu, T.T., Schabus, M., Desseilles, M., Sterpenich, V., Bonjean, M., Maquet, P.: Functional neuroimaging insights into physiology of human sleep. Sleep 33(12), 1589–1603 (2010)

10. Danker-Hopfe, H., Schäfer, M., Dorn, H., Anderer, P., Saletu, B., Gruber, G., Zeitlhofer, J., Kunz, D., Barbanoj, M.-J., Himanen, S., Kemp, B., Penzel, T., Röschke, J., Dorffner, G.: Percentile reference charts for selected sleep parameters for 20- to 80-year-old healthy subjects from the SIESTA database. Somnologie - Schlafforschung und Schlafmedizin 9, 3–14 (2005), doi:10.1111/j.1439-054X.2004.00038.x

11. Dement, W., Kleitman, N.: The relation of eye movements during sleep to dream activity: An objective method for the study of dreaming. Journal of Experimental Psychology 53, 339–346 (1957)

12. Dempster, A.P., Laird, N.M., Rubin, D.B.: Maximum likelihood from incomplete data via the EM algorithm. Journal of the Royal Statistical Society, Series B 39(1), 1–38 (1977)

13. Diekelmann, S., Born, J.: The memory function of sleep. Nat. Rev. Neurosci. 11(2), 114–126 (2010)

14. Fawcett, T.: ROC graphs: Notes and practical considerations for data mining researchers. Hewlett-Packard Labs Technical Report HPL-2003-4 (2003)

15. Guyon, I., Elisseeff, A.: An introduction to variable and feature selection. Journal of Machine Learning Research 3, 1157–1182 (2003)

16. Hall, M., Frank, E., Holmes, G., Pfahringer, B., Reutemann, P., Witten, I.H.: The WEKA data mining software: an update. SIGKDD Explor. 11(1), 10–18 (2009)

17. Hubert, L., Arabie, P.: Comparing partitions. Journal of Classification 2, 193–218 (1985), doi:10.1007/BF01908075

18. Iber, C., Ancoli-Israel, S., Chesson, A.L., Quan, S.F.: The AASM Manual for the Scoring of Sleep and Associated Events: Rules, Terminology, and Technical Specifications. American Academy of Sleep Medicine, Westchester (2007)

19. Johns, M.W.: A new method for measuring daytime sleepiness: the Epworth sleepiness scale. Sleep 14(6), 540–545 (1991)

20. Khasawneh, A., Alvarez, S.A., Ruiz, C., Misra, S., Moonis, M.: EEG and ECG characteristics of human sleep composition types. In: Traver, V., Fred, A., Filipe, J., Gamboa, H. (eds.) Proc. HEALTHINF 2011 (BIOSTEC 2011), pp. 97–106. SciTePress (January 2011)

21. Kishi, A., Struzik, Z.R., Natelson, B.H., Togo, F., Yamamoto, Y.: Dynamics of sleep stage transitions in healthy humans and patients with chronic fatigue syndrome. Am. J. Physiol. Regul. Integr. Comp. Physiol. 294(6), R1980–R1987 (2008)

22. Krueger, J.M., Rector, D.M., Roy, S., Van Dongen, H.P.A., Belenky, G., Panksepp, J.: Sleep as a fundamental property of neuronal assemblies. Nat. Rev. Neurosci. 9(12), 910–919 (2008)

23. Lo, C.-C., Nunes Amaral, L.A., Havlin, S., Ivanov, P.C., Penzel, T., Peter, J.-H., Stanley, H.E.: Dynamics of sleep-wake transitions during sleep. Europhys. Lett. 57(5), 625–631 (2002)

24. Loomis, A.L., Harvey, E.N., Hobart, G.A.: Cerebral states during sleep, as studied by human brain potentials. J. Experimental Psychology 21(2), 127–144 (1937)

25. Moser, D., Anderer, P., Gruber, G., Parapatics, S., Loretz, E., Boeck, M., Kloesch, G., Heller, E., Schmidt, A., Danker-Hopfe, H., Saletu, B., Zeitlhofer, J., Dorffner, G.: Sleep classification according to AASM and Rechtschaffen & Kales: Effects on sleep scoring parameters. Sleep 32(2), 139–149 (2009)

26. Neal, R., Hinton, G.E.: A view of the EM algorithm that justifies incremental, sparse, and other variants. In: Learning in Graphical Models, pp. 355–368. Kluwer (1998)

27. Penzel, T., Kantelhardt, J.W., Lo, C.-C., Voigt, K., Vogelmeier, C.F.: Dynamics of heart rate and sleep stages in normals and patients with sleep apnea. Neuropsychopharmacology 28(S1), S48–S53 (2003)

28. Propper, R.E., Christman, S.D., Olejarz, S.: Home-recorded sleep architecture as a function of handedness II: Consistent right-versus consistent left-handers. J. Nerv. Ment. Dis. 195(8), 689–692 (2007)

29. Rand, W.M.: Objective criteria for the evaluation of clustering methods. Journal of the American Statistical Association 66(336), 846–850 (1971)

30. Rao, M.N., Blackwell, T., Redline, S., Stefanick, M.L., Ancoli-Israel, S., Stone, K.L.: Association between sleep architecture and measures of body composition. Sleep 32(4), 483–490 (2009)

31. Rechtschaffen, A., Kales, A. (eds.): A Manual of Standardized Terminology, Techniques, and Scoring System for Sleep Stages of Human Subjects. US Department of Health, Education, and Welfare Public Health Service – NIH/NIND (1968)

32. Storch, E.A., Roberti, J.W., Roth, D.A.: Factor structure, concurrent validity, and internal consistency of the Beck depression inventory–second edition in a sample of college students. Depression and Anxiety 19(3), 187–189 (2004)

33. Sulekha, S., Thennarasu, K., Vedamurthachar, A., Raju, T.R., Kutty, B.M.: Evaluation of sleep architecture in practitioners of Sudarshan Kriya yoga and Vipassana meditation. Sleep and Biological Rhythms 4(3), 207–214 (2006)

34. Usher, F.W., Wang, C., Alvarez, S.A., Ruiz, C., Misra, S., Moonis, M.: Machine learning of human sleep patterns based on stage bout durations. In: Conchon, E., Correia, C., Fred, A., Gamboa, H. (eds.) Proc. HEALTHINF 2012 (BIOSTEC 2012), pp. 71–80. SciTePress (February 2012)

35. Zhang, L., Samet, J., Caffo, B., Punjabi, N.M.: Cigarette smoking and nocturnal sleep architecture. Am. J. Epidemiol. 164(6), 529–537 (2006)

Touch and Speech: Multimodal Interaction for Elderly Persons

Cui Jian[1], Hui Shi[1], Frank Schafmeister[2], Carsten Rachuy[1], Nadine Sasse[2],
Holger Schmidt[3], Volker Hoemberg[4], and Nicole von Steinbüchel[2]

[1] SFB/TR8 Spatial Cognition, Universität Bremen, Germany
{ken,shi,rachuy}@informatik.uni-bremen.de
[2] Medical Psychology and Medical Sociology,
University Medical Center Göttingen, Germany
{frank-schafmeister,n.sasse,
nvsteinbuechel}@med.uni-goettingen.de
[3] Neurology, University Medical Center Göttingen, Germany
h.schmidt@med.uni-goettingen.de
[4] SRH Health Centre Bad Wimpfen, Germany
mueller-hoemberg@t-online.de

Abstract. This paper reports our work on the development and evaluation of a multimodal interactive guidance system for navigating elderly persons in hospital environments. A list of design guidelines has been proposed and implemented in our system, addressing the needs of designing a multimodal interfaces for elderly persons. Meanwhile, the central component of an interactive system, the dialogue manager, has been developed according to a unified dialogue modelling method, which combines the conventional recursive transition network based generalized dialogue models and the classic agent-based dialogue theory, and supported by a formal language based development toolkit. In order to evaluate the minutely developed multimodal interactive system, the touch and speech input modalities of the current system were evaluated by an elaborated experimental study with altogether 31 elderly. The overall positive results on the effectiveness, efficiency and user satisfaction of both modalities confirm our proposed guidelines, approaches and frameworks on interactive system development. Despite the slightly different results, there is no significant evidence for one preferred modality. Thus, further study of their combination is considered necessary.

Keywords: Multimodal interaction, Elderly-centered system design, Human-computer interaction, Spoken dialogue systems, Formal methods.

1 Introduction

Multimodal interfaces is gaining more and more importance for its promising possibility to achieve a significantly more effective and efficient human computer interaction (cf. [1]), they also increase users' satisfaction and provide a more natural and intuitive way of interaction (cf. [2]). Meanwhile, due to the demographic development

J. Gabriel et al. (Eds.): BIOSTEC 2012, CCIS 357, pp. 385–400, 2013.
© Springer-Verlag Berlin Heidelberg 2013

towards increasingly more elderly persons, there rises a growing research focus on the ultimodal communication technology, which aims at enhancing the quality of interaction by taking age-related decline into account (cf. [3]).

Elderly people often suffer from decline of sensory, perceptual, motor and cognitive abilities. Considering these facts we first present a list of elaborated design guidelines regarding basic design principles of conventional interactive systems and the most common elderly-centered characteristics. Meanwhile, in order to achieve a flexible and context-sensitive, yet formally tractable and controllable interaction, we designed a unified dialogue modelling approach, which combines a finite state based generalized dialogue model and the classic agent based dialogue model, and implemented this by a formal language based development framework. According to the proposed design guidelines and the unified dialogue modelling approach, an interactive guidance system was especially designed and developed for the elderly. To evaluate the touch input and natural spoken language modalities with respect to their feasibility and acceptance by elderly persons, an empirical study was conducted with 31 older participants. The general framework PARADISE [4] has been applied in our evaluation process. The study also aimed at the evaluation of the multimodal interactive guidance system as a whole, while regarding the essential criteria of the following aspects for interaction: the effectiveness of task success, the efficiency of executing tasks and the user satisfaction with the system.

The following text is organized as follows: section 2 presents the proposed general guidelines for designing multimodal interactive system for elderly persons; section 3 introduces the unified dialogue modelling approach which combines the classic agent based approach and the recursive transition network based theory for building the discourse management of the multimodal interaction; section 4 then describes the multimodal interactive guidance system, which is developed based on the unified dialogue modelling approach and the presented set of design guidelines; section 5 describes the experimental study, and the results are analyzed and discussed in section 6. Finally, section 7 concludes and gives an outline of the future work.

2   Design Guidelines of Multimodal Interactive Systems for Elderly Persons

[5] indicated that the decline of elderly persons should be considered while designing interactive systems for the elderly. Therefore, we defined a set of design guidelines for multimodal interaction with respect to the decline of seven very important abilities. They are implemented and integrated into our multimodal interactive guidance system and tested in an empirical pilot study. The results are described in [6] and the improved guidelines are presented as follows:

2.1   Visual Perception

Visual perception declines for most people with age (cf. [7]) in different ways: many people find it more difficult to focus on objects up close and to see fine details; the size of the visual field is decreasing and the peripheral vision is successively declining. Rich colors and complex shapes are making perception difficult. Rapidly moving objects are either causing too much distraction, or becoming less noticeable. To cope with these impairments, the following guidelines should be taken into account:

- Layouts of the user interface should be devised as simple and clear as possible, with few (if any) or no overlapping items.
- All texts should be large enough to be readable on the communicating interfaces.
- Strong contrast should be used with as few colors as possible; this also applies to simple and easily recognizable shape designs.
- Unnecessary and irrelevant visual effects should be avoided.

2.2   Speech Ability

Elderly persons need more time to produce complex words or longer sentences, probably due to reduced motor control of tongue and lips (cf. [8]). Furthermore, speech-related adaptation is necessary to improve the interaction quality to a sufficient level (cf. [9]). Based on these, the following aspects should be taken into account:

- Special acoustic models for the elderly should be used for speech recognizer.
- Vocabulary should be built with more definite articles, auxiliaries, first person pronouns and lexical items related to social interaction. Texts should be as simple as possible.
- Dialogue strategies should be able to cope with elderly specific needs such as repeating, helping and social interaction, etc.

2.3   Auditory Perception

Hearing ability declines at least to 75% after 75 year olds (cf. [10]). High pitched sounds are increasingly lost, as well as long and complex sentences becoming difficult to follow (cf. [11]). Therefore special attention should be paid to the following:

- Text displays can help when information is mis- or not heard, which should not provide conflicting information.
- Synthesized texts should be intensively revised regarding style, vocabulary, length and sentence structures suitable for elderly.
- Low pitched voices are more acceptable for speech synthesis, e.g., female voices are less preferred than male ones.

2.4   Motor Ability

Computer mice are unsuitable for many elderly due to the lack of good hand-eye coordination and decline of fine motor abilities (cf. [12]). Positioning the cursor is difficult if the target is too small or too irregular to locate, and fine movements are harder to control (cf. [13]).Thus, the following procedures are suggested:

- Direct interaction is recommended, e.g., touch screen.
- All GUI items should be accessibly shaped, sized and well spaced from each other.

- Simple movements are recommended, such as clicking instead of dragging.
- Text input should be avoided or replaced with other simpler input actions.
- An undo function is needed to correct errors.
- Simultaneous multimodal input such as the combination of speech and other input should be avoided or replaced.

2.5   Attention and Concentration

Elderly persons are more easily distracted by details or noise (cf. [14]). They show great difficulty maintaining divided attention, where attention must be paid to more than one aspect at a time (cf. [15]). Therefore, the following points are suggested:

- Only relevant images should be used.
- Items should not be displayed simultaneously.
- Unified or similar fonts, colors and sizes of displayed texts are recommended.
- Changes on the user interface should be emphasized in an obvious way.

2.6   Memory Functionalities

Different memory functions decline at different degrees during ageing. Short term memory holds fewer items while ageing and declines earlier; also more time is needed to process information (cf. [16]). Working memory also becomes less efficient (cf. [17]). Semantic information is believed to be preserved in long term memory for a longer period(cf. [18]). To compensate the decline of the different memory functions, the following points are suggested:

- Pure image items should be avoided or placed near relevant key words.
- Presented items in a sequence should not exceed five, the average maximum capacity of short term memory of elderly persons.
- Information should be categorized to assist storage into long term memory.
- Context sensitive information is necessary to facilitate working memory activities.

2.7   Intellectual Ability

Fluid intelligence does decline with ageing, while, crystallized intelligence does not or to a less extent (cf. [19]); it can assist elderly people to perform better in a stable well-known interface environment. Thus, we suggest assuring the following points:

- Unified interface layout, where changes should only happen on data level.
- Semantically intuitive structure, where users should not be too surprised while traversing the interaction levels.
- Consistent interaction style facilitates learning and assist elderly to master interface use.

3   A Formal Unified Dialogue Modelling Approach

As a typical recursive network based approach, generalized dialogue models were developed by constructing dialogue structures at the illocutionary level (cf. [20]). However, it is criticized for its inflexibility of dealing with dynamic information exchange. Meanwhile, information state update based theories were deemed the most successful foundation of agent based dialogue approaches (cf. [21]), which provides a powerful mechanism to handle dynamic information and gains a context sensitive dialogue management. Nevertheless, such models are usually very difficult to manage and extend (cf. [22]).

Thus, a unified dialogue modelling approach was developed. It combines the generalized dialogue models with information state updated based theories. This approach is supported by a formal development toolkit, which is used to implement an effective, flexible, yet formally controllable dialogue management.

3.1   A Unified Dialogue Modelling Approach

Generalized dialogue models can be constructed with the recursive transition networks (RTN). They abstract dialogue models by describing illocutionary acts without reference to direct surface indicators (cf. [23]). Fig. 1 (left) shows a simple generalized dialogue model as a recursive transition network diagram. It is initiated with an assertion from a person A, responded by B with three actions: accept, agree or reject.

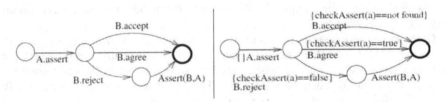

Fig. 1. A generalized dialogue model as a simple recursive transition network (RTN) (left) & a generalized dialogue model as a simple deterministic RTN with conditional transitions (right)

The generalized dialogue model above is a none-deterministic model. To build a feasible interaction model, deterministic behavior should be assured for the interaction flow. Thus, conditional transitions are introduced to improve the above dialogue model (cf. Fig. 1 right). Let *checkAssert* be a method to check whether an assertion holds with B's knowledge and *a* an assertion given by A, if the assertion holds, B can agree with it; otherwise, B rejects it and initiates further discussion; if the assertion is not known by B, then B accepts it. Such conditional transitions can only be activated if the relevant condition is fulfilled. We call it the conditional RTN.

Although the conditional RTN based generalized dialogue model defines a deterministic illocutionary structure, it does not provide the mechanism to integrate discourse information. Thus, information state based theory was integrated into our unified dialogue model by eliminating some typical elements, e.g. AGENDA for planning the next dialogue moves, because such information is already captured by the generalized dialogue model; furthermore it complements illocutionary structure with update rules, which is associated with the information state of current context, and can update the information state respectively if necessary. As a result, a unified dialogue model is constructed as shown in Fig. 2 (left). Four update rules are added, so that the discourse context can always be considered and updated; e.g. the update rule ACCEPT adds a new assertion *a* into B's belief and refer it as known from then on.

Finally, we define a unified dialogue model as a deterministic recursive transition network built at the illocutionary level of interaction processes; its transitions can only be triggered by fulfilled conditions concerning the information state, and with the consequences of information state update according to a set of update rules.

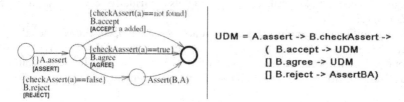

UDM = A.assert -> B.checkAssert ->
    (B.accept -> UDM
    [] B.agree -> UDM
    [] B.reject -> AssertBA)

Fig. 2. A simple unified dialogue model and its CSP specification

3.2   A Formal Language Based Development Toolkit for Dialogue Modelling

Deterministic recursive transition networks can be illustrated as a typical finite state transition diagram (cf. Fig. 2 left), which provides the possibility of specifying the described illocutionary structure with mathematically well-founded formal methods, e.g., with Communicating Sequential Processes (CSP) in the formal methods community of computer science.

CSP can not only be used to specify finite state automata structured patterns with abstract, yet highly readable and easily maintainable logic formalization (cf. [24]), but it is also supported by well-established model checkers to verify the concurrent aspects and increase the tractability (cf. [25]). Thus, CSP is used to specify and verify the unified dialogue models (cf. the example in Fig. 2 (right)).

To support the development of unified dialogue models within interactive systems, we provided the Formal Dialogue Development Toolkit (FormDia cf. Fig. 3).

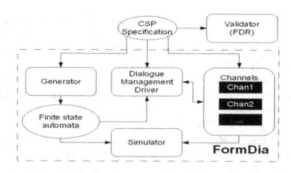

Fig. 3. the Structure of the FormDia Toolkit (cf. [26])

To develop the unified dialogue model based discourse management, FormDia toolkit can be used according to the following steps:

- Validation: the CSP specified structure of a unified dialogue model can be validated by using Failures-Divergence Refinement tool, (FDR (cf. [27])), which is a model checking tool for validating and verifying concurrency of state automata.
- Generation: according to the given CSP specification, finite state automata can then be generated by the FormDia Generator.
- Channels Definition: channels between the dialogue management and application/domain specific components can be defined. These channels are at first black boxes, which will be filled with deterministic behavior of concrete components.
- Simulation: with the generated finite state automata and the communication channels, dialogues scenarios are simulated via a graphical interface, which visualizes dialogue states as a directed graph and provides a set of utilities to trigger events and the dialogue state update for testing and verification.
- Integration: after the dialogue model is validated and verified, it can be integrated into a practical interactive dialogue system via a dialogue management driver.

The FormDia toolkit shows a promising way for developing formally tractable and extensible interaction. It enables an intuitive design of dialogue models with formal language, automatic validation of related functional properties, and it also provides an easy simulation, verification for the specified dialogue models, and the straightforward integration within a practical interactive system. In addition, with the unified dialogue model, FormDia toolkit can even be used in multimodal interactive system.

4   Multimodal Interactive Guidance System for Elderly Persons

The Multimodal Interactive Guidance System for Elderly Persons (MIGSEP) was developed for elderly or handicapped persons to navigate through public spaces. MIGSEP runs on a portable touch screen tablet PC. It serves as the interactive media designed for an autonomous intelligent electronic wheelchair that can automatically carry its users to desired locations within complex environments.

4.1   System Architecture

The architecture of MIGSEP is illustrated in Fig. 4. A Generalized Dialogue Manager is developed using the unified dialogue modelling approach. It functions as the central processing unit and enables a formally controllable and extensible, meanwhile

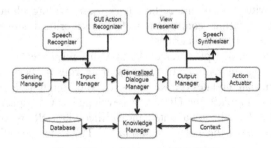

Fig. 4. The architecture of MIGSEP

context-sensitive multimodal interaction. An Input Manager receives and interprets all incoming messages from GUI Action Recognizer for GUI inputs, Speech Recognizer for natural language understanding and Sensing Manager for other sensor data. An Output Manager on the other hand, handles all outgoing commands and distributes them to View Presenter for visual feedbacks, Speech Synthesizer to generate natural language responses and Action Actuator to perform necessary motor actions. Knowledge Manager uses Database to keep the static data of certain environments and Context to process the dynamic information exchanged with users during the interaction.

Although the essential components of MIGSEP are closely connected with each other via predefined XML-based communication mechanism, each of them is treated as an open black box and can be implemented or extended for specific use, without affecting other MIGSEP components. It provides a general platform for both theoretical researches and empirical studies on multimodal interaction.

4.2   The Unified Dialogue Model in MIGSEP

The current unified dialogue model (UDM) consists of four extended state transition diagrams.

Each interaction is initiated with the diagram Dialogue(S, U) (cf. Fig. 5 (left)), by the initialization of the system's start state and a greeting-like request.

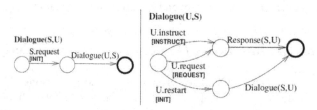

Fig. 5. The Initiate diagram and the transition diagram triggered by user

The dialogue continues with user's instruction to a certain location, request for a certain information or restart action, leading to the system's further response or dialogue restart, respectively, as well as updating the information state with the attached update rules (cf. *Dialogue(U, S)* in Fig. 5 (right)).

After receiving user's input, the system tries to generate an appropriate response according to its current knowledge base and information state (cf. Response(S, U) in Fig. 6 (left)). This can be informing the user with requested data, rejecting an unacceptable request with or without certain reasons, providing choices for multiple options, or asking for further confirmation of taking a critical action, each of which triggers transitions to different diagrams.

Finally, the user can accept or reject the system's response, or even ignore it by simply providing new instructions or requests, triggering further state transitions as well as information state updates (cf. Response(U, S) in Fig. 6 (right)).

Using the FormDia toolkit, the UDM was developed as CSP specifications, and its functional properties have been validated and verified via FDR, as well as its conceptual interaction process using FormDia simulator. The tested specification was then

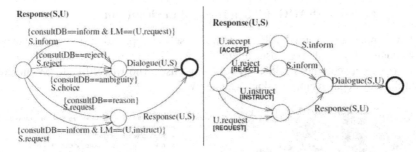

Fig. 6. The model for system's response and the model for user's response

used to generate corresponding machine-readable state transition automata and integrated into the Generalized Dialogue Manager of MIGSEP.

4.3   The Elderly-Friendly Design Elements in MIGSEP

According to the design guidelines in the previous section, a set of elderly-centered design elements were implemented in MIGSEP. Specifically, the most essential elements are listed below:

• Visual perception: simple and clear layout was constructed without overlapping items; 12-14 sized sans-serif fonts were chosen for all displayed texts. Simple and high contrast colors without fancy visual effects were used and placed aside; regularly shaped rectangles and circles were selected, enabling comfortable perception and easy recognition.

• Auditory perception: both text and acoustic output are provided as system responses. Styles, vocabulary, structures of the sentences have been intensively revised. A low-pitched yet vigorous male voice is chosen for the synthesis.

• Motor functions: regularly shaped, sufficiently sized and well separated interface elements were designed for easy access. Clicking was decided to be the only action to avoid otherwise frequently occurring errors caused by decline of motor function. "Start" was provided as the only way of orientating oneself to avoid confusion.

• Attention/Concentration: fancy irrelevant images or decorations were avoided. Unified font, colors, sizes of interface elements were used for the entire interface. Simple animation notifying changes were constructed, giving sufficiently clear yet not distracting feedback to the user.

• Memory functionalities: all items are used with relevant keywords. The number of displayed items is restricted to no more than three, considering the maximum capacity of short term memory, the accessible size as well as the readable amount of information of the interaction items on a table PC. Logically well-structured and sequentially presented items were intensively revised to assist orientation during interaction. Context sensitive clues are given with selected colors.

• Intellectual ability: consistent layout, colors and interaction styles are used. Changes on the interface happen only on data level.

4.4   Interaction with MIGSEP in a Hospital Environment

We have implemented a MIGSEP system and set its application domain to hospital environments. Fig. 7 shows a user interacting with it via speech modality and a sample dialogue between the MIGSEP system and a user who would like to be guided to the cardiology department, to a doctor named Wolf.

S: <shows 3 cards, green for persons, yellow for rooms and
     blue for departments>
   hello, where do you want to go?
U: I want to go to Wolf in the cardiology department.
S: <shows a blue bar texted with cardilogy on the top, and 2
     green cards, with photos, titles and names in the middle>
   I found two persons working in the cardiology department
   with the name wolf, to which do you want to go?
U: to Dr. Wolf please.
S: <enlarges the second green card>
   Do you want to go to Mrs. Dr. Diana Wolf?
U: Yes please.
S: <resizes the second green card and shows the 3 cards in
     the start menu again>
   OK, I have saved this goal. Where else do you want to go?

Fig. 7. A user is interacting with MIGSEP and an Example of a dialogue with MIGSEP

5   The Experimental Study

To evaluate how well an elderly person is assisted by MIGSEP system, an experimental study concerning speech and touch input modalities was conducted.

5.1   Participants

Altogether 31 elderly persons (m/f: 18/13, mean age of 70.7, standard deviation 3.1), all German native speakers, took part in the study, in which 15 participants were using the speech input and 16 were using touch input. They all finished the mini-mental state examination (MMSE), which is a screening test to measure cognitive mental status (cf. [28]). A test value between 28 and 30 indicates slight decline yet sufficiently normal cognitive functioning, therefore, our participants showing 29.0 averagely (std.=.84) were in the acceptable range.

5.2   Stimuli and Apparatus

Except the variation of the input possibilities between touch and speech, where the touch modality was supported by a touchable screen of a laptop and the spoken language instructions were only activated if the button was being pressed and the green lamp was on (cf. Fig. 7), all other stimuli are the same for both modalities, e.g., visual stimuli were given by a green lamp and a graphical user interface; audio stimuli as complementary feedbacks were also generated by the MIGSEP system and presented via two loudspeakers at a well-perceivable volume. All tasks were given as keywords on the pages of a calendar-like system.

The same data set contains virtual information about personnel, rooms and departments in a common hospital, was used for both modalities throughout the experiment.

During the experiment each participant was accompanied by only one investigator, who gave the introduction and well-defined instructions at the beginning, and provided help if necessary.

An automatic internal logger of the MIGSEP system was used to collect the real-time data, while the windows standard audio recorder program kept track of the whole dialogic interaction process.

A questionnaire, which is focusing on the user satisfaction and includes questions of seven categories: system behavior, speech output, textual output, interface presentation, task performing, user-friendliness and user perspective, was filled in by each participant via a five point Likert scale based grading system.

5.3   Procedure

Each participant (both touch and speech) had to undergo four phases:

- Introduction: a brief introduction was given to the participants.

- Standardized learning phase: they were instructed how to interact with the MIGSEP system, either using the touchable screen or using the button device and spoken natural language. After they made no more mistakes with the assigned input modality, a further introduction was given to the verbal and graphical feedbacks the system provides. Then they were asked to perform one task to gather more practical experiences.

- Testing: Each participant had to perform 11 tasks, each of which contains incomplete yet sufficient information about a destination. Each task was ended, if the goal was selected, or the participant gave up trying after six minutes.

- Evaluation: After all tasks were completed, each participant was asked to fill in the questionnaire for subject evaluation.

5.4   Questions and Methods

Altogether, there are three important questions to be answered by the experiment:

- "Can elderly use the MIGSEP system to complete the tasks?"

Besides a general assessment of the task success, a standard measurement method Kappa coefficient ([4]) is also used to detail the evaluation of the effectiveness.

- "Can elderly persons handle the tasks with MIGSEP efficiently?"

This is answered by the automatically logged data of every single interaction.

- "Do elderly find it comfortable to interact with MIGSEP?"

This is answered by the data of the evaluation questionnaires.

6   Results and Discussion

6.1   Effectiveness of MIGSEP

Regarding the effectiveness of the MIGSEP system, 326 out of 341 tasks (10.5 of 11 for each, 95.6%) were correctly performed by all the participants, where 10.6 (96.6) and 10.4 (94.5%) tasks were completed by each participant using touch or speech input modalities respectively. This suggests a generally high effectiveness of the interaction with the MIGSEP system. However, in order to assess the effectiveness at a more detailed level, the standard statistical method Kappa coefficient was used.

In order to apply the kappa method, we needed to define the attribute value matrix (AVM), which contains all information that has to be exchanged between MIGSEP and the participants. E.g. table 1 shows the AVM for the task: "Drive to a person named Michael Frieling." for both touch and speech modalities, where the expected values of this task are also presented.

Table 1. An example AVM for the task "drive to a person name Michael Frieling"

Touch		Speech	
Attribute	Expected value	Attribute	Expected value
Reached Level	L1, L2, L3, L4	FN	Michael
Goal Selection	Michael Frieling	LN	Frieling
Confirm	Yes (to the correct goal)	G	Male
		M	Person

By combining the actual data recorded during the experiment with the expected attribute values in the AVMs, we can construct the confusion matrices for all tasks. Table 2 shows e.g. the confusion matrix for the task "drive to a person named Michael Frieling" with the speech input modality, where "M" and "N" denote whether the actual data match with the expected attribute values in the AVMs, and "SNU" for the system failed-understanding situation. E.g. first name of Michael is wrongly targeted for 4 times and wrongly understood by the system 14 times. Similar construction is done with the AVM of touch input.

Table 2. The confusion matrix fort the task „drive to a person named Michael Frieling"

	FN			LN			G			M			
Data	M	N	SNU	M	N	SNU	M	N	SNU	M	N	SNU	sum
FN	81	4	14										99
LN				82	3	12							97
G							57		4				61
M										3	2	1	6

Given one confusion matrix, the Kappa coefficient can then be calculated with

$$\kappa = \frac{P(A) - P(E)}{1 - P(E)}, \text{(cf. [4])}$$

In our experiment,

$$P(A) = \frac{\sum_{i=1}^{n} M(i, M)}{T}$$

is the proportion of times that the actual data agree with the attribute values, including the system failed-understanding situation as not matched, and

$$P(E) = \sum_{i=1}^{n} (\frac{M(i)}{T})^2$$

is the proportion of times that the actual data are expected to be agreed by chance, where M(i, M) is the value of the matched cell of row i, M(i) the sum of the cells of row i, and T the sum of all cells.

Therefore, we summarized the results of all the tasks and constructed one confusion matrix for all the data, and got: kappa coefficient $\kappa = 0.88(std.=0.10)$ for touch and $\kappa = 0.74\ (std.=0.13)$ for speech modality. This in general still suggests a successful degree of interaction using touch and speech modality. However, touch input is performing more effectively at the detailed interaction level compared to the speech modality, due to the common problem caused by the automatic speech recognizer (294 SNU, 19.6 averagely for each participants).

6.2   Efficiency of MIGSEP

Regarding the efficiency of MIGSEP, the automatically logged quantitative data are summarized in table 3, with respect to user turns, system turns and the elapsed time for each participant and each task.

Table 3. Quantitative results calculated based on the recorded data concerning efficiency

	Touch		Speech	
	Mean	Std.	Mean	Std.
User turns	15.5 (7)	4.1	4.3 (3)	1.7
Sys turns	15.4 (5)	3.9	4.3 (3)	1.6
Elapsed time (s)	88.9	40.2	57.6	24.2

From a general interaction perspective, a very good overall performance efficiency is shown by averagely 4.3 user turns and 4.3 system turns per task for each participant using speech modality, because the average basic turn numbers, inferred by the shortest solution for each task to be filled, are 3 user turns and 3 system turns. This indicates that almost every participant (std. < 2) was able to find the shortest way to complete the tasks while tolerating the problem of automatic speech recognizer. However, 15.5 user turns and 15.4 system turns compared with the shortest solution 7 and 5 respectively were less convincing for the touch modality. Given further insight into the individual data, four participants were having much more turns (averagely 21.8 user turns) than the others, since they were slightly lost finding certain targets and started to do unreasonable brute-force searching.

On the other hand, the elapsed time for each task and each participant for both modalities is considered as satisfying: with averagely 88.9 second for theoretically minimal 12 interaction paces (7+5) with touch input, which is only 7.4 second each, and 57.5 second for minimal 6 interaction paces (3+3), which is only 9.6 second each. However, the standard deviation of 40.2 for touch input is a bit high, due to the interaction context unawareness of the four individuals with 144.8 second averagely consumed time.

6.3   User Satisfaction

Overall, it shows a very good user satisfaction for touch and speech modalities, with averagely 4.7 and 4.3 out of 5. Specifically, the speech and textual outputs are considered appropriately constructed; the interface is intuitive and easy to understand; the process to perform the task is feasible; and the system is considered user-friendly.

However, the scores of system behavior and user perspective for the speech modality were a bit lower than the others. This is mainly due to the problem of the automatic speech recognizer, which could trigger unexpected system responses, and therefore make the future use from the user perspective less attractive. This impression is also reflected to every other aspect for speech modality, with overall lower score compared to touch input.

Table 4. The assessment of subjective user satisfaction

	Touch		Speech	
	Mean	Std.	Mean	Std.
System behavior	4.8	0.3	3.6	0.7
Speech output	4.7	0.5	4.6	0.6
Textual output	4.9	0.3	4.5	0.5
Interface presentation	4.8	0.3	4.7	0.4
Task performing	4.5	0.3	4.3	0.5
User-friendliness	4.7	0.4	4.4	0.6
User perspective	4.3	0.9	3.9	0.8
Overall	4.7	0.2	4.3	0.4

7   Conclusions and Future Work

This paper summarized our work on multimodal interaction for elderly persons, centering the following two essential aspects:

• The design and implementation of a multimodal interactive system according to a number of elderly-friendly guidelines concerning with the basic design principles of conventional interactive interfaces and ageing centered characteristics;

• The modelling and development of multimodal interaction using a tool-supported, formally tractable and extensible unified dialogue modelling approach.

In order to evaluate the minutely designed and developed multimodal interactive system, an experimental study was conducted with 31 elderly persons and concerned with the touch and speech input modality respectively. The evaluation showed high effectiveness, sufficient efficiency and a high satisfaction of the participants with our system for both modalities. Due to the problem caused by the automatic speech recognition, touch modality displays a better performance with respect to effectiveness and is preferred by the elderly. But the speech modality helped the participants in a more efficient way. Thus, the combination of both modalities is motivated.

The presented work served as a continuing step towards building an effective, efficient, adaptive and robust multimodal interactive framework extensively for elderly persons. The result of a further study focusing on the touch and speech combined modality is being analyzed. Corpus-based supervised and reinforcement learning techniques will be applied to improve the current dialogue model and gain more flexible interaction, with the expectation of compensating for the insufficient reliability of automatic speech recognizers.

Acknowledgements. We gratefully acknowledge the support of the Deutsche Forschungsgemeinschaft (DFG) through the Collaborative Research Center SFB/TR8, the department of Medical Psychology and Medical Sociology, the department of Neurology of the University Medical Center Göttingen and the SRH Health Centre Bad Wimpfen, Germany.

References

1. Jaimes, A., Sebe, N.: Multimodal human-computer interaction: A survey. In: Computational Vision and Image Understanding, pp. 116–134. Elsevier Science Inc., New York (2007)
2. Oviatt, S.T.: Ten myths of multimodal interaction. Communications of the ACM 42(11), 74–81 (1999)
3. Holzinger, A., Mukasa, K.S., Nischelwitzer, A.K.: Introduction to the Special Thematic Session: Human–Computer Interaction and Usability for Elderly (HCI4AGING). In: Miesenberger, K., Klaus, J., Zagler, W.L., Karshmer, A.I. (eds.) ICCHP 2008. LNCS, vol. 5105, pp. 18–21. Springer, Heidelberg (2008)
4. Walker, M.A., Litman, D.J., Kamm, C.A., Kamm, A.A., Abella, A.: Paradise: a framework for evaluating spoken dialogue agents. In: Proceedings of the Eighth Conference on European Chapter of Association for Computational Linguistics, NJ, USA, pp. 271–280 (1997)
5. Morris, J.M.: User interface design for older adults. Interacting with Computers 6(4), 373–393 (1994)
6. Jian, C., Scharfmeister, F., Rachuy, C., Sasse, N., Shi, H., Schmidt, H., Steinbüchel-Rheinwll, N.V.: Towards Effective, Efficient and Elderly-friendly Multimodal Interaction. In: PETRA 2011: Proceedings of the 4th International Conference on PErvasive Technologies Related to Assistive Environments. ACM, New York (2011)
7. Fozard, J.L.: Vision and hearing in aging. In: Birren, J., Sloane, R., Cohen, G.D. (eds.) Handbook of Metal Health and Aging, vol. 3, pp. 18–21. Academic Press (1990)
8. Mackay, D., Abrams, L.: Language, memory and aging. In: Birren, J.E., Schaie, K.W. (eds.) Handbook of the Psychology of Aging, vol. 4, pp. 251–265. Academic Press (1996)
9. Moeller, S., Goedde, F., Wolters, M.: Corpus analysis of spoken smart-home interactions with older users. In: Calzolari, N., Choukri, K., Maegaard, B., Mariani, J., Odjik, J., Piperidis, S., Tapias, D. (eds.) Proceedings of the Sixth International Conference on Language Resources Association. ELRA (2008)
10. Kline, D.W., Scialfa, C.T.: Sensory and Perceptual Functioning: basic research and human factors implications. In: Fisk, A.D., Rogers, W.A. (eds.) Handbook of Human Factors and the Older Adult. Academic Press (1996)
11. Schieber, F.: Aging and the senses. In: Birren, J.E., Sloane, R.B., Cohen, G.D. (eds.) Handbook of Mental Health and Aging, vol. 2. Academic Press (1992)
12. Walkder, N., Philbin, D.A., Fisk, A.D.: Age-related differences in movement control: adjust submovement structure to optimize performance. Journal of Gerontology: Psychological Sciences 52B, 40–52 (1997)
13. Charness, N., Bosman, E.: Human Factors and Design. In: Birren, J.E., Schaie, K.W. (eds.) Handbook of the Psychology of Aging, vol. 3, pp. 446–463. Academic Press (1990)

14. Kotary, L., Hoyer, W.J.: Age and the ability to inhibit distractor information in visual selective attention. Experimental Aging Research 21(2) (1995)
15. McDowd, J.M., Craik, F.: Effects of aging and task difficulty on divided attention performance. Journal of Experimental Psychology: Human Perception and Performance 14, 267–280 (1988)
16. Hoyer, W.J., Rybash, J.M.: Age and visual field differences in computing visual spatial relations. Psychology and Aging 7, 339–342 (1992)
17. Salthouse, T.A.: The aging of working memory. Neuropsychology 8, 535–543 (1994)
18. Craik, F., Jennings, J.: Human memory. In: Craik, F., Salthouse, T.A. (eds.) The Handbook of Aging and Cognition, pp. 51–110. Erlbaum (1992)
19. Shaie, K.W.: Intellectual development in adulthood. In: Birren, J.E., Shaie, K.W. (eds.) Handbook of the Psychology of Aging, vol. 4. Academic Press (1996)
20. Sitter, S., Stein, A.: Modelling the illocutionary aspects of information-seeking dialogues. Journal of Information Processing and Management 28(2), 165–180 (1992)
21. Traum, D., Larsson, S.: The information state approach to dialogue management. In: van Kuppevelt, J., Smith, R. (eds.) Current and New Directions in Discourse and Dialogue, pp. 325–354. Kluwer (2003)
22. Ross, J.R., Bateman, J., Shi, H.: Using Generalized Dialogue Models to Constrain Information State Based Dialogue Systems. In: Symposium on Dialogue Modelling and Generation (2005)
23. Alston, P.W.: Illocutionary acts and sentence meaning. Cornell University Press (2000)
24. Roscoe, A.W.: The Theory and Practice of Concurrency. Prentice Hall (1997)
25. Hall, A., Chapman, R.: Correctness by construction: Developing a commercial secure system. IEEE Software 19(1), 18–25 (2002)
26. Shi, H., Bateman, J.: Developing human-robot dialogue management formally. In: Proceedings of Symposium on Dialogue Modelling and Generation, Amsterdam, Netherlands (2005)
27. Broadfoot, P., Roscoe, B.: Tutorial on FDR and Its Applications. In: Havelund, K., Penix, J., Visser, W. (eds.) SPIN 2000. LNCS, vol. 1885, p. 322. Springer, Heidelberg (2000)
28. Folstein, M., Folstein, S., Mchugh, P.: "Mini-mental state", a practical method for grading the cognitive state of patients for clinician. Journal of Psychiatric Research 12(3), 189–198 (1975)

Data Integration Solution for Organ-Specific Studies: An Application for Oral Biology

José Melo[1], Joel P. Arrais[1], Edgar Coelho[1], Pedro Lopes[1], Nuno Rosa[2],
Maria José Correia[2], Marlene Barros[2], and José Luís Oliveira[1]

[1] Department of Electronics, Telecommunications and Informatics (DETI),
Institute of Electronics and Telematics Engineering of Aveiro (IEETA),
University of Aveiro, 3810-193 Aveiro, Portugal
{jmsmelo,jpa,eduarte,pedrolopes,jlo}@ua.pt
[2] Health Sciences Department, Portuguese Catholic University, 3504-505 Viseu, Portugal
{nrosa,mcorreia,mbarros}@crb.ucp.pt

Abstract. The human oral cavity is a complex ecosystem where multiple interactions occur and whose comprehension is critical in understanding several disease mechanisms. In order to comprehend the composition of the oral cavity at a molecular level, it is necessary to compile and integrate the biological information resulting from specific techniques, especially from proteomic studies of saliva. The objective of this work was to compile and curate a specific group of proteins related to the oral cavity, providing a tool to conduct further studies of the salivary proteome. In this paper we present a platform that integrates in a single endpoint all available information for proteins associated with the oral cavity. The proposed tool allows researchers in biomedical sciences to explore microorganisms, proteins and diseases, constituting a unique tool to analyse meaningful interactions for oral health.

Keywords: Oral health, Data integration, Proteins, Diseases, Web services.

1 Introduction

Information available online is increasing rapidly. In order to be processed, this increased amount of data requires the constant development of computer applications that must adapt to increasingly complex requirements, particularly those related to the integration of heterogeneous data and composition of distributed services.

Oral health is an area of research where these problems are particularly relevant. Being a very specific area of study, researchers are faced with many problems in obtaining clinically relevant information concerning the oral cavity in an easy and transparent way. This information must be stored and managed using tools that should provide the user with functionalities to retrieve, store and search for this data.

Usually, databases for molecular biology are centred either on a specific organism, such as SGD for Saccharomyces [1], or on a specific research topic, such as STRING for protein-protein interaction [2]. In addition, databases like Entrez [3] or the Universal Protein Resource (UniProt) [4] play a major role as hubs of biomolecular information, storing data from multiple topics and several organisms.

J. Gabriel et al. (Eds.): BIOSTEC 2012, CCIS 357, pp. 401–412, 2013.

Although this effort to create long-lasting hubs of biomedical data has been very successful, we should not ignore the major contribution that many, more specific, databases make to the current state of science. They are of special interest for small communities that share common research interests. The UMD-DMD database [5], specialized in Duchenne Muscular Dystrophy, is one example.

Being aware of the redundancy of features shared by many of those databases, and of the lack of technical expertise regarding the curators, several frameworks have been proposed to ease the task of deploying new databases. Examples include LOVD [6], specialized in annotating locus-specific databases, GMODWeb [7] for organism-specific databases, or Molgenis [8] which allows deployment of more generic bio-medical databases. Despite the validity of those frameworks, there is none focused on simulating the behaviour of a single human organ or set of adjacent organs. This need to partition the data of the "whole" human system is relevant because, on the one hand, it reduces the time and resources involved in searching, processing and curating information, and on the other, it facilitates the use of algorithms to retrieve biological-ly meaningful results.

The main objective of this project was the development of a web information sys-tem that can collect genotypic information about oral health and that can be useful both for researchers and dentists. A comprehensive integrated resource of the saliva proteins, currently missing in the field of oral biology, would enable researchers to understand the basic constituents, diversity and variability of the salivary proteome, allowing definition and characterization of the human oral physiome. This goal was achieved at two levels: (1) Oralome for the application developer, which consists of a proprietary database, and a set of tools to retrieve biomolecular information from the major platforms like NCBI (National Center for Biotechnology Information) and UniProt; (2) for the end-user, a web portal (OralCard) directed to researchers and dentists, with a set of tools for searching and filtering data from the database, and the possibility to add new information.

Through OralCard web portal, users are able to perform their queries and search among a list of provided results. For each entity, users will be able to consult and ana-lyse a list of dependencies and information retrieved from other major databases. To demonstrate the usefulness of this project, we also present its application in the oral cavity research domain. A platform designed to integrate protein data related to this field will be implemented. This will include salivary proteins obtained in proteomic studies by different research groups, as well as proteins potentially produced and ex-creted by microorganisms assigned to the oral cavity. The ultimate goal is to present a tool for the community that contains accurate, manually curated and updated data re-garding the oral cavity, to enable interaction studies, categorization and exploration.

We expect this work to be a valuable resource for investigators aiming to clarify oral biology, identify molecular disease markers, develop diagnostic tests and im-prove prognosis, as well as providing information for the design of biological path-ways preparing the ground for the discovery of new therapeutic agents.

2   Motivation

The oral cavity consists of a complex ecosystem where a variety of proteins from numerous origins are present. Being able to estimate the impact of the interactions among those proteins is crucial in the understanding of underlying disease mechanisms and hopefully in developing new treatment methods.

Saliva is the watery and usually frothy substance produced in the oral cavity of humans and most other animals. It is an unique clear fluid, composed of a complex mixture of electrolytes and proteins, and represented by enzymes, immunoglobulins and other antimicrobial factors, such as mucosal glycoproteins, traces of albumin and some polypeptides and oligopeptides, of importance for oral health [9].

Whole saliva is secreted mainly from three pairs of major salivary glands: the parotid, the submandibular and the sublingual glands. Approximately 90% of total salivary volume results from the activity of these three pairs of glands, with the bulk of the remainder from minor salivary glands located at various oral mucosal sites [10]. Whole saliva also contains proteins from gingival crevicular fluid, oral mucosa and oral microbiota. The various components of saliva from these sources, together with the plasma proteins that appear in saliva, define the physiological behaviour of the oral cavity, the oral physiome (Oralome).

Saliva is an ideal translational research tool and diagnostic medium and is being used in novel ways to provide molecular biomarkers for a variety of oral conditions, such as oral cancer [11, 12], dental caries [13] and periodontitis [13, 14], as well as systemic disorders such as breast cancer [15], Sjögren's syndrome [16], diabetes mellitus [17], cystic fibrosis [18] and diffuse systemic sclerosis [19]. The ability to analyse saliva to monitor health and disease is a highly desirable goal for oral health promotion and research [20, 21]. The most important advantage in collecting saliva is that it is obtained in a non-invasive way and is easily accessible.

Over the past thirty years, there have been many efforts to determine and identify the main salivary proteins and peptides. Nevertheless, the fluctuating nature of saliva from different individuals, huge dynamic protein concentration ranges and the protein detection limits of most proteomic techniques have made the saliva proteome difficult to define [22]. Even when a healthy individual's saliva is considered, with multidimensional separations and advanced bioinformatics search software tools, proteins identified in different saliva proteomics experiments are often inconsistent with each other except for the most abundant proteins. To overcome the poor coverage, potential bias and complementary nature of each experimental measurement of the human saliva proteome, it is necessary for biomedical researchers to collect and evaluate all reliable publicly available saliva protein data sets generated from different analytical and computational platforms for healthy individuals as well as in disease conditions.

A comprehensive integrated resource of the saliva proteins would provide a great amount of comparative power for interpreting proteomics profile changes in patients' saliva, and may supplement or compensate the limitations and biases associated with the set of controls for a given study. It would also improve the ability to find protein biomarkers that are known to occur in healthy human saliva, for instance where a protein is differentially expressed in a patient sample related to the quantities observed in the study control.

Oralome will have as a vital component an integrated database, by compiling and manually reviewing all the existing experimental data performed on healthy individual samples as well as in several oral and systemic diseases. It will include a collection of microbial proteins expected to be present in saliva due to their presence in the genomes of the oral microbiota [23, 24] and a subset of microbial proteins determined experimentally [25].

We think OralCard will be a fundamental resource to clarify human oral biology and to establish protein biomarkers for salivary diagnostic processes based on the analysis of saliva samples both in health and in disease, exploring the Oralome functionalities. With OralCard, clinical samples from patients' saliva may be better analysed, contributing to improved diagnostic methods and to the development of more effective therapies.

3   Methods

This solution for biomedical data integration is based upon two major components, Oralome and OralCard, illustrated in Fig. 1.

Oralome, our backend pipeline, comprises four functional phases: data curation, modelling, information extraction and publication.

For the OralCard portal frontend, we have selected two iterations: establishment of a relationship between the previously created database and the enterprise tier, and the design of views for the entities using web application frameworks.

3.1   Compilation and Curation of the Saliva Proteome

Regarding specific domains, such as the oral cavity, manual data curation is the key iteration where system restrictions are imposed. These are essential to correctly focus the system and to establish a common platform for further iterations. Here three main activities are performed: bibliography review, target identification and requirements analysis. Reviewing the bibliography consists of analysing state-of-the-art work and deciding where the final system's uniqueness will be. Next, the target identification process involves exploring both relevant bibliography and databases, and filtering which data and/or features should be available in the final system. Once data and features are selected, a careful requirements analysis process must be conducted. This implies getting a first idea of the technological requirements related to the desired features and integrated data.

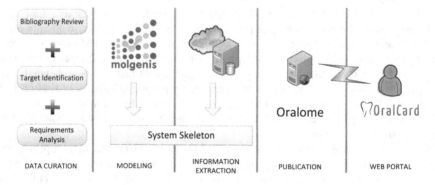

Fig. 1. Pipeline for biomedical data integration

By the time this work was done, to our knowledge, there was no database joining the proteomes of major and minor salivary glands. For this reason, the first step was to compile this information from different sources. The proteome data of major salivary glands (parotid, submandibular/sublingual) were obtained from the Salivary Proteome Knowledge Base and from Yates Lab, The Scripps Research Institute. The proteome of human minor salivary gland secretion was obtained from Oppenheim Laboratory, Henry M. Goldman School of Dental Medicine, Boston University. The proteins identified in different studies were compared and repeated entries eliminated.

Biological information is constantly being updated. Since the first publication of saliva proteomes, many of the originally identified proteins, catalogued as different entries in biological databases, have been merged with others and some deleted due to misidentification. Therefore, all information concerning the identified proteins was manually curated and updated. The update of the IPI (International Protein Index) entries was carried out with the "IPI History Search" (www.ebi.ac.uk/IPI) tool. All other updates have been made using the UniProt database.

3.2   Oral Cavity Data Integration

The orthogonal nature and innate heterogeneity associated with life science resources have always hampered easier developments regarding the integration of distributed data. Furthermore, research in this field has entered a cycle where computational solutions lag one step behind technological requirements in biology. This brought about a growing disparity regarding bioinformatics software, where a few well-known and widely used resources, such as UniProt or NCBI, co-exist with hundreds of smaller independent tools.

Although the oral cavity presents a narrower scope, it involves assorted life science fields, from microorganisms to proteins or diseases. Establishing new connections amongst these diverse entities creates a high degree of complexity, thus requiring the development of new ad-hoc data integration software solutions. On the one hand, large warehouses that might contain this domain-specific information also contain many other resources. Consequently, researchers are overwhelmed by huge datasets, making their data of interest impossible to find. For instance, discovering oral cavity information amongst UniProt is a nightmarish task.

From a technological perspective, there are miscellaneous strategies for solving data integration problems. However, they all rely on three elementary concepts: warehousing, middleware and link integration. Warehouse approaches intend to support an efficient decision-making process, requiring the aggregation of all desired data in a huge central dataset [26]. Middleware-based solutions rely on the development of specific wrappers to mediate connections between users' requests and original data servers [27]. Finally, link-based integration attempts to connect heterogeneous data types by creating graphs or networks based on pointers between distinct data units [28]. These approaches can be distinguished by the way they treat aggregated data. Warehouses replicate entire resources, creating a truly integrated environment, whereas middleware or link-based solutions only provide streamlined access to data, resulting in virtual integration.

Although many examples can be found for each integration strategy, the most common solutions involve developing a hybrid architecture, where some data is replicated whilst other data are simply connected through links and identifiers. This approach was integrated into the Oralome project.

Once target resources and data were identified, the modelling iteration started. This task consisted of designing a common information model to support oral cavity data from distinct resources.

Before the actual data integration, a system skeleton needed to be deployed. As mentioned above in this article, there are several frameworks designed for rapid prototyping of data portals for life science projects, such as LOVD or GMODWeb. For this specific task, we have chosen the Molgenis framework for its agility in creating a database and application, complete with data exploration web workspace, REST and SOAP web services, and R interface out of the box. For the data integration process, Molgenis provides easy and direct data input, whether through the web interface, through any of the available services, or through a provided database API. Therefore, custom data wrappers, collecting data from miscellaneous resources, can be easily implemented. Oralome required the deployment of general-purpose wrappers, combining external data in the newly deployed Molgenis instance. These wrappers allow for systematic information extraction from resources such as UniProt, NCBI or STRING, amongst others. These resources provide several ways to retrieve information, such as REST interfaces or APIs for Java development.

Executing this streamlined data integration workflow, curated oral cavity data is collected and re-organized in a publicly available web framework.

3.3   Oralome Development

Oralome consist of a set of tools and a database that provide access to information related to several entities, such as microorganisms, proteins, diseases and pathways, integrating crucial data regarding the oral cavity.

The upper entity is a microorganism which has several associated proteins. A protein itself has other identifiers linked to it, such as OMIM (Online Mendelian Inheritance in Man), KEGG (Kyoto Encyclopedia of Genes and Genomes), PDB (Protein Data Bank) and GO (Gene Ontology) terms. The main subject for this tool consists of two groups of proteins: (1) a subset of microbial proteins determined experimentally, and (2) microbial proteins expected to be present in saliva. Regarding the first group, besides the information retrieved from UniProt, Oralome will integrate information related to the environment where a protein was identified (health or disease, regulation, age group, and the particular source where it resides, for instance, mucosa or tongue).

For Oralome tool development we chose the Molgenis framework for generating all the necessary tools and features needed to start compiling our database and to view this data in an easy and rapid way.

Molgenis consists of a framework written in Java, which accepts two XML files as input: a database and a user interface descriptor file. Using the first file, users can specify how the database will be structured, its entities and relations; the second file specifies the layout for the web interface. Molgenis generates a Java model and a database API which are used to deploy the related SQL tables, web services and web interface into a web server (Fig. 2).

In order to start using this framework we needed to have preinstalled a MySQL server to store its database, an Apache Tomcat web server for deploying web services and a simple user interface, and the Java Development Kit for generation of SQL and HTML code.

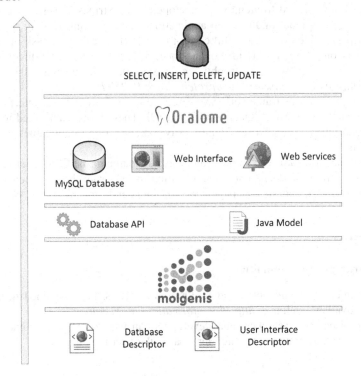

Fig. 2. Oralome architecture

Along with the Oralome application, we developed tools and wrappers to obtain specific information on each of the elements that build up the system (proteins, diseases, pathways and others). For this, we carried out a first survey of sources where this information would be available, and built a runnable script to update the Oralome database.

This data fetch is made easier using Molgenis. It bundles a Database API that has the advantage of hiding complex SQL commands.

To import and filter the information needed in our database, we used Java as the programming language because it is highly compatible with most APIs provided by the major external services resources (UniProt, KEGG, and NCBI Entrez Utilities).

3.4   Oralcard Development

In order to take advantage of the Oralome functionalities, we propose a tool that enables searching over the oral cavity database and show different and customized views for each entity. This led to the OralCard web application, a fundamental resource for salivary diagnostic processes of protein biomarker studies of health and disease based on the analysis of saliva samples.

For information retrieval from Oralome, we decided to use Hibernate, an object-relational mapping tool for Java. This architecture is illustrated in the following diagram (Fig. 3). OralCard frontend was developed using Stripes, a web framework that makes the development of Java web applications easier, by introducing some useful tools. Stripes enabled us to take full control over URLs, easing the task of accessing an entity by only knowing its id. For instance, researchers can have direct access to the protein P22894 (Neutrophil collagenase), introducing the address *http://bioinformatics.ua.pt/oralcard/proteins/view/P22894*.

By using the Stripes framework, we were able to improve the user interface, introducing frameworks such as jQuery and jQueryUI. These tools contributed to presenting information in a more user-friendly way, taking control over tables and AJAX interactions.

Finally, the OralCard web application takes advantage of the CSS benefits. It is designed to separate the document content written in JSP from the document presentation, including elements such as page layout, colours and fonts used.

4   Results

4.1   Oralome Functionalities

The web interface feature provided by Oralome reflects data contained in the database. It is available online at *http://bioinformatics.ua.pt/oralome*.

Both oral cavity researchers and developers can use this tool. Oralome provides two distinct entry points, each matching a particular user type needs.

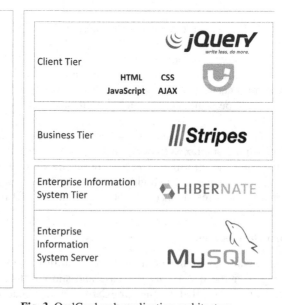

Fig. 3. OralCard web application architecture

This workspace enables searching, filtering, browsing and viewing all collected data in a typical web interface. Moreover, researchers can combine data and download it for personal usage.

Along with the web interface for researchers, Oralome encompasses a set of web services for programmatic data access. Oralome's API is available for R, HTTP, REST and SOAP. These interfaces enable any developer to build custom solutions using any development framework, using Oralome's curated data related to the oral cavity.

In addition to the described oral cavity data exploration scenarios, Oralome allows the deployment of an unlimited number of customized applications. Using any of the remote API interfaces, developers can use Oralome data to enrich already existing applications or to develop new ones.

This key Oralome feature envisages the use of collected curated oral cavity information in education scenarios or by medical dentists. Oralome web framework's openness will enable the creation of an entire application ecosystem built around expertly curated information, aiming for the delivery of improved dental health care.

4.2   Oralcard Interface

OralCard web application is available online at http://bioinformatics.ua.pt/oralcard. The researcher is first presented with a home page where he can insert search items. These can be related to proteins, diseases or microorganisms. Each of these entities has dedicated views, where the user can find answers to several questions.

The OralCard web application can provide the user with information about:

- Diseases involved in a specific saliva protein;
- Salivary proteins involved in a specific disease;
- Proteins whose levels are altered in a specific disease;
- Proteins obtained from a particular source (which ones and how many);
- Microorganisms involved in a particular disease;
- Diseases in which a particular microorganism is evolved;
- Proteins produced by a given microorganism and present in a certain disease;

Fig. 4 shows the dedicated view for the *Neutrophil collagenase* protein. The user is presented with a list of choices, where he can consult related diseases, Protein Data Bank structures, PubMed references and the related sources. We also embed some online tools, like UniProt, Panther or BRENDA, in order to facilitate access to specific information provided by these services.

5   Evaluation

As an example of application, we used the Oralome tool to evaluate how saliva proteins contribute to or reflect impaired healing of oral cavity tissues in diabetic patients [29]. Then we identified all the salivary proteins whose amount is changed in patients with diabetes mellitus.

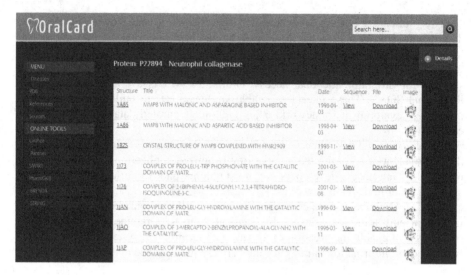

Fig. 4. Structure view for the Neutrophil collagenase protein

Subsequently, in order to understand which salivary protein molecular functions were altered in diabetes, we found that binding and catalytic activity functions changed more evidently. That led to the search for pathways in which the binding molecular function could be acting, in other words, which signalling pathways could be involved. Our results indicate that the blood coagulation pathway (directly related to healing) was the one in which the proteins altered in diabetes mellitus patients' saliva were involved. We then traced molecular networks reflecting the interactions between proteins involved in this pathway. We found that two key molecules in this network are two blood coagulation cascade inhibitors.

This methodology has allowed us to identify molecular reasons for the impaired healing of oral tissues in diabetic patients, as well as some key molecules in this process, which may be good molecular markers, useful for diagnosis and even good targets for therapeutic agents.

6   Conclusions

In this paper we described a pipeline for creating web-based databases specialized in a set of adjacent human organs. We believe that the proposed work is of major importance for research projects that need an easy and agile solution to share the results of studies.

Information on the oral cavity is dispersed through different databases focused on more general systems. In addition, data is not always standardized, which makes their integration and comprehensive study a colossal task. This work resulted in development of an integrated database which comprises a comprehensive catalogue and characterization of the human oral proteome. It aims to become a fundamental resource for clarifying human oral biology and establishing a protein biomarker for salivary diagnostic processes.

We presented Oralome as a set of tools combining a database, web services and user interface, useful for joining specific results from several major databases, such as UniProt or NCBI. This framework generated a database, web services and a web application where users can access the downloaded data.

Finally, we proposed OralCard as an example of a web application that takes advantages of the Oralome functionalities. It works as a search engine where the researcher can input any searched-for item, concerning the field of oral biology. It uses the Oralome database and some of the new web standards to present specific information regarding a protein (JavaScript, AJAX and CSS).

Some of the advantages of the Oralome approach are reduction of the time and resources involved in searching, processing and curating information, as well as facilitating the use of algorithms to retrieve biologically meaningful results. As a particular example of Oralome use, OralCard offers the advantage of gathering several major external tools in one single web application, making it easier and more comfortable for researchers to access important information regarding one specific protein. Using the oral cavity as a particular case study, we have shown how it can be used to obtain a fully functional tool that enables both interactive categorization and exploration.

We believe this project will be a valuable resource for investigators to clarify the oral cavity biology, identify molecular disease markers, develop diagnostic tests and improve prognosis, as well as providing invaluable help in discovering new therapeutic agents.

Acknowledgements. This work was supported by the European Community's Seventh Framework Programme (FP7/2007-2013), under grant agreement no. 200754 (GEN2PHEN project), and from Fundação para a Ciência e Tecnologia, FCT, under grant agreement PTDC/EIA-CCO/100541/2008. Joel P. Arrais is funded by FCT grant SFRH/BPD/79044/2011.

References

1. Cherry, J.M., et al.: SGD: Saccharomyces genome database. Nucleic Acids Research 26(1), 73 (1998)
2. Mering, C., et al.: STRING: a database of predicted functional associations between proteins. Nucleic Acids Research 31(1), 258 (2003)
3. Maglott, D., et al.: Entrez Gene: gene-centered information at NCBI. Nucleic Acids Research 33(suppl. 1), D54 (2005)
4. Bairoch, A., et al.: The universal protein resource (UniProt). Nucleic Acids Research 33(suppl. 1), D154 (2005)
5. Humbertclaude, V., et al.: G.P.9.10 Clinical development of the French UMD-DMD database. Neuromuscular Disorders 17(9-10), 817–818 (2007)
6. Fokkema, I.F.A.C., den Dunnen, J.T., Taschner, P.E.M.: LOVD: Easy creation of a locus specific sequence variation database using an "LSDB in a box" approach. Human Mutation 26(2), 63–68 (2005)
7. D O'Connor, B., et al.: GMODWeb: a web framework for the Generic Model Organism Database. Genome Biology 9(6), R102 (2008)
8. Swertz, M., et al.: The MOLGENIS toolkit: rapid prototyping of biosoftware at the push of a button. BMC Bioinformatics 11(suppl. 12), S12 (2010)

9. de Almeida, P.V., et al.: Saliva composition and functions: a comprehensive review. J. Contemp. Dent. Pract. 9(3), 72–80 (2008)

10. Greabu, M., et al.: Saliva—a diagnostic window to the body, both in health and in disease. J. Med. Life 2, 124–132 (2009)

11. Nagler, R.M.: Saliva as a tool for oral cancer diagnosis and prognosis. Oral Oncology 45(12), 1006–1010 (2009)

12. Shpitzer, T., et al.: Salivary analysis of oral cancer biomarkers. British Journal of Cancer 101(7), 1194–1198 (2009)

13. Rudney, J., Staikov, R., Johnson, J.: Potential biomarkers of human salivary function: a modified proteomic approach. Archives of Oral Biology 54(1), 91–100 (2009)

14. Gonçalves, L.D.R., et al.: Comparative proteomic analysis of whole saliva from chronic periodontitis patients. Journal of Proteomics 73(7), 1334–1341 (2010)

15. Streckfus, C.F., et al.: Breast cancer related proteins are present in saliva and are modulated secondary to ductal carcinoma in situ of the breast. Cancer Investigation 26(2), 159–167 (2008)

16. Hu, S., et al.: Salivary proteomic and genomic biomarkers for primary Sjögren's syndrome. Arthritis & Rheumatism 56(11), 3588–3600 (2007)

17. Rao, P.V., et al.: Proteomic identification of salivary biomarkers of type-2 diabetes. Journal of Proteome Research 8(1), 239–245 (2009)

18. Livnat, G., et al.: Salivary profile and oxidative stress in children and adolescents with cystic fibrosis. Journal of Oral Pathology & Medicine 39(1), 16–21 (2010)

19. Giusti, L., et al.: Specific proteins identified in whole saliva from patients with diffuse systemic sclerosis. The Journal of Rheumatology 34(10), 2063 (2007)

20. Seymour, G.J., Cullinan, M.P., Heng, N.C.: Oral Biology: Molecular Techniques and Applications (Methods in Molecular Biology). Humana Press (2010)

21. Wong, D.T.: Salivary diagnostics powered by nanotechnologies, proteomics and genomics. The Journal of the American Dental Association 137(3), 313 (2006)

22. Helmerhorst, E., Oppenheim, F.: Saliva: a dynamic proteome. Journal of Dental Research 86(8), 680 (2007)

23. Chen, T., et al.: The Human Oral Microbiome Database: a web accessible resource for investigating oral microbe taxonomic and genomic information. Database: the Journal of Biological Databases and Curation 2010 (2010)

24. Nelson, K.E., et al.: A catalog of reference genomes from the human microbiome. Science 328(5981), 994–999 (2010)

25. Xie, H., et al.: Proteomics analysis of cells in whole saliva from oral cancer patients via value-added three-dimensional peptide fractionation and tandem mass spectrometry. Molecular & Cellular Proteomics 7(3), 486 (2008)

26. Santos, R.J., Bernardino, J.: Real-time data warehouse loading methodology. In: Proceedings of the 2008 International Symposium on Database Engineering & Applications, pp. 49–58. ACM, Coimbra (2008)

27. Barbosa, A.C.P., Porto, F.A.M., Melo, R.N.: Configurable data integration middleware system. Journal of the Brazilian Computer Society 8(2), 12–19 (2002)

28. Lopes, P., Dalgleish, R., Oliveira, J.L.: WAVe: web analysis of the variome. Human Mutation (2011)

29. Lamster, I.B., et al.: The relationship between oral health and diabetes mellitus. The Journal of the American Dental Association 139(suppl. 5), 19S (2008)

Author Index